The Internet and
the
New Biology

Tools
for Genomic and Molecular
Research

The Internet and the New Biology

Tools for Genomic and Molecular Research

Leonard F. Peruski Jr.
U.S. Naval Medical Research Unit No. 3
Cairo, Egypt

Anne Harwood Peruski
American University in Cairo
Cairo, Egypt

Library of Congress Cataloging-in-Publication Data

Peruski, Leonard F.
    The Internet and the new biology : tools for genomic and molecular
research / Leonard F. Peruski, Jr., Anne Harwood Peruski.
        p.   cm.
    Includes bibliographical references and index.
    ISBN 1-55581-119-1
    1. Molecular biology–Computer network resources.   2. Nucleotide
sequence–Computer network resources.   3. Amino acid sequence–
Computer network resources.   I. Peruski, Anne Harwood.   II. Title.
QH506.P44   1997
025.06'5728–dc21   96-46766
                                                                CIP

Contents

Preface

This book grew out of a "self-defense" laboratory manual that was put together to teach fellow post-docs how to analyze sequence data using the resources of the Internet. Basically, we got tired of answering the same set of questions over and over:

- How do I submit sequence data to the public sequence databases?
- How do I search a sequence against the public databases?
- How can I look for sequence motifs such as promoters, splice sites, etc.?
- How do I retrieve sequences from the public databases?

This small and terse guide became quite popular within our local community, and out of this in-house laboratory "cookbook" came the idea for a more formalized text for a broader scientific community. The focus of this text still remains the same: to give simple, straightforward answers to the above four questions.

To answer these questions this book concentrates on the simplest component of the Internet, e-mail servers, and their application to molecular and genomic biology. There are a lot of resources relevant to the biological sciences on the Internet today. Many require high-speed connections, complicated software, or much time and effort to use. In contrast, the e-mail servers are powerful, fast, and cheap to access and use. Any laboratory, school, or home office with an Internet connection can make use of these tools with a minimum of time and effort. The same cannot be said of any others of the Internet resources.

The Internet is in transition, perhaps the greatest in its history. Originally the Internet was concerned with text-based interfaces and keyboard control of remote computers, because of the limited informational capability of the network. It has evolved into a robust and sophisticated system with an informational capacity and computational power that its originators could have scarcely dreamed of. As a result, the Internet and its molecular and genomic biology tools are increasingly becoming graphical in nature. High-resolution graphics, formatted text, hyper-links, and mouse-driven interfaces are rapidly becoming the standard in much of the scientific community. Recognizing that this transition is happening, but is by no means complete, this book also highlights and demonstrates the use of two other practical aspects of the Internet: Gopher and the World Wide Web.

Please note that nearly all of the services or resources mentioned in this book are the direct result of the voluntary contributions of scientists and students from all types of institutions throughout the world. Any service or resource on the Internet should be considered by the scientific community as formally published, and properly cited as such. Consequently, if any of these services and resources are used in a publication or public presentation, their authors should be cited to acknowledge their efforts.

The book is divided into ten chapters. The first is an introduction to the Internet and why it is of increasing importance to the genomic and molecular biology community specifically and the biological community as a whole. The second chapter takes a broad look at e-mail and e-mail servers, Gopher, and the Web. Chapter 3 sets the foundation for using the Internet for sequence analysis by explaining how to use a sampling of the different resources. This chapter describes the structure of an e-mail message, the directives used by the servers, and the databases that can be accessed.

Chapter 4 describes the major public primary and specialized sequence databases that most of the Internet-based resources use as foundations. The primary databases are the principal repositories of DNA and predicted protein sequences, while the specialized databases are just that, serving as repositories of information on protein structure, restriction enzymes and methylases, and sequence motifs.

The next three chapters describe the servers themselves in greater detail. Chapter 5 covers resources that are designed to compare and retrieve sequence data. Chapter 6 looks at resources designed to predict structure and function of DNA and protein sequences. Chapter 7 highlights servers that are archives of software, news, and information for the biological sciences community.

Chapter 8 describes how to use the analysis tools described in Chapters 5 though 7 to analyze a query sequence. This analysis process takes you through the identification of an unknown sequence using BLAST; the downloading of related sequences using Retrieve; the analysis of translated protein products using Blocks, Domain, and MotifFinder; the prediction of secondary structure with nnPredict and PredictProtein; and finally the determination of phylogenetic relationships using MAlign and CBRG.

Chapter 9 examines the relational and graphical side of the Internet: Gopher servers and Web browsers that reflect the cutting edge of technology and ease of use. The final chapter takes a peek into the future of these tools.

Finally, a series of appendices provide a summary of all of the servers described in the book, a guide to domains and usage, suggested computer hardware and software needs, information on nucleic acids and proteins, and finally a glossary of terms common in this field.

The services described in this book are by no means the sum total of the e-mail servers, Gopher servers, or Web sites that are available on the Internet. What we have tried to describe is a set of electronic tools which are reliable and robust systems that we have used in our own projects. Much as everyone has their favorite "tools" in their personal laboratory methodology, these are our favorites and the ones that appear to be the most broadly useful to the molecular and genomic biology community at large. So, as such, this selection of resources is by no means complete, nor is it meant to be. While we have made an extensive effort to sift through the information and make this listing as accurate as possible, there will be some mistakes. For these errors we apologize. Please feel free to point them out to us, so that subsequent editions of this guide can be more useful. If we have missed a resource that is felt to be particularly useful, please pass that information along to us as well. It is our hope to ultimately make this guide a community tool kit for genomic and molecular biology, rather than a personal one.

Recognizing the rapid pace of development of the Internet in general and the Web in particular, ASM Press has established a Web site as a complement to this text. The URL is

http://www.asmpress.org/isbn/1555811191/

The password is 9646766.

This Web site will serve many functions relative to the book:

- a compendium of active links to the resources described in the text

- a mechanism to highlight and report on recent advances in existing Internet-based tools and the creation of new ones

- a vehicle for the reader to offer comments and suggestions in a timely and convenient fashion

With this Web site, we hope to keep *Internet Tools for the New Biology* up to date and accurate, but above all a resource that reflects the needs of the user community.

In closing, this book is not meant to be an advanced text on sequence analysis or on the Internet, but instead to show the reader some of the tools of computational biology and how they can be applied to problems in genomic and molecular biology. It is our hope that this book will be useful both as an introduction to this area for the true novice and as a laboratory resource for scientists who are looking for simple answers to the four basic questions outlined above.

Leonard F. Peruski Jr.
U.S. Naval Medical Research Unit No. 3
PSC 452, Box 110
FPO, AE 09835-0007 U.S.A.

Anne Harwood Peruski
American University in Cairo
113 Kasr El Aini Street
Cairo, Egypt

Telephone: +011-202-352-8688
Facsimile: +011-202-284-1382
E-mail: annep@acs.auc.eun.eg (Internet)

Acknowledgments

Any book is the result of more than the efforts of the authors listed on the cover. Over the course of completing this text, we became deeply indebted to several individuals. First and foremost among this group of unsung heroes is Ellie Tupper, Senior Production Editor at ASM Press. Without her friendly reminders and careful reading of our text, we could not have completed this book. She took the writing of a couple of neophyte scientists and turned it into readable prose. For all of her work in the editing process she should be considered an author as well. Patrick Fitzgerald, former director of ASM Press, was a constant source of encouragement and constructive criticism during the formative stages of the book. Kenn Rudd, University of Miami School of Medicine, did an outstanding job as the scientific editor of the original draft. He spotted numerous errors and inconsistencies, refined details and concepts, and greatly improved the content and structure.

We would also like to thank Karen Gray, Marketing Manager at ASM Press, for her work on the development of the companion Web site as well as for her insight into the business side of publishing. Susan Graham and Gerry Quinn did a wonderful job in designing and formatting the text. Also, we thank our children Elizabeth, Kyle, Jennifer, and Scott. They kept us on track and allowed us the time and space we needed to complete this project.

Finally, we would like to thank the American University in Cairo for the unrestricted use of an Internet account. It was the essential tool without which this book would not have been born.

## CHAPTER 1

Introduction to the Internet

- What the Internet is (and is not)
- A brief history of the Internet
- How the Internet works
- Components and resources of the Internet
- The importance of the Internet in biology
- Scope and limits of this book

What the Internet is (and is not)

What is the Internet? Pose that question to the general public and you would get a wide range of answers. Ask that same question of a scientist and you would get a divergent and only slightly more focused range of answers. **In the most general terms, the Internet is an international network of computational hardware and software that can be accessed and shared by any other computer linked to the network.** It thus serves as a repository of data files, images, and text. It functions as a system by which users can access other more powerful computers in a shared, community-based fashion to ask (and hopefully answer) complex or computationally intensive questions.

In brief, the Internet is a worldwide network that links networks of computers to one another. This loosely organized federation of networks is interconnected and shares a common underlying communication protocol. This communication protocol, Transmission Control Protocol/Internet Protocol (TCP/IP), acts as a sort of universal language translator, permitting different types of computers to communicate with each other. All networks that use TCP/IP and that are linked together use this protocol as the foundation to perform common tasks, such as e-mail messaging.

In practical terms, the Internet at the simplest level is a personal computer that is linked to other computers and computer networks around the world. By being connected to other computers and associated networks, it becomes possible to access other computers and services in a community-based or shared fashion. Specialized analysis techniques are no longer limited by the need for expensive and dedicated hardware and software on site. Instead, one can simply access such facilities remotely through the Internet.

While the Internet offers a range of information distribution services, each of these services has strengths and weaknesses. Originally, each aspect of the Internet was controlled by a single, unique software protocol specific to a given Internet component, such as e-mail, FTP, Telnet, or Gopher. In contrast, the more recent and sophisticated services, such as the World Wide Web, offer a highly integrated environment that blurs the boundaries between the original services of the Internet while offering graphical interfaces to services and content. As a result, the traditional text-based interfaces are being replaced with graphical interfaces in either the Macintosh or Windows environments or through other graphic interface environments. However, the text-based interfaces and services still offer advantages over the graphical interfaces and will be a significant factor in Internet access for some time.

There are two major reasons for the survival of the text-based interface. First, many if not most Internet users worldwide have access only to text-based terminals. Second, text-based interfaces have a lower bandwidth requirement, and when users must pay by the volume of data moved, text interfaces will be much cheaper. Further, over slower transmission links to the Internet, text-based interfaces, with the much smaller amount of data to be transmitted, are much more efficient and time-effective. On the downside, text-based interfaces are generally more intimidating to most computer users and, as a result, require more effort to master. Additionally, text-based interfaces cannot access or display information in graphical form, which is becoming increasingly common both on the Internet in general and in science specifically.

A brief history of the Internet

The beginnings of the Internet can be traced back to the early 1960s and the height of the Cold War period. Computers were expensive commodities, but vital in the sciences even then. One way to share limited computer resources was to link or network these machines to permit wider access. Further, when individual computers were linked, they could be distributed over a wide area, which reduced the risk that all of the computers would be destroyed in the event that the Cold War became hot.

During this period, networks were designed in such a fashion that if one computer on the network failed, the entire network failed. The U.S. Department of Defense began a research project designed, in part, to develop the software and hardware for a network that could survive the failure of at least some computers that made up a network. By the late 1960s, the Defense Advanced Research Projects Agency (DARPA or ARPA) created the first such network, ARPAnet, the forerunner of the Internet.

By 1969, the first primitive routers were installed at the University of California at Los Angeles. By the end of that year, a total of four sites were connected to the fledgling ARPAnet. In 1972, the ARPAnet concept was demonstrated publicly: approximately 40 computers were connected at that time. After that, the pace and expansion of the ARPAnet from a so-called "Cold War network" increased. In 1979, Usenet, the first news distribution protocol, was introduced. In 1982, TCP/IP (Transmission Control Protocol/Internet Protocol) was established as the standard data transfer protocol for the ARPAnet. TCP/IP was important for two reasons. First, this protocol has the ability to decide which way data should be transmitted on the network. This routing capability permits data to be efficiently and rapidly transmitted even in the event of failed computers on the network, so one failed computer doesn't affect the function of the entire network. TCP/IP proved to be such a robust and successful protocol that it was adopted as a standard for other networks that were beginning to emerge around the world.

In 1987, the National Science Foundation of the United States created a TCP/IP-based network designed to connect university supercomputing centers. This public network became known as the NSFnet. Over the next few years, the NSFnet absorbed the functional role of the ARPAnet and evolved into what is known as the Internet today. With the end of the Cold War, the Internet was no longer seen as a defense computer network, and its growth mushroomed in the private and public sectors worldwide.

With the growth of commercial links to the Internet and the introduction and popularization of the World Wide Web in the early 1990s, the Internet became increasingly popular not just in science and technology-related fields, but with the general public. Since that time, the growth of the Internet is estimated at about 10% per month. Currently, it is estimated that the Internet has about 30 to 50 million users worldwide.

How the Internet works

As we mentioned, the Internet is a computer network of smaller computer networks. In order to transmit information between computers on the Internet, each computer must have a unique identifier; therefore, each computer is assigned a number. Because this number is difficult for most people to remember, this identifying number is linked to a unique name by the DNS or Domain Naming System. Names under the DNS are composed of words, each separated by a period.

To permit fast and error-free data transmission between these smaller networks, a specialized computer, called a router, and a software protocol are used to manage data transmission. TCP/IP is the software protocol used to transfer data throughout the Internet. TCP stands for Transmission Control Protocol; its function is to transfer data between two computers connected to the Internet. IP stands for Internet Protocol; IP handles computer addresses and data packet fragmentation. Whenever data needs to be transferred between computers on the Internet, it is split up into small packages, called packets, which are sent over the network. Working in concert, TCP passes data packets along the Internet while IP directs the data packets to the proper Internet address and reassembles the packets in the correct order at the final destination. Routers manage the flow of the data packets between networks and decide which way on the network the data will be sent. Figure 1.1 shows a highly simplified and abstract view of the Internet. In this schematic, several computers and their networks are connected with routers.

Transmitting data from computer 1 to computer 2 could be as simple as sending the information along the path from computer 1 through routers A and B to computer 2. But imagine that some part of the network between routers A and B is damaged and no longer functions. In this case, the data goes from computer 1 to router A. Router A now transmits the data via routers C and D to router B, bypassing the damaged part of the network. Router B then sends the data to computer 2.

As we mentioned, data is split into packets when it is transmitted over the network. Using the previous example, assume that the data was split up into four packets. If the line between routers A and B fails after the first two packets, then the remaining two packets

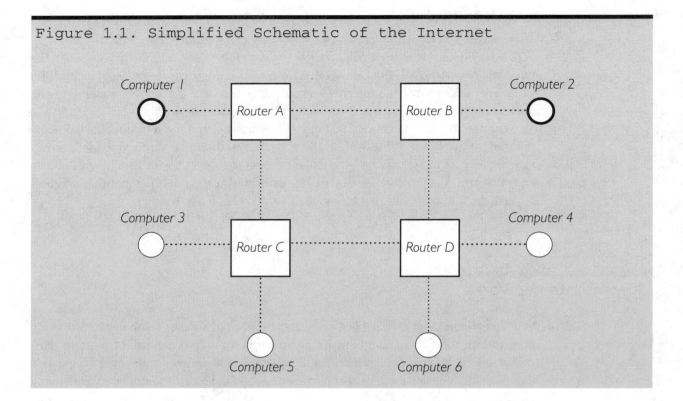

Figure 1.1. Simplified Schematic of the Internet

will take another path to reach computer 5. This mechanism of transmitting data is called packet switching. Packet switching makes the Internet error-tolerant and adaptable to network problems.

Because of packet switching in data transmission, the data packets might not arrive at the destination computer in the same order as they were sent from the originating computer. At this point, the IP part of the TCP/IP system is responsible for reassembling the independent data packets back into the original transmitted data.

Components and resources of the Internet

■ Internet access tools

One of the wonders of the Internet is its flexibility and adaptability. A particular brand of computer or operating system is not required in order to be connected and use its resources. The Internet offers seven distinct access tools or services that can be utilized by nearly any computer system: electronic mail (e-mail); File Transfer Protocol (FTP); remote computer operation or Telnet; news and discussion groups or Usenet; Internet Relay Chat (IRC) for online conferencing; Gopher, a primarily text-based, interactive information retrieval system; and the World Wide Web. All of these Internet tools share a common underlying structure in that they are composed of three distinct components:

a server (or daemon) program that runs all the time on a host computer and accepts requests over the Internet from other computers, **a client program** that is used by the client computer to connect to or access these servers, and **a standard communications protocol** that allows clients and servers to exchange information.

■ Electronic mail

Electronic mail, or e-mail, is used to communicate with other Internet users, regardless of where they are on the Internet. E-mail is fast and inexpensive. While one of the oldest components of the Internet, it is the most widely used and has amazing flexibility. With e-mail, it is possible not only to send messages, but also to execute programs and analyses on remote computers, retrieve data and files, and subscribe to news and informational services.

■ File Transfer Protocol

File Transfer Protocol or FTP is used to transfer files between computers. There are many FTP computer archives around the world that store documents, software, images, and audio files. Most of these FTP computer archives permit other computers to connect in an anonymous fashion, thus allowing the general public to retrieve files from the archives. By convention, anonymous FTP users log on to the host computer using ANONYMOUS as the user identification (userid) requested by the host computer and then provide their e-mail addresses when asked for a password. In this fashion, the archive managers can keep track of access to the FTP host servers. Two modes of file transfer exist: an ASCII mode, which is used to transfer plain text, and a binary mode to transfer all other files.

Because of the vast number of files that are stored in anonymous FTP archives around the world, additional tools have been developed to help locate and sort files. One of these tools, **Archie**, can be used to locate specific files stored in any one of the thousands of anonymous FTP archives around the world. Archie can be used interactively or via e-mail. To learn how to use Archie by e-mail, send a message containing the single word "HELP" to archie@ans.net. A more powerful tool for locating files in FTP archives is the Wide Area Information Server or WAIS. The concept behind WAIS is simple: to make anonymous FTP archives more accessible by indexing their contents for searching and browsing. WAIS servers are often linked to Gopher servers, permitting the use of Gopher to retrieve files located in FTP archives.

■ Usenet

Usenet Newsgroups are a discussion forum organized in a quasi-hierarchical way. Usenet is one of the oldest services of the Internet and has over 20,000 different discussion forums, encompassing a wide range of topics in the sciences, literature, current events, and social issues. Usenet is e-mail-based, allowing the transmission of news and information to a wide range of computers. Through Usenet, posted articles can be read and replied to, either in a broad fashion to the entire discussion forum or specifically to the author of the article.

■ Telnet

Telnet is based on UNIX, a multi-user/multitasking operating system often used on computer workstations. Through Telnet, a remote computer can be used to operate a distant computer almost transparently, as if the distant computer and the remote computer were one and the same. This allows the user of the remote computer to interactively control the distant computer to perform analyses of data or images or to execute programs.

■ Internet Relay Chat

Internet Relay Chat or IRC is simply an interactive (and live), online discussion group. When using IRC, real-time conferencing is possible. IRC is broken down by channels that have specific topics for discussion. As IRC is a flexible component of the Internet, it is possible to create specific channels for focused discussions on a topic, or to create broader, more open channels.

■ Gopher

Gopher was created in 1991 at the University of Minnesota as a campus-based information distribution system. The name Gopher is a double pun: first, Gopher is designed to "go for" information, and second, the University of Minnesota mascot is the "Golden Gopher." As the ability to search for, locate, and retrieve information on the Internet had not kept pace with the sheer volume of this information and its richness, Gopher quickly became a popular tool. Therein lies the basic principle behind Gopher: it is a system that is designed to make information both widely available and retrievable over a network.

To simplify and speed searches for files and information in Gopher servers, the search tool **Veronica** was developed. Veronica functions in a fashion analogous to Archie and FTP archives. Veronica searches through Gopher servers looking for files that match a keyword supplied by the user. It then assembles a list of Gopher servers that contain items of interest. It should be noted that Veronica searches the titles of files stored in Gopher servers, not the file contents. Like the name Gopher, the name Veronica is a double pun: first, playing on the concept of Gopher, the name means "Very Easy Rodent-Oriented Net-wide Index to Computerized Archives," and second, the name is a play on the comic book couple of Archie and Veronica.

Gopher is sometimes viewed as an ancestor of the World Wide Web (WWW or Web). Like the Web, Gopher offers a menu to browse in order to locate information. As with the Web, Gopher almost seamlessly links a wide range of Internet services. The ability of Gopher to access many different files, such as text, audio, and images, parallels many aspects of the Web. Three major differences exist between Gopher and the Web. First, Gopher is primarily a text-based interface to the Internet, while the Web is generally a graphic-based interface. Second, except for text, the information located using Gopher cannot be viewed on line in an interactive fashion as is the case with the Web. Third, and most importantly, Gopher organizes and retrieves information in a hierarchical fashion as opposed to a web-like fashion.

■ **The Web**

The Web is the newest of the Internet components and is rapidly becoming the most popular. Resources and information displayed on graphical-based Web browsers appear much like a newspaper or magazine page. Text, images, video, and sound can be presented. Information is linked together, much like a web (hence the name). This web-like linkage of information makes it possible to browse through vast amounts of information without specifying a starting point or a destination.

The importance of the Internet in biology

With its background in mind, note that while the Internet is broadly based and popularly slanted, it offers the biology community at large many useful if not indispensable resources. The Internet provides the ability to communicate with colleagues using e-mail, mailing lists, and discussion groups. Public databases can be browsed, and the information they contain can be accessed and retrieved for subsequent use. Files can be transferred between distant locations using any one of a number of resources on the Internet. Finally, data can be analyzed on remote computers from the convenience of a personal computer.

Thus, for biologists in general and scientists specifically working in genomic and molecular biology, the Internet has become a tool of critical importance. The reason for this is simple: the current pace of genomic sequencing and the nearly overwhelming volume of information that has resulted. An explosion of information is coming to the biological research community as a result of the Human Genome Project, the worldwide effort to determine the complete nucleotide sequence of all 26 chromosomes found in humans, as well as dozens of other related projects to determine the complete genomic sequences of organisms as divergent as bacteria, plants, and other animals. Unfortunately, much of the biological research community is not aware of the sophistication and depth of the sequence information and the analysis software that are available on the Internet. Instead, most labs spend significant sums of money purchasing dedicated software and hardware instead of relying on these readily accessible tools. The resources of the Internet that are directly applicable to molecular and genomic biology research have increased in parallel with the knowledge base of the field of genomic biology. No longer is sequence analysis the sole dominion of a small group of interested scientists. It is now possible, using the resources of the Internet in conjunction with the wealth of genomic data freely available, to perform sophisticated analyses of DNA, RNA, and protein sequences. All that is required is a personal computer and an e-mail connection to the Internet.

Because of this fundamental shift in access to informational technologies, along with improvements in the electronic tools for sequence analysis, it is possible that in the not too distant future much of the analysis of protein structure, sequence homologies, evolutionary relationships, and general annotation of the sequence data contained in repositories could be done by interested students and their instructors independently of the traditional research institutions. As a result, smaller universities, community colleges, and even high schools can enter the world of genomic biology, with the price of admission being simply a personal computer and a gateway to the Internet.

The Internet-based tools available for pursuing this new biology include sequence repository databases such as GenBank, literature resources such as Medline, and sequence analysis software such as BLAST, GenQuest, and NetServ. How capable and usable are these electronic tools? The answer is that these tools, even at the present, are incredibly powerful and relatively simple to access and use. You can ask relatively complex questions, do sophisticated analyses, and get answers back simply by sending e-mail to these servers.

On the downside, some tools are still missing and some of the currently implemented tools are less than ideal. But the beauty of these tools is that they are available now at negligible cost to an increasing community of scientists and lay people. As this community of new biologists grows and evolves, as the tools grow and evolve, the questions that can be asked by molecular and genomic biologists and by the educational and lay community will be limited only by imagination. In a very real sense, a new era in biology is dawning.

Scope and limits of this book

This book is not meant to be a guide to the history of the Internet, the story of computational biology, the history of sequence analysis and databases, or even the conceptual theory behind computational sequence analysis. This book is a guide to some of the resources available on the Internet in the broad area of molecular and genomic biology. It is focused on e-mail servers, Gopher servers, and the emerging World Wide Web. The reason for this is twofold. First, in an effort to maintain the focus of this document and achieve a manageable size, some sacrifices had to be made with respect to scope. Second, these three components of the Internet have the broadest selection of electronic tools for genomic and molecular biology; at the same time, they offer practicality to scientists rather than just computer specialists.

This book assumes that the reader either has an Internet connection or has access to one. Internet access can be obtained in a variety of ways: through a corporate or university-sponsored account, through a direct connection via an Internet services provider (ISP) company, or simply through an online service such as America Online (AOL) or Compuserve.

Further, this book is meant to serve a specific audience in the genomic and molecular biology community: the students and staff at schools, colleges, and universities that have (or can get) connections to the Internet and would like to get simple and straightforward answers to sequence analysis questions in a simple and straightforward manner. Only two things are needed to pursue sequence analysis via the Internet: a personal computer such as an Apple Macintosh or a PC-compatible system and a connection to the Internet via a modem, local area network, or direct connection.

Selected readings

Journal Articles

Cohen, D., I. Chumakov, and J. Weissenbach. 1993. A first-generation physical map of the human genome. *Nature* (London) **366**:698-701.

Collins, F., and D. Galas. 1993. A new five-year plan for the U.S. human genome project. *Science* **262**:43-46.

Gilbert, W. 1991. Towards a paradigm shift in biology. *Nature* (London) **349**:99.

Olson, M. V. 1995. A time to sequence. *Science* **270**:394-396.

Pennisi, E. 1996. From genes to genome biology. *Science* **272**:1736-1738.

Zehetner, G., and H. Lehrach. 1994. The reference library system: sharing biological material and experimental data. *Nature* (London) **367**:489-491.

Books and Monographs

Engst, A. C. 1993. *Internet Starter Kit for Macintosh.* Hayden Books, a division of Prentice Hall Computer Publishing, Indianapolis.

Falk, B. 1994. *The Internet Roadmap.* SYBEX, Alameda, Calif.

Gribskov, M., and J. Devereux (ed.). 1991. *Sequence Analysis Primer.* Stockton Press, New York.

Hahn, H. 1996. *The Internet Complete Reference,* 2nd ed. Osbourne McGraw-Hill, Berkeley, Calif.

Swindell, S. R., R. R. Miller, and G. S. A. Myers (ed.). 1995. *Internet for the Molecular Biologist.* Horizon Scientific Press, Norfolk, England.

U.S. Department of Health and Human Services and Department of Energy. 1990. *Understanding Our Genetic Inheritance: The U.S. Human Genome Project: The First Five Years.* National Technical Information Service, U.S. Department of Commerce, Springfield, Va.

Internet Publications

Smith, U. R. 1993. *A Biologist's Guide to Internet Resources.* Yale University, New Haven, Conn.

E-mail, Gopher, and the Web

## CHAPTER 2

- Comparison of the three most critical Internet tools for genomic biology
- Electronic mail (e-mail)
- Gopher
- The Web (WWW)

Comparison of the three most critical Internet tools for genomic biology: E-mail, Gopher, and the Web

While the vast Internet has seven major components, three are critical for molecular and genomic biology: e-mail, Gopher, and the World Wide Web. E-mail began with the creation of the Internet. Gopher appeared just as the Internet became a global reality. The Web represents the graphical interactive future of the Internet. Each has distinct advantages and disadvantages for scientists, depending on the type of Internet connection that is available and how much data or bandwidth that connection can support. E-mail is the most widespread of the Internet tools, and with it a surprising depth of information and range of services can be accessed. Gopher was the first true attempt to create a common, unified interface to the variety of information, resources, and services found on the Internet. The Web is an innovative tool that seamlessly integrates all of the services of the Internet under one common interface.

While other components of the Internet can be of use within the scientific community, these three have emerged as the most useful for the research community. What follows in this chapter is a brief overview of each of these three components and what they can offer the scientific community.

Electronic mail (e-mail)

Everyone is familiar with postal mail. Simply write a note on paper, place it in an envelope, seal it, address it, stamp it, and drop it in the mailbox. The Internet also offers a mail service. In contrast to the physical nature of postal mail, mail on the Internet is sent electronically, so it is called electronic mail or e-mail for short.

Conceptually, e-mail is like conventional postal mail except that it is delivered with much greater speed at a much lower cost. For example, postal mail can take several days to a couple of weeks to travel between countries; e-mail messages take seconds to minutes. As such, e-mail is the foundation of the Internet. Despite the fact that it is one of the oldest components of the Internet and that more exotic protocols such as the Web have appeared over recent years, e-mail remains the most widely used Internet-based tool. Further, anyone who is connected to or has access to the Internet will have e-mail capability, making it possible to access a wide range of Internet-based tools simply by sending a message. With only an e-mail account on the Internet, it is a simple matter to download files, analyze sequence data, query databases, or search for information (as we will demonstrate in succeeding chapters).

An e-mail message is composed of two parts: addresses and the message itself. All e-mail addresses use a similar structure:

```
name@domain
```

where *name* is the name of the Internet user and *domain* is the address of the mail server where the user has an account. The user name and domain or domains are separated by the "at" character. Mail servers support multiple users, from a few individuals to entire organizations. For example, the authors' e-mail account is:

```
annep@acs.auc.eun.eg
```

where annep is the name of the Internet user and acs, .auc, .eun, and .eg are domains of increasing hierarchy. The second part of an e-mail is the message itself. This can be any alpha-numeric character from the ASCII set (Appendix C). Putting it together, the structure of an e-mail is straightforward. An example is shown in Fig. 2.1.

Figure 2.1	An E-mail Message

```
From:      annep@acs.auc.eun.eg
To:        lenperuski@aol.com
Subject:   Travel plans
The flight reservations are confirmed. We will leave for the
States on 21 July and return the following week. I'll have the
tickets at the lab by the end of the day.

           Anne
```

Once posted by the sender, the message is sent to the recipient. If the recipient is logged on to the Internet when the message arrives, a notification of a new message will be displayed on the recipient computer. If the recipient is not logged on, then the notification of a new message will be held until the recipient logs on to the Internet.

As we stated above, Internet e-mail addresses are composed of at least two distinct parts: a user name and one or more hierarchical domains. The user name refers to a specific computer or, more specifically, to the software located on that computer. The hierarchical domains serve to locate the computer on the Internet, with the domains becoming more specific going from right to left in the Internet address. For example, in the e-mail message in Fig. 2.1, annep is the user name for the sender, followed by an @, then the domain names acs, .auc, .eun, and finally, .eg. The last domain name, .eg, refers to the country (in this case, Egypt). The next domain, .eun, refers to the site within Egypt (in this case, the Egyptian University Network), while the two domains immediately following the @ sign, acs and .auc, refer to the Academic Computing Service in the American University in Cairo. For the recipient of the above e-mail message, the domains are even simpler: aol and .com. The domain .com refers to a commercial site, while aol refers to a commercial on-line service. Some examples of domains are given in Table 2.1, but for a more complete listing of domains and their translations, turn to Appendix B.

Table 2.1.	Examples of Domain Names and Internet Addresses		
Countries		**Organizations (international)**	
Australia	.au	Academic	.ac
France	.fr	Commercial firms	.co
Germany	.de	General	.gen
New Zealand	.nz	Organizations	.org
South Africa	.za		
Spain	.es	**Organizations (specific to the U.S.)**	
Switzerland	.ch	Academic/Educational	.edu
United Kingdom	.uk	Commercial firms	.com
United States	.us	Government	.gov
		Military	.mil
		Organization	.org

Example Internet addresses

Specific to the U.S.:	retrieve@ncbi.nlm.nih.gov
International:	malign@nig.ac.jp

Normally, the recipient of an e-mail message is not on line when a message is sent. This means that the e-mail has to be stored until it can be read. A computer at the recipient's e-mail site, called a server, stores the e-mail. Generally, each e-mail user has a fixed amount of storage space on the server for messages. Once this space is full, no more messages can be stored. Thus, it is important to check an e-mail box on a regular basis, much like checking a postal mailbox. Failure to check for mail and retrieve it from the server will result in storage space full to capacity. When that event occurs, any subsequent e-mail received by the server is returned to the sender or "bounced back."

As with any tool, electronic or otherwise, learning how to properly use the tool permits more efficient work. When using e-mail, try to adhere to the following guidelines:

- **Use a text editor (like a word processor) to compose messages.** This permits the preparation of messages off line, which saves the resources of the Internet and allows you to compose a better message.

- **Exercise caution with attached files.** Internet e-mail is designed for simple text, not complicated word processing documents. A formatted file, graphics files, or database file will generally not travel well via Internet e-mail. While a growing number of e-mail servers can successfully process nontext files, do not assume that the recipient of a complex e-mail message is capable of receiving it.

- **Retain copies of important messages.** This is important for two reasons. First, it helps to keep track of details, and second, if it is necessary to resend the same message or a similar one, an electronic copy makes the task simple.

- **Delete old messages.** Do not waste resources. If it is necessary to save an old e-mail message, save it to a disk or even keep a printed copy. Do not leave it on the mail server to waste system resources.

- **Receive all e-mail at one account.** Avoid the temptation to have several different Internet accounts, or if this is unavoidable, have e-mail forwarded to one account. This makes it a simple matter to check one computer (or account) for messages.

- **Remember that e-mail is not necessarily private**. An e-mail message is usually read only by the person writing the message and the person receiving it. But it is possible for other parties to eavesdrop on or even intercept e-mail messages sent to other people.

Despite the advent of fancier and more diversified tools on the Internet, e-mail will remain a vital tool for genomic biologists. Beyond its obvious usefulness as a tool for personal communication, it can be used to access sophisticated computational resources with a minimum of network requirements. It has proven itself to be flexible and adaptable, limited only by the user's imagination. For example, while originally designed simply to transmit text-based data built around the ASCII code, it is capable of handling binary data as well. So, while e-mail is the most basic of all of the tools that form the Internet, it is by no means the least, nor is it in any danger of becoming obsolete.

Unlike e-mail, Gopher and the Web are menu-driven Internet applications. Further, Gopher and the Web are Internet applications that were developed by the user communities to simplify the accessing of information on the Internet. With that background in mind, let's look at the origins and differences of these two applications.

Gopher

The second of the critical Internet tools for molecular and genomic biologists is Gopher. Gopher offers the Internet community a number of advantages over other Internet protocols such as FTP or Telnet. Gopher is similar in some aspects to FTP in that it is designed to locate and retrieve information over a network; however, Gopher offers significant advantages. First, it places a friendlier, menu-driven interface over the complexities of the Internet services. Even with a text-based Internet client, Gopher makes it a much less intimidating process to search for and retrieve information. Second, Gopher encompasses a much wider variety of information than does FTP. FTP only allows the retrieval of files. With Gopher, it is possible to access catalogs, databases of information, e-mail directories, news, and files, accompanied by images, audio, and video clips. Further, all of this information is integrated under one interface on the computer and is accessible via the standardized menu-driven system of the Gopher client software. Using this menu-driven system, Gopher allows the user to transparently browse throughout all Gopher servers around the world in search of information of interest. Thus, the physical location of the information makes little difference in Gopherspace. All you need to know is what information you want, rather than what information you want and where it might be. Gopher then operates in a hierarchical or "tree-like" fashion; the user's computer is the trunk of the tree or its base. In Gopher, users can search through all of the branches of the tree until they find the "leaf" or specific information they need.

When retrieving a file via FTP, if another site is referenced besides the site being used, then the user has to connect to that additional site to retrieve any of its files. In contrast, when connected to a Gopher server, it is a simple matter to wander out into the domains of other Gopher servers around the world, which makes it quick and simple to locate and retrieve any information desired. Browsing in Gopher basically ignores the constraints of physical location inherent with FTP and Telnet. Further, Gopher uses less bandwidth and overall networking resources than does FTP or Telnet. With FTP or Telnet, the client computer must remain connected to the host throughout the operation, whereas with Gopher, the connections are brief — just long enough to retrieve the information — and then the client and server disconnect until the next request or task.

Several Gopher client software packages exist for the Macintosh and Intel-PC platforms. Among the best are those designed by the University of Minnesota Gopher programmers, the original developers of Gopher. They are fast and clean and make use of the normal operating system interface. Gopher can also be accessed by Telnet and the VT-100 interface, making Gopher a useful option for command line systems or networks with very limited resources. The drawbacks with this approach are that Gopher operating in this text-based environment is slower, not multi-task capable, and images cannot be viewed online.

When using a Gopher client software package, simply type in GOPHER for a text-based interface or click on the Gopher icon for a graphic interface. A menu will appear that can then be navigated by the use of arrow keys or a mouse. In this menu structure will be an assortment of information that can be searched and retrieved such as text files, directories, document collections, and multimedia items. Typically, four additional choices will also be present: a search directory, a Telnet session, a phone book directory, and a Gopher link. The **search directory** accepts keyword terms that will be used to locate items that match the search criteria. The **Telnet session** permits the Gopher client to connect to another computer system host at a remote location. This permits the Gopher client to browse other computer systems for information such as an on-line library catalog. The **phone book directory** searches for specific names, addresses, and e-mail addresses. The **Gopher link** is just that, a link to other Gophers around the world, making it possible to retrieve information from widely dispersed sources.

This ability to use Gopher to traverse the Internet worldwide in search of information is both powerful and confusing, in the sense that it is easy to get lost in the "sea" of information. In response to this limitation of Gopher, more powerful tools have evolved. It was not unexpected that Gopher would be supplanted by another Internet-based informational tool, the Web. What was surprising was how quickly this change happened. Even so, Gopher will remain a powerful search and retrieval tool on the Internet for the molecular and genomic biology communities, at least on text-based terminals.

The Web (WWW)

The Internet tool that supplanted Gopher for the browsing and retrieval of information is the Web (the World Wide Web or WWW). The Web had its beginnings in the early 1990s at the European Centre of Particle Physics (CERN) in Geneva, Switzerland, at almost the same time as the advent of Gopher. CERN is a scientific research institute that specializes in high-energy physics. The function of the Web for the physics community was as an electronic medium to make a wide range of information available in a linked fashion to other researchers. The key foundations of the Web are the range of accessible information and the linkage of the information. On the Web, information can range from text to sounds to images to line art.

In contrast to Gopher, which operates in a hierarchical or tree-like search fashion, the Web operates in a "spiderweb" fashion. All information is interconnected or linked via hypertext, which permits browsing in both specific and general directions and allows relational searches for information. Linking via hypertext is one of the wonders and frustrations of the Web. It permits the browsing of vast archives of information in a completely user-selected fashion. One can simply follow interesting links to wherever they lead. It is frustrating because it can be time-intensive, completely unstructured, and depending on the resources of the Internet user — namely the data capacity and speed of the Internet access — painfully slow.

Because of its graphical nature and ease of use, the Web has completely restructured the Internet and, in essence, opened the Internet to the masses. This makes the Web the future of the Internet for the general public and the research community. All of the pieces that would make the Web a truly ubiquitous Internet tool are not yet in place. These pieces include mature sites that are not transient in nature and sites that have limited scope and function. The primary limitation of the Web has been and remains the demands that it places on the infrastructure of the Internet. The Web is a resource-intensive tool. To be both graphical in nature and effective for use, it requires a fast and reliable connection to the Internet: 28 kilobaud is what most serious users would consider to be the minimum for an adequate connection when using a graphical browser such as Netscape Navigator or Microsoft Internet Explorer. Even when using a text-based browser such as

Lynx, 14 kilobaud is barely adequate. In most of the United States and much of Europe and Japan, networks are available that readily support this access speed. For most of the rest of the world, reliable network speed of this level is wishful thinking.

Navigating the Web

■ Web pages

All Web pages or sites have an address known as a Uniform Resource Locator (URL). This allows the page to be located on the Internet. Additionally, all Web pages are made up of a special language, called HyperText Markup Language (HTML). This language is made up of commands called tags, which define how the text and images look when displayed on a browser. When looking at a Web page with a graphics-based browser, some of the text appears in a different color or is underlined. Some of the images may have a colored frame around them. These components of the Web page indicate hypertext or a hyperlink. Hypertext or hyperlinks are text or images that are connected or linked to other Web pages or Web sites. If a hyperlink is selected by clicking on it with a mouse, then a different Web page containing information related to that hyperlink is displayed.

■ Clients and browsers

As with most of the other Internet access tools, use of the Web requires specialized client software, called a browser. Browsers are programs that read specially formatted documents on the Web and display them on the computer screen. Browsers are available for nearly every personal computer operating system. There are two basic types of browsers, graphics based (such as Netscape Navigator, Microsoft Internet Explorer, and NCSA Mosaic) or text based (such as Lynx).

A browser navigates the Web and permits connections to nearly all other types of Internet access services. To accomplish this, browsers make use of URLs to locate sites and resources. In this manner, by using URLs as a kind of "road map" to the Internet, it becomes possible to use a Web browser not just to surf the Web but to read Usenet news, send e-mail, access a Gopher site, or conduct an FTP session, all from the convenience of one client software program. Web users can control this navigation process by either accessing a specific web site through typing in the site-specific URL or by moving to different locations through selecting a hyperlink. Thus users can jump to other pages and sites via hyperlinks or URLs to follow their own intellectual curiosities.

■ URL addresses

Except for a few Internet services, the general form of a URL address is

```
URL://domain/directory:port/document
```

where URL is the service being accessed (WWW, FTP, e-mail etc.), domain is the address of the server to connect to, directory is the directory to access on the server, port is the port number to connect to, and document is the full name and extension of the document to view. Note that directory, port, and document may be omitted if the system has default settings. There are several standardized URL descriptions. Table 2.2 lists the currently accepted versions for Internet services.

Table 2.2 URL Descriptions for Accessing Internet Services Using a Web Browser

URL	Description
http://	**HyperText Transfer Protocol.** This description is used for accessing Web sites and is the default protocol for transferring the HTML documents that make up Web pages. An example of a valid URL is: `http://www.somehost.somewhere`
gopher://	**Gopher protocol.** This description is used for connecting to a Gopher service using a browser client. An example of a valid URL is: `gopher://gopher.service`

ftp://	**File Transfer Protocol.** This description is used for connecting to an FTP service using a browser client. Examples of valid URLs are:
	`ftp://ftp.a.host`
	`ftp://ftp.a.host/filename.txt`
	
	(retrieves the specified file)

news:	**Usenet protocol (without the //).** This description is used for reading a Usenet Newsgroup using a browser client. An example of a valid URL is: `news:alt.binaries.windows.shareware`
telnet:	**Telnet protocol (without the //).** This description is used for starting a Telnet session with a remote computer using a browser client. An example of a valid URL is: `telnet:192.56.23.2`
mailto:	**E-mail protocol (without the //).** This description is used for sending e-mail using a browser client. An example of a valid URL is: `mailto: annep@acs.auc.eun.eg`

■ Search engines

During the early years of the Web, there was no way to search it for specific sites or information. In response to this need, individual users of the Web began to collect addresses of homepages and display them as hyperlinks on homepages within the Web, making a primitive directory system. When this system became too large to be manageable, the first Web search engines were developed. Search engines are simply large databases of Web sites and forms for entering search criteria.

On current Web browsers, search engines and associated directories can be found in the search engine section of the Starting Points Command. Some of the more popular tools in this category are Lycos, WebCrawler, and Yahoo (described in more detail in Chapter 9). To use these tools, simply select a search engine or directory, type in the search criteria, and press the search button to begin.

As a result of its power and simplicity, most people now think of the Web as the Internet. It is a remarkable tool that allows the retrieval of limitless amounts of information. However, because of its vastness, it may be surpassing the capabilities of the infrastructure of the Internet. It will be interesting to see how this mode of information service will evolve.

Selected readings

Journal Articles

Germain, E. 1996. Fast lanes on the Internet. *Science* **273**:585-588.

Books and Monographs

Engst, A. C. 1993. *Internet Starter Kit for Macintosh*. Hayden Books (a division of Prentice Hall Computer Publishing), Indianapolis.

Falk, B. 1994. *The Internet Roadmap*. SYBEX, Alameda, Calif.

Hahn, H. 1996. *The Internet Complete Reference,* 2nd ed. Osbourne McGraw-Hill, Berkeley, Calif.

Internet Publications

Eggenberger, F. 1995. *WWW Introduction*. EMBnet, University of Basel, Switzerland.

Hayden, D. 1994. *Guide to Molecular Biology Databases*. School of Library and Information Studies, University of Alberta, Calgary, Alberta, Canada.

Hughes, K. 1994. *Entering the World-Wide Web: A Guide to Cyberspace*. Enterprise Integration Technologies, Palo Alto, Calif.

Rankin, R. 1996. *Accessing the Internet by E-Mail: Doctor Bob's Guide to Offline Internet Access*. Tillson, New York.

Smith, U. R. 1993. *A Biologist's Guide to Internet Resources*. Yale University, New Haven, Conn.

CHAPTER 3 An Internet server primer

- ■ Introduction to sequence analysis on the Internet
- ■ Server directives and parameters
- ■ Structure of a query to an analysis server

Introduction to sequence analysis on the Internet

We have arranged the chapters in this book to provide a protocol for sequence identification and structure/function analysis using a combination of servers. The information on how to use these tools is adapted, in part, from their original electronic help files. We strongly recommend that the original help files be retrieved and read prior to using any of these tools. There are two practical reasons for this recommendation. First, the resources and their help files are updated periodically. The help files will report how to use new features and will supply other pertinent information. Second, by retrieving the help file first, the Internet path to and from the e-mail server is tested. If there is a problem in accessing a server via the Internet, it is better to find out while retrieving a help file than while waiting for a reply to an important query. Unless otherwise stated, a help file can be retrieved for each of these servers by sending an otherwise blank message containing a single line with the word "HELP."

This chapter is an introduction to the theory and practice of using the Internet server for sequence analysis. For more detailed information on specific servers, refer to the appropriate sections in the following chapters and to the appendices. Of particular use to more experienced users is the summary listing of servers in Appendix A. It may be useful to pin a photocopy of this listing next to the computer terminal.

Sequence analysis on the Internet can be readily done via Web-based servers or e-mail servers. Sequence analysis by e-mail servers is useful for those scientists who do not have access to the Web or who do not have a fast enough connection to the Internet to make practical use of the Web. Most scientists have access to Internet e-mail and via this simple connection, they have ready access to a range of computational tools and databases. The only disadvantage — and a minor one at that — is that at busy times, the request will be queued and the user may have to wait a few hours for the results of an analysis.

It appears that sequence analysis through the Web or at least through Web client software is the future of this type of work. While e-mail servers will remain, they will in all likelihood be absorbed into or, at the very least, be linked to one or more Web sites. There are four main tasks that can be performed using the Internet-based analysis servers currently available: (1) comparing a sequence against the sequences in public databases; (2) analyzing a sequence in detail to search for motifs and clues to function; (3) retrieving sequence data from the public sequence databases; and (4) obtaining news and information from the scientific community.

Some of the servers that are described in the following sections are able to provide a wide variety of sequence analysis functions, ranging from simple database search and retrieval to powerful sequence analysis and comparison. Others offer resources for the determination of structure and function of either a nucleotide or an amino acid sequence. This means locating genes within a sequence, determining the protein expressed, identifying the motifs and domains in sequences, and when possible, assigning them a tentative function. There has been a great deal of progress in these methods over the last few years. For some species and groups of organisms, such as *Escherichia coli, Caenorhabditis elegans*, yeasts, and mammals, the methodologies used by some servers have progressed to the point that they not only can locate similar sequences or identify coding regions, but also can suggest the overall structure of genes and proteins. They all can be accessed via e-mail or, in most cases, via the Web. This means that a home or lab user equipped with a personal computer and a modem can have access to and control a worldwide assortment of extremely sophisticated sequence analysis tools.

There is a significant limitation attached to this power. The e-mail servers are computers programmed to respond in a specific and automated way to text messages received in a rigidly defined format. The message sent to the server is, in essence, a set of commands in a specific format defined by each server. Any deviation from this format can cause the server to send back a general help file to the user, return the original e-mail message unprocessed, send back an error message, or do nothing at all.

The first step in sequence analysis usually involves extracting the sequence in a computer-readable format for a variety of purposes including restriction analysis, primer, and oligonucleotide designs. Published sequences are usually submitted to a particular database and the sequence is given a unique identifier — the accession number — which makes it possible to extract the sequence by this identifier alone. Without an accession number, the user is compelled to search by using a keyword, author's name, citation, etc.

Overall, computational methods are sufficiently accurate that they can refine subsequent experimental approaches and give significant practical help in many projects of biological and medical import. However, the decision of which server to use and how to use it is not simple. The choice of a server or servers is dependent on many interrelated factors, including what organism is being studied, the type of sequence data (protein, genomic DNA, or cDNA), and how much effort the user is willing to expend. The ideal situation would be to have one server handle all of these needs, but that is unlikely in the immediate future.

Server directives and parameters

All servers are operated through the use of commands called **directives** which usually require user-entered modifiers (called parameters) in order to function. Directives are simply instructions telling a server what sequence to analyze, how to analyze it, and how and where to present the results. Directives are the foundation of Internet-based sequence analysis. They make it possible to automate the entire search, analysis, and reporting process, thus permitting great computational efficiency. Directives can be of several distinct types and have a variety of parameters: they can be mandatory or optional, require numerical or textual arguments, or be Boolean or non-Boolean in nature. Some directives are required in every search request, while others are optional and can alter the default behavior of the server.

Unfortunately, not all Internet servers use the same directive language. In many instances, different servers have directives that perform the same function, but are called by different names. The scientist who is the end user of these tools must be "multilingual" to use all the different directive dialects.

Fortunately, there are some underlying principles that govern the use of directives and parameters. The first principle is that a directive can be mandatory or optional. A mandatory directive must appear in every search request. Most e-mail similarity search servers have only two mandatory directives: PROGRAM and DATALIB. Sometimes one of two additional directives, either QUERY or BEGIN, but not both, must be specified. An optional directive does not have to be used in a query to a server in order to perform an analysis.

The second principle is that the parameters for a directive, mandatory or optional, fall into three classes: numeric, text, or Boolean. The parameter for a numeric-based directive must have a number value to complete the directive. An example is the EXPECT directive, which details the expected number of alignments that would occur by chance alone. After this directive, a number would be entered, such as 10 or 100. A text directive uses a word or abbreviation to complete the directive. An example is the DATALIB directive, which indicates which database to search. After this directive, an abbreviation for the desired database would be entered, such as NR for a Non-Redundant database. A Boolean directive requires a YES, NO, TRUE, FALSE, 1, or 0 to be present after the directive in order to complete the statement.

The following list gives some of the most commonly used server directives. They have been culled from the major servers such as BLAST or Retrieve. This list defines the purpose of the directive and explains the types of parameters needed to complete the directive. Some of these directives must be present in every search request; others are optional, but can be used to alter the default behavior of the server. Unless specifically stated as a mandatory directive, the listed directives are optional in query messages. Most servers will not use all of these commands. So, the list below is general in nature and provides an overview of the directives that can be part of a typical analysis query. For the specific directives that are used by each server, refer to the individual chapter sections.

ACKNOWLEDGE (Numeric) The server will acknowledge the receipt of a request only if the backlog of queries to be processed is longer than a specified time limit. The default value for this directive is generally 120 minutes. The time limit can be varied by entering the desired time in minutes after the directive. For example, to have the server acknowledge the receipt of a query, set the time limit to zero as in ACKNOWLEDGE 0. If this directive is used in a query, a parameter value for the time limit must be entered.

ALIGNMENT (Numeric) This directive restricts the number of matching sequences reported by a server. The usual default number of sequences reported is 50. If more sequences than this value are considered matches by the server, then only those that are the most statistically significant are reported. For example, to have the server report the best ten alignments, use ALIGNMENT 10. If this directive is used in a query, a parameter value for the number of sequences to be reported must be entered.

BEGIN

(None) This mandatory directive is one of the very few that has no associated parameter. It must be the last directive before the actual query sequence or text. The server uses this directive to identify the end of the directives and the start of an actual query or request. For some servers, the directives QUERY and BEGIN are interchangeable. In those cases, one of the directives, but not both, must be used.

CUTOFF

(Numeric) This is the cutoff score for reporting high-scoring segment pairs (HSP). The default value is calculated from the EXPECT value. HSPs are reported for a sequence only if their statistical significance is at least as high as that of a lone HSP having a score equal to the CUTOFF value. Higher CUTOFF values are more stringent, leading to fewer chance matches being reported. Typically, these thresholds can be more intuitively managed using the EXPECT directive.

DATALIB

(Text) The mandatory DATALIB directive is used to indicate which database should be searched. Only one database can be searched per query. Databases are generally searched in their entirety. For example, with the BLAST server, the following databases are available for analysis with a query sequence:

Peptide Sequence Databases

nr	Non-redundant protein database; it includes sequences from the PDB, SWISS-PROT, PIR, and translations from GenBank
month	All new or revised sequences from the PDB, SWISS-PROT, PIR, and translations from GenBank released in the last 30 days
swissprot	The last major release of the SWISS-PROT protein sequence database
pdb	Sequences derived from the three-dimensional structure Brookhaven Protein Data Bank
kabatpro	Kabat's database of sequences of immunological interest
acr	Ancient Conserved Region subset of SWISS-PROT
alu	Translations of Alu repeats. This database is designed for masking Alu repeats from query sequences.

Nucleotide Sequence Databases

nr	Nonredundant nucleotide sequence database; it includes sequences from PDB, GenBank, DDBJ, EMBL, but not the EST or STS databases.
month	All new or revised sequences from the PDB, GenBank, DDBJ, EMBL, but not the EST or STS databases, released in the last 30 days
pdb	Sequences derived from the three-dimensional structure Brookhaven Protein Data Bank
alu	Translations of Alu repeats. This database is designed for masking Alu repeats from query sequences.
vector	Vector subset of the GenBank database
kabatnuc	Kabat's database of sequences of immunological interest
est	Database of Expressed Sequence Tags (ESTs)
sts	Database of Sequence Tagged Sites (STSs)
epd	Eukaryotic Promoter database

DESCRIPTIONS	(Numeric) Restricts the number of short descriptions of matching sequences reported; default limit is 100 descriptions. If this directive is used in a query, a numeric value must follow the directive. For example, to restrict the number of sequence descriptions to 50, use DESCRIPTIONS 50.
END or //	(None) For a few servers, the mandatory END directive is critical because it indicates the end of a sequence to be analyzed. For those servers that use this directive, it is mandatory and has no parameter.
EXPECT	(Numeric) The EXPECT directive is the number of gap-free alignments expected by chance when searching the database with a query sequence. The default value for this directive is 10. If it is used in a query, a numeric value must follow the directive. For example, EXPECT 200 means that 200 matches satisfying the cutoff score would be expected to occur by chance alone. Both whole number and fractional values such as 0.1, 0.01, and 0.001 are usually acceptable.
FILTER	(Text) This optional directive filters a query sequence for regions of low complexity such as proline-rich sections of proteins, the poly(A) tails of DNA, or short internal repeats. The filtered query sequence retains the more complex regions. In this way, filtering can

eliminate the statistically significant but biologically uninteresting sequences that the server reports; filtering then allows the server to return what are hopefully more biologically interesting matches with sequences in the databases. If filtering is not wanted, simply put the directive FILTER NONE into the query.

GCODE   (Numeric) This directive can be used to select an alternate genetic code for translation of sequence data by the programs BLASTX, TBLASTN, and TBLASTX. The standard or universal code is 1. A complete list of codes for use with the BLAST server is given in Table 5.4. The code specified here is used to translate the query sequence in BLASTX and TBLASTX; for TBLASTN requests, the code is used to translate the database sequences. If this directive is used in a query, a parameter must be specified. For example, the directive GCODE 1 uses the universal code. See also Appendix D.

HISTOGRAM   (Boolean) Display a histogram of scores for each search. The default value is YES. To prevent the display of a histogram of scores in the results reported by a server, enter the directive HISTOGRAM NO.

KTUP   (Numeric) With some servers, this optional directive is used to set the degree of sensitivity of the search. The default value is generally 4 for searching DNA databases and 1 for protein databases. If using other values, set the value for KTUP in a range from 3 to 6 for DNA or RNA sequences and from 1 to 2 for protein sequences. For example, to set the KTUP to 5 in a query, enter KTUP 5.

MATRIX   (Text) This optional directive specifies the scoring matrix that will be used by the program in the analysis of the query sequence. Matrices are simply tables of accepted substitutions, insertions, and deletions within a sequence and the penalty values for permitting such events in a sequence comparison. Default matrices are always specified by the program, and for first-time server users it is best to use the default settings for this directive. Further, most users will not need to adjust the default settings to obtain excellent results. Details on alternative permitted matrices can be found in each of the respective server sections.

MAXDOCS   (Numeric) This optional directive is used in servers like Retrieve to set the maximum number of sequence documents to send back in response to a query. For the Retrieve server, the default value is 20 and the maximum value is 2,400.

MAXLINES   (Numeric) As with the directive MAXDOCS, this optional directive is used in servers like Retrieve. Its purpose is to set the maximum number of lines of output in a reply message. For the Retrieve server, the default value is 1,000 and the maximum value is 50,000.

PATH   (Text) This directive is used to provide a return Internet e-mail address if the server seems unable to correctly parse the e-mail address from the message header of a search request. If the server is not responding to an e-mail message request, try explicitly telling it the return e-mail address with this directive, for example, PATH ANNEP@ACS.AUC.EUN.EG. If used in a query, an argument is required for this directive.

PROGRAM

(Text) This mandatory directive specifies which type of search program to execute when comparing a query sequence to a database. For example, the BLAST server includes the following five programs; each program performs specific tasks:

- BLASTP: compares an amino acid query sequence against a protein sequence database

- BLASTN: compares a nucleotide query sequence against a nucleotide sequence database

- BLASTX: compares the six-frame conceptual translation products of a nucleotide query sequence (both strands) against a protein sequence database

- TBLASTN: compares a protein query sequence against a nucleotide sequence database dynamically translated in all six reading frames (both strands)

- TBLASTX: compares the six-frame translations of a nucleotide query sequence against the six-frame translations of a nucleotide sequence database

QUERY

(Text) This is an alternative directive used by some servers to specify the query sequences by locus name or accession number. This directive helps reduce Internet traffic and simplifies creation of search request messages using sequences already found in the databases. The QUERY directive can appear anywhere in the request. It is very important to note that either the QUERY or BEGIN directive, but not both, must be specified.

REPORT

(Text) This optional directive is used by some servers to modify the report returned by a server. The acceptable parameters are ALIGN, COMPRESS, NONE, SEGMENT, and SIMPLE. The default setting is generally SEGMENT. ALIGN gives a report that contains search parameters and the number of matches found, a one-line description of each match, and the multiple-sequence alignment of the query sequence against the databases matches. COMPRESS gives a compressed version of the ALIGN report. NONE returns a report that has the search parameters and the number of matches found and a one-line description of each match. SEGMENT generates a report similar to ALIGN, but with the query sequence aligned independently with each database match (unlike the multiple-sequence alignment of ALIGN or COMPRESS). Finally, SIMPLE gives a report that has a one-line summary of search parameters and the number of matches found, along with a one-line description of each match. If this directive is used in a query, then the parameter must be specified. For example, to have a simple report returned by the server, use the directive REPORT SIMPLE.

SCORES

(Numeric) A few servers use this optional directive to set the lower limit of similarity scores for a server to report. So, if the score for a sequence is greater than this value, it is reported; otherwise the sequence is not reported. The default value is usually 100. To change the default setting to another limit, such as 200, use the directive SCORES 200.

SPLIT

(Numeric) This optional directive is used to break large replies from the server into smaller messages. The SPLIT directive can thus be very useful if a computer network will not accept large messages. The use of a parameter for the SPLIT directive is optional because the default value for the directive is preset to 1,000 lines. For example, to use this directive to break a reply into messages of no more than 1,000 lines, simply put SPLIT into the query message. To change the size of the reply messages to a length of 500 lines, put SPLIT 500 into the query message.

STARTDOC

(Numeric) This optional directive is used by servers like Retrieve to set the starting database record for a server to return in response to a query. The default value is 1. This directive can be used in successive e-mail messages to retrieve blocks of records when a query to the server generates data that exceeds the limits of the MAXDOCS or MAX-LINES directives.

STRAND

(Text) This optional directive can either limit a TBLASTN search to just the top or bottom strand of the sequences in a database, or limit a BLASTN, BLASTX, or TBLASTX search to only the reading frames on the top or bottom strand of the query sequence. The default setting is to search both strands. The following parameters are acceptable for this directive: TOP, BOTTOM, PLUS, MINUS, +, -, COMPLEMENTARY, SINGLE, DOUBLE, or BOTH. The directive STRAND SINGLE means the same as STRAND TOP or STRAND PLUS. The directive STRAND DOUBLE is the default setting.

TITLES

(None) This is another of the very few directives that have no associated parameter. If this optional directive is included in a query message, only the titles of the sequence records will be sent in the reply e-mail message.

Structure of a query to an analysis server

Now that you understand the different e-mail components, we will move on to the practical aspects of preparing an e-mail query message; we will send a query to the BLAST server as an example. All of the examples we list in this book have the search directives and parameters in uppercase letters. Please note that this is only for clarity in the text. Servers are generally case insensitive; it does not matter whether upper or lowercase letters or both are used in any part of the query.

A search request consists of an e-mail message with a set of search parameters identifying the program, such as BLASTP for proteins or BLASTN for nucleic acids, the database to be searched, values related to the search parameters, and the query sequence to be used in the search. Generally, components of the message must be provided in this order: two mandatory directives, usually PROGRAM and DATALIB; any optional directives and their associated parameters; another mandatory directive, such as BEGIN or QUERY, to denote that the next lines of the message are sequence data to be analyzed; and finally, the query sequence on the subsequent lines. Blank lines within a query message are usually not permitted by servers because blank lines signal the end of a message. However, it is a good idea to have a blank line at the end of a query. This is because some e-mail programs automatically affix a signature to an outgoing message and the server that receives the query e-mail message needs to distinguish the end of a query from the signature.

In Fig. 3.1, the first four lines of the e-mail message are a mail header that is automatically created when addressing an e-mail message to a server. Usually nothing needs to be entered in the Subject line, but many people use this line of text to include identifying words or a sequence number. Some servers use the information in the Subject line as part of the reply message. This can be useful to sort out replies when sending queries to multiple servers.

Figure 3.1. Example of a Query Sent to the BLAST Server at NCBI

```
From:        annep@acs.auc.eun.eg
Date:        28 Jul 1992 21:29:02-
To:          blast@ncbi.nlm.nih.gov
Subject:     Analysis of unknown phosphatase subunit Q

PROGRAM BLASTN
DATALIB NR
EXPECT 0.7
BEGIN
>partial DNA sequence of phosphatase subunit Q
tgcttggctgaggagccataggacgagagcttcctggtgaagtgtgtttcttgaaatcat
caccaccatggacagcaaaagtcgttgcagtcgatgccagtcgaccgtagccagtatatt
```

The actual search request begins with the mandatory parameter PROGRAM in the first line followed by the value BLASTN (the name of the program) for searching nucleic acids. The next line contains the mandatory search parameter DATALIB with the value NR for the combined nucleic acids database. The third line contains an optional EXPECT parameter and the value for it. The fourth line contains the mandatory BEGIN directive, followed by the query sequence. Each line of information must be less than 80 characters in length. Note that each directive and its parameter are specified on a separate line.

Most server programs require a particular sequence format, have limits on the length and number of sequences that may be submitted for analysis, and — except in the case of database searches — are designed for only one or a few specific organisms. Most search servers require that the directive BEGIN, or its equivalent, be present in a query.

The query sequence should immediately follow the BEGIN directive and must appear in FastA format. A sequence in FastA format begins with a single-line description. The description line (usually required) begins with a greater-than (>) symbol, which distinguishes it from the lines of sequence data that follow it. In order to successfully pass through all computers that may need to relay the search request to a server, all lines of the sequence should be kept to less than 80 characters. Figure 3.2 shows an example sequence in FastA format.

Figure 3.2. Example of a Sequence in FastA Format

```
>Human Chromosome Y DNA, unknown function
ggttaagggaatggtgattttttatgctaaaaaaaaaaaattgtgtattttaccacttttt
tttttttaaggcagattcctttcaatcatctgagtgagcccagtgcgatctgaagggtcc
ctacaggtggaagaggcagtggccaggatcgttttttttttcccccccccccccccggt
```

E-mail and Web browser clients generally import a file containing a sequence into the query or insert by "cutting and pasting." Assuming the sequence is already in FastA format, the sequence file should be imported or pasted into the query on the line after the BEGIN directive. A blank line after the last line of the query sequence is recommended for inclusion with each request, since some clients automatically append a signature block to the messages they send; without the intervening blank line, a server could treat the signature block as part of the sequence itself.

In addition to adhering to FastA format, sequence data used in a query is expected to be represented in the standard IUB/IUPAC amino acid and nucleic acid codes. As we mentioned previously, most servers are case insensitive; the sequence data can be in either upper or lowercase or a mixture. If using a sequence that has been downloaded from a database as part of a query, then any numerical digits in the sequence should be removed. Appendix D lists the accepted nucleotide and amino acid codes supported by most servers.

Selected readings

Journal Articles

Altschul, S. F., W. Gish, W. Miller, E. W. Myers, and D. J. Lipman. 1990. Basic local alignment search tool. *J. Mol. Biol.* **219**:403-410.

Altschul, S. F. 1993. A protein alignment scoring system sensitive at all evolutionary distances. *J. Mol. Evol.* **36**:290-300.

Appel, R. D., A. Bairoch, and D. F. Hochstrasser. 1994. A new generation of information retrieval tools for biologists: the example of the ExPASy WWW server. *Trends Biochem. Sci.* **19**:258-260.

Pearson, W. R., and D. J. Lipman. 1988. Improved tools for biological sequence comparison. *Proc. Natl. Acad. Sci. USA* **85**:2444-2448.

Wilbur, W. J., and D. J. Lipman. 1983. Rapid similarity searches of nucleic acid and protein data banks. *Proc. Natl. Acad. Sci. USA* **80**:726-730.

Books and Monographs

Gribskov, M., and J. Devereux (ed.). 1991. *Sequence Analysis Primer.* Stockton Press, New York.

Swindell, S. R., R. R. Miller, and G. S. A. Myers (ed.). 1995. *Internet for the Molecular Biologist.* Horizon Scientific Press, Norfolk, England.

Internet Publications

BioSCAN Help Document. 1996. University of North Carolina, Chapel Hill.

BLAST FAQ List. 1996. National Center for Biotechnology Information, National Library of Medicine, National Institutes of Health, Bethesda, Md.

BLAST Help Document. 1996. National Center for Biotechnology Information, National Library of Medicine, National Institutes of Health, Bethesda, Md.

BLAST Manual. 1994. National Center for Biotechnology Information, National Library of Medicine, National Institutes of Health, Bethesda, Md.

Molecular and Genomic Databases

■ Primary sequence databases

| DDBJ | EBI | GenBank | GSDB |

■ Specialized databases

| Blocks | GDB | PDB | PIR |
| PROSITE | REBASE | SWISS-PROT | |

The numerous repositories for the archiving of primary nucleotide and amino acid sequence data can be somewhat confusing when it comes time to submit a sequence to a public database. Fortunately, all of the major primary databases cooperate with each other; that is, a sequence accepted at one database is shared freely with all of the other databases. **Accession numbers**, which identify submissions, are unique and used uniformly across these four major sequence databases: the DNA Data Bank of Japan (DDBJ), the European Bioinformatics Institute (EBI; formerly the European Molecular Biology Laboratory [EMBL] Data Library), GenBank (National Center for Biotechnology Information, U.S.A.), and the Genome Sequence Database (GSDB; National Center for Genome Resources, U.S.A.). In this chapter we will briefly describe the four major sequence databases, the services that they offer, and how to contact them. Later in the chapter, we will list and describe five principal specialized databases and how to contact them.

Primary sequence databases

Primary sequence databases can be loosely defined as those databases that store sequence data derived from the direct experimental characterization of a nucleic acid or protein. In other words, if a laboratory determined the sequence of a nucleic acid clone or a purified protein, then the resulting data would be submitted to and stored in one of these databases.

The primary databases can also contain additional sequence information (such as the predicted translated sequence of a DNA clone), putative sequence motifs (such as phosphorylation sites), promoter regions, or other such annotations within a submitted sequence. However, the principal thrust of these databases — or more accurately, databanks — is the cataloging, storage, and dissemination of the experimentally derived sequence data.

Most scientific journals now expect that the DNA and amino acid sequences that appear in articles have been submitted to the sequence databases prior to publication. Most journals, in fact, require an accession number for sequence data as a condition of publication. To simplify the submission of sequence data and the assigning of accession numbers, the sequence databases have established a number of submission channels, ranging from postal mail to e-mail. The fastest and most reliable route to submit sequence data to any of the databases is through e-mail or via the Web.

To facilitate authors' direct submission of data in an electronic format, the GenBank project developed the computer program **Authorin**. This program (for PC or Macintosh) is designed to help researchers prepare their sequence and associated comments or annotations for computer-readable submission to the sequence databases at EBI, DDBJ, NCBI, or GSDB. Authors may enter their data in any order and may make revisions at any time prior to submission. Partially completed entries may be saved and finished in a later session. In many fields, menus are provided to standardize terminology and reduce typing. Files generated by Authorin are simple text files that can be mailed electronically to a sequence database via the Internet or saved to a disk and sent via postal mail. Authorin software can be obtained free of charge from EBI or NCBI at the postal or Internet addresses for these organizations as listed in the sections below. Simply specify whether a PC or Macintosh version of the software is needed.

Each sequence submitted to a primary database receives a unique identification number (this is the accession number) issued by the database staff. Accession numbers are usually assigned within 24 hours of the receipt of the sequence data, before any processing of the submission begins. Processing of the submitted data can involve an extensive review of the sequence and its annotation, which can result in a lag between submission and the release of the data to the user community. This lag time varies between the databases, with delays of two weeks not uncommon. High-priority submissions may be processed in a single day, but this is rare.

Processed sequence data is generally released to the scientific community immediately. Many submitters of sequence data are concerned that release of this data before the appearance of a manuscript may compromise ongoing studies. For that reason, (upon request) the staff of the primary databases will withhold the release of sequence data until the publication of a manuscript.

The DDBJ, EMBL, and GenBank sequence databases are the foundation for the analysis power of the e-mail servers on the Internet. By serving as repositories and clearinghouses for most publicly available sequence information, these databases make it possible to rapidly compare query sequences against known sequence information and to retrieve sequence information of interest.

DDBJ

The DNA Data Bank of Japan (DDBJ) is a Japanese government-sponsored and maintained computerized repository of all reported nucleic acid sequences. DDBJ began service in 1986 in collaboration with the Japanese molecular biology community and the support of the Ministry of Education, Science, Sports, and Culture. The database is managed by the DDBJ, which collaborates with the other major sequence database repositories.

Internet addresses

E-mail:	ddbj@ddbj.nig.ac.jp	(Information)
	ddbjsub@ddbj.nig.ac.jp	(Data submission)
	ddbjupdt@ddbj.nig.ac.jp	(Data updates)
	fasta@nig.ac.jp	(FastA server)
	blast@nig.ac.jp	(BLAST server)
	malign@nig.ac.jp	(Multiple alignment server)
Web:	http://www.ddbj.nig.ac.jp	(DDBJ home page)
Organization:	DNA Data Bank of Japan (DDBJ), Center for Information Biology, National Institute of Genetics, 1111 Yata, Mishima, 411, Japan	
Telephone:	+81-559-81-6853	
Facsimile:	+81-559-81-6849	

Services

Sequence data can be analyzed for similarity to other sequences using the DDBJ e-mail servers BLAST and FastA. The DDBJ BLAST and FastA servers work the BLAST server at NCBI and the Mail-FastA server at EBI. In addition, DDBJ offers a multiple sequence alignment server, MAlign. (Use of these tools is described in detail in Chapter 5.)

Submission of sequence data

Sequence data can be submitted to the DDBJ database in several ways. Electronic files created by the program Authorin can be submitted by postal or electronic mail. More recently, a Web-based tool called **Sakara** was put into operation. Sakara is a forms-based sequence entry and annotation tool that can be accessed through the DDBJ home page.

EBI

The European Bioinformatics Institute (EBI) replaced the European Molecular Biology Laboratory (EMBL) Data Library in the fall of 1994 (Rodriguez-Tome et al. 1996). Like its predecessor, EBI is an EMBL-sponsored and maintained computerized repository of all reported sequences and a variety of software tools for their analysis (Emmert et al. 1994).

EBI's primary function is the development, curation, and distribution of both a comprehensive nucleotide sequence database (the EMBL Nucleotide Sequence database) and, in collaboration with Dr. Amos Bairoch of the University of Geneva, a companion protein sequence database, SWISS-PROT. In addition to these two sequence databases, EBI maintains over fifty specialty databases.

 Internet addresses

E-mail:	datalib@ebi.ac.uk	(Information)
	datasubs@ebi.ac.uk	(Data submission)
	netserv@ebi.ac.uk	(E-mail fileserver)
	nethelp@ebi.ac.uk	(Help system)
	update@ebi.ac.uk	(Data updates and corrections)
	blitz@ebi.ac.uk	(BLITZ e-mail server)
	fasta@ebi.ac.uk	(FastA e-mail server)
	prosite@ebi.ac.uk	(PROSITE e-mail server)
	retrieve@ebi.ac.uk	(Retrieve e-mail server)
Gopher:	gopher.ebi.ac.uk	(EBI Gopher server)
Web:	http://www.ebi.ac.uk	(EBI home page)
Organization:	EMBL Outstation, the EBI, Hinxton Hall, Hinxton, Cambridge CB10 1RQ, United Kingdom	
Telephone:	+44-1223-494400	
Facsimile:	+44-1223-494468	

■ Services

EBI also maintains an extensive, free series of network services. These include access to the primary and specialized databases maintained by EBI, collections of computer software and documentation, and sequence analysis tools that can be utilized by an e-mail server, Gopher, or the Web. The EBI network fileserver, NetServ, permits access via an e-mail message to all of the databases, software, and documentation maintained by EBI. (Use of the NetServ system is described in detail in Chapter 7.) Sequence data can be analyzed for similarity to other sequences by using the EBI e-mail servers BLITZ, Mail-

FastA, and Mail-PROSITE. (These tools are described in detail in Chapter 5.) To retrieve sequence records from databases, EBI offers two options. First, the Sequence Retrieval System (SRS) is a tool to query the EBI/EMBL databases using keywords to retrieve specific sequences. SRS is accessed via the Web through the EBI home page. (Use of this service is described in more detail in Chapter 9.) The second option is a satellite site for the Retrieve e-mail server, which mirrors the one offered by NCBI. (Use of this server is described in more detail in Chapter 5.)

The EBI Gopher server uses a simple graphical interface to streamline access to the available network services; it also offers links to other genomic biology resources throughout the world. The EBI Web server (described in Chapter 9) is the most wide-reaching of the network services offered by EBI. It permits access to all of EBI's resources including submission of sequence data, database access, similarity analysis, and database query and retrieval, along with links to other Web resources worldwide.

■ Submission of sequence data

Sequence data can be submitted to the EBI databases in several ways. First, printed forms can be filled out and mailed back to EBI. (The forms are available from EBI upon request or can be found in the first issue each year of the journal *Nucleic Acids Research*.) Second, electronic copies of the forms can be obtained from the EBI fileserver or from the EBI Web site. These forms can be filled out on a computer and sent back to EBI via e-mail or by postal mail. Third, the program Authorin can be used to submit data by disk or by e-mail. Finally, sequence data can be submitted via the Web using electronic forms available on the EBI home page.

GenBank

GenBank is an NIH-sponsored and maintained computerized repository of all reported nucleic acid sequences (Benson et al. 1996). Since 1992, the database has been managed by the National Center for Biotechnology Information (NCBI), which collaborates with the other sequence database repositories. GenBank is a registered trademark of the National Institutes of Health. The GenBank database and its associated files are offered to the public at no charge.

 ### Internet addresses

E-mail:	info@ncbi.nlm.nih.gov	(Information)
	gb-sub@ncbi.nlm.nih.gov	(Data submission)
	update@ncbi.nlm.nih.gov	(Data update)
	blast@ncbi.nlm.nih.gov	(BLAST server)
	retrieve@ncbi.nlm.nih.gov	(Retrieve server)
Web:	http://www.ncbi.nlm.nih.gov/	(NCBI Web site)
	http://www3.ncbi.nlm.nih.gov/omim	(OMIM Web site)
Organization:	National Center for Biotechnology Information, National Library of Medicine, Building 38A, Room 8S-803, 8600 Rockville Pike, Bethesda, MD 20894, U.S.A.	
Telephone:	+1-301-496-2475	
Facsimile:	+1-301-480-9241	

■ Services

NCBI also maintains an extensive, free series of services that are GenBank-based. Sequence data can be analyzed for similarity to other sequences using the BLAST e-mail server (described in detail in Chapter 5). To retrieve specific sequences from databases, the Retrieve e-mail server permits searches by accession number or text terms (this server is also described in Chapter 5). On the Web, NCBI offers access to the BLAST and Retrieve servers, submission of sequence data, links to other related Web sites, and access to the integrated database retrieval system **Entrez**. Entrez links the nucleotide and protein sequence databases with literature references from MEDLINE, three-dimensional structure data, and complete genome data.

Another valuable Web-based resource, **OMIM**, the Online Mendelian Inheritance in Man database, can be accessed through a link on the NCBI Web home page. OMIM is a catalog of human genes and genetic disorders originated by V. A. McKusick of Johns Hopkins School of Medicine. The Web version of this pioneering database contains text, pictures, and reference information with extensive links to NCBI's database of MEDLINE articles and sequence information.

In addition to these electronic services, a free subscription to *NCBI News*, a periodical describing the services and resources provided by the NCBI, may be obtained by sending an e-mail request, giving your name and a postal mailing address, to info@ncbi.nlm.nih.gov.

■ Submission of sequence data

As with other primary databases, sequence data should be submitted to GenBank in an electronic format. There are three ways to do this. First, NCBI has a Web site tool, called **BankIt**, for submission of sequence data. With BankIt, sequence data can be entered directly onto electronic forms that can be annotated, reviewed, and revised prior to submission to GenBank. Second, **Sequin**, NCBI's new stand-alone submission software for Macintosh, PC, and UNIX platforms, is also available. This cross-platform computer tool was developed to replace Authorin and to run on individual computers or over networks. When using Sequin, the output files for direct submission should be sent to GenBank by electronic mail. Third, the data files may be copied to a floppy disk and mailed to NCBI.

As previously mentioned, the program Authorin can still be used to format your submission, although submitters are encouraged to switch to either BankIt or Sequin. Authorin is compatible with both Macintosh and Intel-PC computers. Sequence data can be entered into electronic forms generated by this program, which in turn can be submitted to GenBank via e-mail or by sending a computer disk via postal mail. To obtain any of these software tools, send a request through the NCBI home page on the Web, send an e-mail request to the general information address listed above, or download them from the NCBI FTP site, which is accessible from the Web.

GSDB

The newest of the major international nucleotide sequence databases, GSDB (the Genome Sequence Database), is a publicly accessible relational database of DNA sequence and annotation maintained by the National Center for Genome Resources (NCGR) under an agreement with the U.S. Department of Energy (DOE) (Keen et al. 1996). NCGR is a private, not-for-profit corporation established in 1994 to manage resources and information developed by genome projects and related research.

 Internet addresses

E-mail:	ncgr@ncgr.org	(Information on NCGR)
	gsdbhelp@ncgr.org	(Information on GSDB)
	datasubs@gsdb.ncgr.org	(Data submission)
	update@gsdb.ncgr.org	(Data updates)
	websub@gsdb.ncgr.org	(Help with Web submission)
Web:	http://www.ncgr.org/gsdb/	(GSDB Web server)
	http://www.ncgr.org/gsdb/websub.html	
	(Data submission)	
	http://www.ncgr.org/gsdb/update.html	
	(Data updates)	
Organization:	National Center for Genome Resources (NCGR), 1800 Old Pecos Trail, Santa Fe, NM 87505, U.S.A.	
Telephone:	+1-505-982-7840	
Facsimile:	+1-505-982-7690	

■ **Services**

In cooperation with the other major DNA sequence databases, GSDB collects data directly from authors in many forms, including submissions via the program Authorin. GSDB also supports three new methods of data access, tailored to the needs of large-scale genome sequencing. First, the Web-based sequence and entry retrieval form locates entries by accession number or locus name. Entries retrieved through this method contain hyperlinks to other data collections. Second, an on-line data access program is available to registered users. This program permits 24-hour access to the database at GSDB for submission and retrieval of sequence information. Anyone connected to the Internet can obtain an account on the GSDB database and run the client program, Annotator's WorkBench, to enter and edit data. The third method currently being implemented is the use of the standard query language SQL as an interface to GSDB. Both example queries used by researchers as templates and customized queries will be permitted.

Specialized databases

Besides the four primary sequence databases, there are several other repositories that are important resources within the molecular and genomic biology community. The most important are, currently, the Blocks, GDB, PDB, PIR, PROSITE, REBASE, and SWISS-PROT databases. Each of these databases is more specialized than the primary sequence databases; they contain information that is not present or readily available from other sources. Each of these databases is derived from the information contained in the primary databases. Some of these specialized databases contain protein translations of the nucleic acid sequences contained in the primary databases. Others offer sets of patterns and motifs derived from sequence homologs in the primary databases that offer clues to structure and function. Because of this derived nature, it is not possible to submit new sequence data to a few of these databases, which does not mean that the specialized databases are less important or useful than the primary databases. Indeed, these specialized databases have taken on an increasingly important role in the determination of structure, function, and identification of unknown clones. (Most of the servers we describe in Chapters 5 and 6 make use of at least some of the information contained in these repositories.) With these comments in mind, we want to devote the balance of this chapter to a brief description of the five principal specialized databases, the services they offer, and how to contact them.

Each of these seven specialized databases is an adjunct to the four primary databases we described earlier in this chapter. However, the roles of Blocks, GDB, PDB, PIR, PROSITE, REBASE, and SWISS-PROT are crucial to the annotation and identification of query sequences and the determination of structure-function relationships. As the correlation between experimental analysis and theoretical analysis of sequences improves, the complexity and utility of these databases and their offspring will increase dramatically.

Blocks

The Blocks database is derived from the conserved domains found in individual entries of the protein families that make up the PROSITE database (Pietrokovski et al. 1996). The theory behind Blocks is that this database is a set of "distilled" sequence homologies from multiple-aligned sequences that reduce background noise and enhance the detection of distant sequence relationships. As such, the Blocks database aids in determining structure and function.

The entries that form the Blocks database are made in an automated three-step process. In the first step, the most highly conserved regions of protein families documented in the PROSITE database are identified. In the second step, these preliminary protein blocks are further refined to generate a set of protein blocks that is routinely present in the original protein family from which the blocks were derived. The protein blocks are then "calibrated" against the SWISS-PROT database to determine the level of matches that occur by chance alone. The Blocks database is made up of this completed set of protein blocks.

Internet Addresses

E-mail:	henikoff@howard.fhcrc.org	(General help/information)
	blocks@howard.fhcrc.org	(Server address)
	blockmaker@howard.fhcrc.org	(Block construction server)
Web:	http://blocks.fhcrc.org	
Organization:	Fred Hutchinson Cancer Research Center, 1124 Columbia Street, Seattle, WA 98104, U.S.A.	

Services

Blocks can be accessed by an e-mail server and via the Web, both of which permit pattern matching against the Blocks database and the retrieval of Blocks and PROSITE sequences and references. (Details on the use of the Blocks server and its associated tools can be found in Chapter 6.)

Recently a new service was added to the Blocks server, called Block Maker. In Block Maker, the user submits a set of protein sequences and generates a set of blocks based on those sequences to use in pattern searches. Block Maker generates protein blocks in the same automated fashion as those in the Blocks database. The only difference is that the reported protein blocks are not calibrated. The Block Maker server can be accessed by e-mail or via the Web.

GDB

Established at Johns Hopkins University in 1990, the Genome Database (GDB) is the official central repository for genomic mapping data from the Human Genome Project. GDB attempts to present the stored data in a manner that captures the complete biological picture. Due to the ease of the Web, GDB uses a generic Web browser to query the database. Most significantly, this system allows the scientific community to enter and edit the database directly and to annotate but not change the data submitted by others. This feature makes GDB the first public database to allow community curation and third-party annotation.

 Internet addresses

E-mail:	help@gdb.org	(General help/ information)
	data@gdb.org	(Data submission)
	mailserv@gdb.org	(E-mail server)
Web:	http://gdbwww.gdb.org/	(GDB Web server)
Organization:	Division of Biomedical Information Sciences, Johns Hopkins University School of Medicine, Baltimore, MD 21205, U.S.A.	
Telephone:	+1-410-955-9705	
Facsimile:	+1-410-614-0434	

PDB

The Protein Data Bank (PDB) is an archive of experimentally determined three-dimensional structural information about proteins, nucleic acids, and other biological macromolecules that serves a global community of scientists. The database contains a broad range of structural information including atomic coordinates, bibliographic citations, primary and secondary structure information, crystallographic structure factors, and nuclear magnetic resonance experimental data. With the expansion of the database to include the archiving of other biomolecules in addition to proteins, PDB will soon have a new name. Although not finalized, the database is expected to be called the Three-Dimensional Database of Biomacromolecules (3DB).

Internet addresses

E-mail:	pdbhelp@bnl.gov	(Information)
	pdb@bnl.gov	(Data submission)
	orders@pdb.pdb.bnl.gov	(Orders)
	errata@pdb.pdb.bnl.gov	(Data updates)
Gopher:	gopher://pdb.pdb.bnl.gov	(Gopher site)
Web:	http://www.pdb.bnl.gov	(PDB Web site)
Organization:	Protein Data Bank, Biology Department, Bldg. 463, Brookhaven National Laboratory, P.O. Box 5000, Upton, NY 11973-5000, U.S.A.	
Telephone:	+1-516-344-3629	(Main telephone)
	+1-516-344-6356	(Help desk)
Facsimile:	+1-516-344-5751	

Services

Because the exact three-dimensional arrangement of biomolecules dictates biological function, the PDB has become a major resource for molecular biological and genomic biology research. As a result, the PDB has significantly improved access to its information, primarily through an aggressive use of Internet access tools including e-mail and both a Gopher and Web site.

To simplify the submission of data and to improve the accuracy and annotation of the data, a Web-based submission tool, AutoDep, has been developed. AutoDep is designed to guide the user through the steps involved in data submission. Other services include Web-based database browsing tools to look at protein structures and to download structural data for analysis.

PIR

The PIR (the Protein Information Resource: an International Protein Sequence Database) traces its origin back to 1965 and the pioneering efforts of M. O. Dayhoff. Three centers maintain the database: PIR, at the National Biomedical Research Foundation (NBRF) in the U.S.A., the Martinsried Institute for Protein Sequences (MIPS) in Germany, and the Japan International Protein Information Database (JIPID) in Japan (George et al. 1996). PIR accepts direct submissions of protein-only sequence data and it comprehensively scans the literature for published conceptual and actual protein sequences.

The database is a repository of information on naturally occurring proteins whose amino acid sequence has been determined. Information in the database includes literature references; the name, classification, and origin of the protein; and the biological characteristics of the protein, such as expression and processing, domains and motifs within the protein, functional aspects, and primary sequence data. Given the wealth of information contained in the database, it is incorporated, at least in part, into many other databases such as SWISS-PROT and those maintained by NCBI and EBI.

Internet addresses

Americas

E-mail:	pirmail@nbrf.georgetown.edu	(General help/information)
	pirsub@nbrf.georgetown.edu	(Data submission)
	fileserv@nbrf.georgetown.edu	(E-mail server)
Web:	http://www-nbrf.georgetown.edu	
Organization:	PIR Technical Services Coordinator, National Biomedical Research Foundation, 3900 Reservoir Road, N.W., Washington, DC 20007, U.S.A.	
Telephone:	+1-202-687-2121	
Facsimile:	+1-202-687-1662	

Europe and Africa

E-mail:	mewes@mips.embnet.org	(General help/information)
Web:	http://www.mips.biochem.mpg.de	
Organization:	Martinsried Institute for Protein Sequences, Max Planck Institute for Biochemistry, D-82152 Martinsried, Germany	
Telephone:	+49-89-8578-2657	
Facsimile:	+49-89-8578-2655	

Asia and Australia

E-mail: tsugita@jipdalph.rb.noda.sut.ac.jp   (Information)

Organization: Japan International Protein Information Database, Science
                              University of Tokyo, 2669 Yamazaki, Noda 278, Japan

Telephone: +81-471-239778

Facsimile: +81-471-221544

PIR-International maintains Internet access to the database both by an e-mail server and via the Web. Both the e-mail server and the Web site have database searching and retrieval tools along with some analysis tools. (These services are described in more detail in Chapters 5 and 9.)

PROSITE

PROSITE is a compendium of protein profiles and patterns which can be useful in assigning functions or annotating protein sequences (Bairoch et al. 1996). The database is maintained by Dr. Amos Bairoch at the University of Geneva, Switzerland. PROSITE can be obtained on CD-ROM from the EBI or via FTP from various servers, including those maintained by NCBI, EBI, and DDBJ, as well as through the ExPASy server on the Web. For additional information, contact Dr. Bairoch at the address listed below or send an e-mail message to the EBI fileserver, NetServ. The body of the message should contain a line of text with the directive HELP, followed by a second line with the directive HELP PROSITE.

PROSITE is the most complete of all of the protein pattern and profile databases. As such, it is the de facto "gold standard" for the analysis of a protein sequence to determine functional domains or relationship to protein families. The strength and utility of PROSITE comes from the completeness and depth of its entries, the high specificity of the profiles and patterns in the dataset, the thorough documentation of the entries, and the level of curation of the dataset.

Further evidence for the stature of PROSITE within the research community comes from the fact that so many other Internet-based servers use it and so many related databases have been developed from its foundations. The list includes the BLOCKS server in the U.S., MotifFinder in Japan, and the EBI PROSITE Server. Every major protein motif, profile, or pattern database is derived, at least in part, from PROSITE. All of these aspects of the PROSITE database make it an essential tool in the structural and functional analysis of a protein sequence.

 Internet addresses

E-mail:	bairoch@cmu.unige.ch	(General help/information)
	prosite@ebi.ac.uk	(PROSITE server)
Web:	http://expasy.ch	(ExPASy server)
Organization:	Dr. Amos Bairoch, Department of Medical Biochemistry, University of Geneva, 1 rue Michel Servet, 1211 Geneva 4, Switzerland	
Telephone:	+41-22-784-4082	
Facsimile:	+41-22-702-5502	

REBASE

REBASE, the Restriction Enzyme Database (Roberts and Macelis 1996), contains detailed information about restriction enzymes, methylases, the microorganisms from which they have been isolated, recognition sequences, cleavage sites, methylation specificity, the commercial availability of the enzymes, and references — both published and unpublished observations — dating back to 1952. The database and its associated files are offered to the public at no charge, but are under copyright protection assigned to New England BioLabs.

Each month an updated set of data files is released publicly and distributed to the scientific community at no charge via a monthly e-mail message list. For placement on this list, send name, postal address, telephone number, fax number, and an e-mail address to the e-mail address listed below. These data files are flat ASCII text files, many of which are designed for use with a variety of commercial software packages such as the GCG suite, DNA Strider, GENPRO, Staden, PC/Gene, SEQAIDII, and others. In addition to lists of restriction enzymes in formats compatible with most DNA sequence analysis software, a variety of other information and data files are available from REBASE via the Web.

Internet addresses

E-mail:	roberts@neb.com	(Information/mailing lists)
	macelis@neb.com	(Information/mailing lists)
Web:	http://www.neb.com/rebase	(Home page)
Organization:	New England BioLabs, 32 Tozer Road, Beverly, MA 01915, U.S.A.	
Telephone:	+1-508-927-5054	
Facsimile:	+1-508-921-1527	

SWISS-PROT

The SWISS-PROT database is an annotated, cross-referenced protein sequence database that contains entries from a variety of sources including translations of nucleic acid entries from the major international sequence databases (Bairoch and Apweiler 1996). The database was established in 1985; since 1987 it has been maintained jointly by the University of Geneva, Switzerland, and the EBI Data Library. SWISS-PROT can be obtained on CD-ROM from EBI or via FTP from various servers, including those maintained by NCBI, EBI, and DDBJ, as well as through the ExPASy server on the Web. For additional information, contact Dr. Bairoch at the address listed below or send an e-mail message to the EBI fileserver, NetServ. The body of the message should contain a line of text with the directive HELP, followed by a second line with the directive HELP PROT.

Information in SWISS-PROT comes from three main sources: translations of DNA sequences from the EMBL Nucleotide Sequence Database, direct submissions from the research community, and from the scientific literature. The database distinguishes itself from other protein sequence databases by four distinct criteria: annotation, minimal redundancy, integration with other databases, and documentation. These features make SWISS-PROT a highly useful tool to look at structure-function relationships, homology and similarity, and the evolution of protein families. Many other protein databases, such as PROSITE, Blocks, and PRINTS, are derived from SWISS-PROT. The richness and depth of annotation of the annotations in SWISS-PROT requires time and care to assemble and results in a database that lacks the currency of the EMBL Nucleotide Sequence Database or GenBank, for example. To address this limitation, SWISS-PROT has recently developed a new database subset, TrEMBL (Translation of EMBL), as a supplement. TrEMBL is a direct, unannotated translation of the EMBL Nucleotide Sequence Database. The combination of SWISS-PROT and TrEMBL gives a more complete protein sequence set, much like the GenPept-GenBank partnership.

For purposes of standardization and close integration with the other EBI-maintained databases, the format of SWISS-PROT entries closely adheres to that for the EMBL Nucleotide Sequence Database.

 Internet addresses

E-mail:	bairoch@cmu.unige.ch	(Information)
Web:	http://expasy.ch	(ExPASy server)
Organization:	Dr. Amos Bairoch, Department of Medical Biochemistry, University of Geneva, 1 rue Michel Servet, 1211 Geneva 4, Switzerland	
Telephone:	+41-22-784-4082	
Facsimile:	+41-22-702-5502	

Selected readings

Journal Articles

Bairoch, A., and R. Apweiler. 1996. The SWISS-PROT protein sequence data bank and its new supplement TrEMBL. *Nucleic Acids Res.* **24**:21-25.

Bairoch, A., P. Bucher, and K. Hofman. 1996. The PROSITE database, its status in 1995. *Nucleic Acids Res.* **24**:189-196.

Benson, D. A., M. Boguski, D. J. Lipman, and J. Ostell. 1996. GenBank. *Nucleic Acids Res.* **24**:1-5.

Emmert, D. B., P. J. Stoehr, G. Stoeser, and G. N. Cameron. 1994. The European Bioinformatics Institute (EBI) databases. *Nucleic Acids Res.* **22**:3445-3449.

Fasman, K. H., S. I. Letovsky, R. W. Cottingham, and D. T. Kingsbury. 1996. Improvements to the GDB Human Genome Data Base. *Nucleic Acids Res.* **24**:57-63.

George, D. G., W. C. Barker, H.-W. Mewes, F. Pfeiffer, and A. Tsugita. 1996. The PIR-international protein sequence database. *Nucleic Acids Res.* **24**:17-20.

Henikoff, S., and J. G. Henikoff. 1991. Automated assembly of protein blocks for database searching. *Nucleic Acids Res.* **19**:6565-6572.

Henikoff, S., and J. G. Henikoff. 1992. Amino acid substitution matrices from protein blocks. *Proc. Natl. Acad. Sci. USA* **89**:10915-10919.

Keen, G., J. Burton, D. Crowley, E. Dickinson, A. Espinosa-Lujan, E. Franks, C. Harger, M. Manning, S. March, M. McLeod, J. O'Neill, A. Power, M. Pumilia, R. Reinert, D. Rider, J. Rohrlich, J. Schwertfeger, L. Symth, N. Thayer, C. Troup, and C. Fields. 1996. The Genome Sequence DataBase (GSDB): meeting the challenge of genomic sequencing. *Nucleic Acids Res.* **24**:13-16.

Pietrokovski, S., J. G. Henikoff, and S. Henikoff. 1996. The Blocks database — a system for protein classification. *Nucleic Acids Res.* **24**:197-200.

Roberts, R. J., and D. Macelis. 1996. REBASE-restriction enzymes and methylases. *Nucleic Acids Res.* **24**:223-235.

Rodriguez-Tome, P., P. J. Stoehr, G. N. Cameron, and T. P. Flores. 1996. The European Bioinformatics Institute (EBI) databases. *Nucleic Acids Res.* **24**:6-12.

Books and Monographs

Abola, E. E., F. C. Bernstein, S. H. Bryant, T. F. Koetzle, and J. Weng. 1987. Protein data bank, p. 107-132. *In* F. H. Allen, G. Begerhoff, and R. Sievers (ed.), *Crystallographic Databases Information Content, Software Systems, Scientific Applications.* Data Commission of the International Union of Crystallography, Cambridge.

McKusick, V. A. 1994. *Mendelian Inheritance in Man. Catalogs of Human Genes and Genetic Disorders*, 11th ed. Johns Hopkins University Press, Baltimore, Md.

CHAPTER 5 Search and Retrieval

■ Overview

■ **Search servers:** for comparison of nucleotide and protein sequence data against international sequence databases

BLAST	BioSCAN	Blitz	DDBJ
GenQuest	Mail-FastA	Mail-QUICKSEARCH	

■ **Retrieval servers:** for retrieval of sequence data and other relevant files from international sequence and genomic databases

GENIUSnet/NetServ	Query	Retrieve	STS and EST

Overview

In Chapter 5 we discuss servers. In the first half of the chapter, entitled "Search," we describe the servers designed to analyze query sequences against the international sequence databases. We will discuss several electronic analysis tools that use a range of algorithms to perform analyses, including the BLAST server at NCBI, the multi-purpose GenQuest, the DDBJ servers for FastA, BLAST, and multiple sequence alignment analysis, and the EBI/EMBL servers (BLITZ, Mail-FastA, and Mail-QUICKSEARCH), along with the highly useful server BioSCAN. These servers were designed to answer the question, "Does this sequence exist within a database and if so, how many other sequences related to it are present?" To answer this question, the servers take a nucleotide or protein sequence and search it against the major international sequence databases (or some subset of them) to determine if there are any potential matches. Of these servers, the BLAST server is arguably the best known and most widely used, but the other servers we list also have unique strengths.

In the second half of the chapter, "Retrieve," we describe the sequence retrieval servers. The function of these servers is complementary to that of the search servers; they were designed to locate specific sequence data records and return them to the requester. We will describe four different servers in this category: the flat-file retrieval servers, typified by the GENIUSnet and NetServ sequence servers; the Query and Retrieve servers, which were designed to locate sequence records as well as other sequence-relevant information such as literature references; and the STS and EST report servers, which use the Retrieve server to access sequence information from the growing EST and STS databases.

We list multiple Internet addresses for each of the servers. Each server can have up to four types of addresses. We list the address for the e-mail server first; this address is for an automated system that handles both analysis queries and help requests (it can automatically return a set of documentation on using the server, called a help file). As this system is automated, it can understand only a limited set of directives and associated parameters. The second e-mail address is designed to handle requests for additional information on the use of the server or specific questions that are not covered in the help file. Queries to this address are processed by a staff member of the organization that operates the server. (When sending a query or request to the server staff, try to explain the request in clear and specific terms.) The third server address, if it is available, is the Gopher site. Not many of the servers have Gopher sites, but they are a useful and interactive way to use the tools that the servers provide. The final server address is the URL for the Web site. Nearly all of the servers now have a Web site that offers the functionality of the e-mail version of the server and makes the analysis process a more user-friendly and interactive one.

Because of the increasing importance of Web-based tools in sequence analysis, it may seem unnecessary to learn the directives and parameters required to send queries to the e-mail versions of these servers. That assumption is short-sighted and limiting. Besides the utility of being able to use e-mail for sequence analysis and retrieval, learning the directives and parameters for the servers then allows you to learn how to construct queries for submission to the e-mail version of the server and to modify the search directives and parameters for the Gopher and Web versions of the servers. To aid in this learning process, the information for each server is adapted both from the help documentation and from actual use of the systems. Many of these servers use the same or a very similar set of directives and parameters to execute the analysis of submitted queries, but have some subtle, but critical differences. Therefore, before you decide to use a certain server, read its section carefully to avoid errors.

Search

With the exception of the BLAST server, we list the search servers in alphabetical order, starting with BioSCAN and ending with Mail-QUICKSEARCH. The BLAST server leads the chapter because so many of the other servers are dependent on it for their function and utility. Hence, understanding how to use the BLAST server gives the foundation for a much broader knowledge of analysis tools. BioSCAN offers a solid set of analysis tools that are particularly appropriate for the analysis of large tracts of genomic sequence as well as smaller regions. This server has been used extensively in the analysis of some of the first complete bacterial genomes elucidated. In contrast, the BLITZ server offered by EBI is designed for a single task: the high-speed analysis of sequence data using the Smith-Waterman algorithm. DDBJ offers three different servers: one for FastA analysis, one that uses BLAST, and a third that performs multiple sequence alignments. (The first two servers are similar to the EBI Mail-FastA and NCBI BLAST servers, but have a somewhat different directive structure.) The multiple sequence alignment server MAlign is a simple and cleanly designed tool to generate multiple sequence alignments of protein or nucleic acid sequences, along with evolutionary trees and putative ancestral and consensus sequences. GenQuest, an analysis from the Oak Ridge National Laboratory, is one of the few servers that can search multiple databases in a single query. It can search databases using algorithms such as Smith-Waterman, FLASH, BLAST, and FastA or perform pattern and profile analysis with Blocks. EBI offers a couple of other servers for sequence analysis: Mail-FastA and Mail-QUICKSEARCH. These two servers are based on well-established algorithms and have proven to be consistent and reliable tools.

Each of these search servers has different strengths, and they should be viewed and used as complementary tools. With this in mind, one hypothetical approach to the analysis of a new piece of sequence data would be to first submit the data to either the BLAST server at NCBI or to BioSCAN for a quick screening against a nonredundant database, then follow with GenQuest, BLITZ, or the Mail-FastA server to confirm the results of the first analysis and look in more detail. Newcomers to Internet-based sequence analysis tools should try each of the servers to learn not only how to use them but also the particular strengths of each one. A few of these servers will become favorite tools, but all of them are simple and powerful methods for the comparison of sequence data against the major public sequence databases.

BLAST

The BLAST (Basic Local Alignment Search Tool) server performs a similarity search of a sequence against public databases using one of five variants of the BLAST algorithm (Altschul et al. 1990). The server accepts queries containing a nucleotide or protein query sequence. A search of the sequence data is then performed against the specified database using the BLAST algorithm, and the results are returned in an e-mail message.

The BLAST algorithm was developed by the National Center for Biotechnology Information (NCBI) at the National Library of Medicine; it is designed for finding un-gapped, locally optimal sequence alignments. A wide range of e-mail servers available to the molecular and genomic biology community are based on or make use of the BLAST algorithm. The BLAST family of programs employs this algorithm to compare an amino acid query sequence against a protein sequence database, or a nucleotide query sequence against a nucleotide sequence database, as well as other combinations of protein and nucleic acid comparisons. The BLAST programs were tailored for sequence similarity searching — or, simply put, to identify homologs to a query sequence. The programs are not generally useful for motif-style searching.

There are several entities that support the BLAST server. In this section we describe EBI and NCBI versions; we will describe the version offered by DDBJ later in the chapter. The BLAST servers that are supported by EBI and NCBI are identical in function and use the same parameters. The BLAST server supported by DDBJ, while it uses the same algorithm as EBI's and NCBI's BLAST server, uses a slightly different set of directives to format a query, which is why the DDBJ server is described later. The reason for multiple servers at different locations is due to their demand for use; if one server is busy or slow, you can use a server at a different location.

Because the BLAST family of algorithms has such widespread use within the sequence analysis community, the "Search" section of this chapter is key to learning how to make effective use of many of the other servers described in this book. Servers ranging from those used to search sequence databases (like BioSCAN, DDBJ's servers, and GenQuest) to those used to determine structure and function relationships (like Domain, GRAIL, PSORT, and SBASE) all make use of BLAST-based routines to accomplish at least some of their analysis functions. So, even if you have no intention of using the BLAST server, it is important that you review this section.

 Internet Addresses

E-mail:	blast@ncbi.nlm.nih.gov	(NCBI server)
	blast@ebi.ac.uk	(EBI server)
	blast-help@ncbi.nlm.nih.gov	(Personal help)
Web:	http://www.ncbi.nlm.nih.gov	(NCBI home page)

Organization:	National Center for Biotechnology Information, National Library of Medicine, Building 38A, Room 8S-803, 8600 Rockville Pike, Bethesda, MD 20894, U.S.A.
Telephone:	+1-301-496-2475
Facsimile:	+1-301-480-9241

■ Accessing the BLAST e-mail server

To access the server, send a properly formatted e-mail message to one of the server addresses listed above. To receive the current set of instructions on using the server, send a help message to the server. The message must contain the word HELP on a single line in the body of the message. For answers to questions not adequately addressed in the help text, send an e-mail message with a detailed question or description of a problem to the NCBI staff at the Internet personal help address listed above.

■ Server directives and parameters

As we discussed in Chapter 3, a variety of directives govern the operation of the BLAST server. They can be considered either mandatory or optional, with numeric, text, or Boolean parameters. Table 5.1 lists the directives and their associated parameters supported by the BLAST server.

Table 5.1. BLAST Directives and Parameters

Directive	Attributes	Explanation
ACKNOWLEDGE	Optional, numeric	The default value is 120 minutes.
ALIGNMENTS	Optional, numeric	The default limit is 50.
BEGIN	Mandatory, none	It must appear after all other parameters and immediately before the FastA-formatted query sequence.
CUTOFF	Optional, numeric	The default value is calculated from the EXPECT value.
DATALIB	Mandatory, text	Available databases are listed in Tables 5.2 and 5.3.
DESCRIPTIONS	Optional, numeric	The default limit is 100 descriptions.
EXPECT	Optional, numeric	The default value is 10. Fractional values are acceptable.

continued

Table 5.1 continued

Directive	Attributes	Explanation
FILTER	Optional, text	Filtering is applied to the query sequence by default. To turn off filtering use the parameter NONE. For BLASTN the default filter is DUST; no other filters are allowed. For all other programs the default filter is SEG; other filter options are XNU, SEG+XNU, and XNU+SEG.
GCODE	Optional, numeric	The default code is 1. A complete list of alternate codes is given in Table 5.4.
HISTOGRAM	Optional, Boolean	The default is YES.
HTML	Optional, Boolean	Puts the results from the server in HTML format, suitable for viewing by a Web browser. The default setting is NO.
MATRIX	Optional, text	The default matrix is BLOSUM62. The valid alternative choices include PAM40, PAM120, PAM250 and IDENTITY. No alternate matrices are available for BLASTN; specifying the MATRIX directive in BLASTN requests returns an error response.
NCBI_GI	Optional, Boolean	Shows the unique GI sequence identifier assigned by the NCBI. Default setting is NO.
PATH	Optional, text	The default value is the e-mail address of the sender of the query. Any alternate e-mail address can be specified.
PROGRAM	Mandatory, text	The parameters supported are BLASTP, BLASTN, BLASTX, TBLASTN, or TBLASTX.
SPLIT	Optional, numeric	The default value is 1,000 lines per message.
STRAND	Optional, text	The default argument is DOUBLE. This directive is ignored for protein searches. The argument to this directive should be chosen from the vocabulary TOP, BOTTOM, PLUS, MINUS, +, -, COMPLEMENTARY, SINGLE, DOUBLE, or BOTH. Specifying STRAND SINGLE is equivalent to STRAND TOP or STRAND PLUS, while STRAND DOUBLE is equivalent to searching both strands.

■ **Formatting a query**

Components of an e-mail message to the BLAST server must be provided in this order: first, the two mandatory directives, PROGRAM and DATALIB, should be listed. This is followed by the desired optional directives and parameters. Finally the mandatory directive BEGIN is inserted, followed by the query sequence itself. Each directive must be specified on a separate line of the query and only one sequence is allowed in a query message. Each line of information in the query message must be less than 80 characters in length.

There are five BLAST parameters that can be chosen by the PROGRAM directive, as follows.

- **BLASTP:** compares an amino acid query sequence against a protein sequence database

- **BLASTN:** compares a nucleotide query sequence against a nucleotide sequence database

- **BLASTX:** compares the six-frame conceptual translation products of a nucleotide query sequence (both strands) against a protein sequence database

- **TBLASTN:** compares a protein query sequence against a nucleotide sequence database dynamically translated in all six reading frames (both strands)

- **TBLASTX:** compares the six-frame translations of a nucleotide query sequence against the six-frame translations of a nucleotide sequence database

The DATALIB directive defines which sequence database will be used for comparison against the query sequence. The protein sequence databases that can be used with the NCBI BLAST server are shown in Table 5.2, while Table 5.3 shows the nucleotide sequence databases that can be used with the NCBI BLAST server. Note that the NR databases are composites of multiple databases, produced by merging the annotations of identical sequences from the different database sources into single entries. It is recommended that the NR database be used for both protein and nucleic acid database searching because the search is more comprehensive and it provides more concise reports.

After the DATALIB directive, any of the optional directives listed in Table 5.1 may be used in the query message. If an optional directive is used, it must be followed immediately by the associated parameter value. If an optional parameter is not used in the query message, then the server will use the default settings in the analysis of the query sequence. (We discussed mandatory and optional directives in greater detail in Chapter 3.) After any optional directives comes the last directive, BEGIN. The BEGIN directive must be followed on the next line of the query message with the text of the query sequence in FastA format.

■ **Examples of queries**

Two examples of queries to the BLAST server are shown in Fig. 5.1 and 5.2. Figure 5.1 shows a simple query using only the default settings for the BLAST server. The first four

Table 5.2. BLAST Peptide Sequence Databases

Database Name	Description
NR	Nonredundant protein database; it includes sequences from the PDB, SWISS-PROT, PIR, and translations from GenBank.
MONTH	All new or revised sequences from the NR database released in the last 30 days.
SWISSPROT [sp]	The last major release of the SWISS-PROT protein sequence database.
YEAST	*Saccharomyces cerevisiae* protein sequences.
PDB [pdb]	Sequences from the Protein Data Bank.
KABATPRO [kabat]	Kabat's database of sequences of immunological interest.
ACR [acr]	Ancient Conserved Region subset of SWISS-PROT.
ALU [alu]	Database of translations of select Alu repeats.

Table 5.3. BLAST Nucleotide Sequence Databases

Database Name	Description
NR	Nonredundant nucleotide sequence database; it includes sequences from GenBank, EMBL, DDBJ, and PDB, but not the STS or EST databases.
MONTH	All new or revised sequences from the NR database released in the last 30 days.
YEAST	*Saccharomyces cerevisiae* nucleotide sequences.
PDB [pdb]	Sequences from the Protein Data Bank.
ALU [alu]	Database of translations of select Alu repeats.
VECTOR [vector]	Vector subset of GenBank.
KABATNUC [kabat]	Kabat's database of sequences of immunological interest.
EST	Nonredundant EST database; it includes sequences from GenBank, EMBL, and DDBJ.
STS	Nonredundant STS database; it includes sequences from GenBank, EMBL, and DDBJ.
EPD [epd]	Eukaryotic Promoter Database.

lines in the example form an e-mail message header automatically created by a mail program and bundled with the message when it is sent. Nothing needs to be entered for the subject of a BLAST e-mail request, but if text is entered in the Subject field of the mail header, the server will use it in the e-mail message reply. The actual search request begins with the mandatory parameter PROGRAM in the first column followed by the program BLASTN for searching nucleic acids. The next line contains the mandatory diretive DATALIB with the value NR for the combined nucleic acids database. The third line contains the mandatory BEGIN directive, followed by the query sequence in FastA format.

Figure 5.1. Example of a Simple Query to the BLAST Server

```
From:       annep@acs.auc.eun.eg
Date:       Tue 24 Sep 1996 - 09:58.16 - GMT
To:         BLAST@ncbi.nlm.nih.gov
Subject:

PROGRAM   BLASTN
DATALIB   NR
BEGIN
>UKNOWN GENE FRAGMENT
AGCTGTGGGTGATGCGCCCGATGCGATGCGATCGATGCTAGGCATGAGCTAGGCT
ACGGATCGATCGATCGATCGGACCACCCAGATGAGATCG
```

Figure 5.2. Example of a Complex Query to the BLAST Server

```
From:       annep@acs.auc.eun.eg
Date:       Tue 24 Sep 1996 - 09:58.16 - GMT
To:         BLAST@ncbi.nlm.nih.gov
Subject:

PROGRAM   BLASTP
DATALIB   NR
EXPECT 10
FILTER SEG+XNU
DESCRIPTIONS 20
ALIGNMENTS 10
HISTOGRAM NO
BEGIN
>UKNOWN PROTEIN TRANSLATED FROM CLONE 45-89
ASDLIPYTRWCCHIPLIESSEATILLMCVIYTADSRLSLIPENLISADEFGGIL
```

The BLAST server requires the use of the single-letter amino acid codes, ABCDE FGHIKLMNPQRSTVWXYZ, along with an asterisk, *, and the hyphen, - (refer to

Appendix D). An asterisk signifies a stop codon, and a hyphen signifies a gap of indeterminate length through which BLAST alignments are never permitted to extend. Any letter that is not a member of this alphabet will be stripped from an amino acid query sequence on submission to the server and will not be included as part of the analysis process.

If instead a nucleotide sequence is used for the query, the alphabet used by the BLAST server consists of the IUB/IUPAC nucleotide codes (ACGTRYMKWSBDHVNU), plus a hyphen, -, to signify a gap of indeterminate length. The code letter U (uracil) is treated like a T (thymidine) in this case (refer to Appendix D). Any letter that is not a member of this alphabet causes the program to report the presence of an unknown character and halt processing of the sequence. Remember, each line of information must be less than 80 characters in length.

The second example, shown in Fig. 5.2, analyzes a protein sequence using the BLASTP program and the NR library. In this example, a variety of optional parameters were used. The EXPECT directive was set to 10, meaning that 10 sequences with matching regions are expected to occur by chance alone. The FILTER directive was set to use the SEG+XNU combination instead of the default SEG. The number of sequence descriptions to report in the results was set to 20 with the best 10 alignments to be displayed. No histogram will be displayed in the results.

■ Restrictions and hints

As the programs BLASTX and TBLASTN consume substantial computer time, requests sent to the BLAST e-mail server are partitioned into two queues depending on the particular BLAST program requested. BLASTN and BLASTP requests are entered in a fast queue. BLASTX and TBLASTN requests are placed in a slow queue. Because of the extremely high computational demands of the TBLASTX algorithm, queries that use it are limited to only a few databases. We discuss this in more detail below.

Both queues operate concurrently. However, the fast queue is typically shorter than the slow queue, so individual BLASTN and BLASTP requests are usually processed sooner than BLASTX and TBLASTN requests. BLASTN was written to achieve high-speed nucleotide versus nucleotide sequence comparisons, but at the expense of decreased sensitivity, particularly for low-scoring or even moderately diverged homologs. The number of alignments observed with this program is frequently less than the number predicted from the value of the EXPECT parameter. Except for closely related homologs, better sensitivity in detecting homology between coding regions is usually found from protein level comparisons using BLASTX, TBLASTN, or TBLASTX, due to the combination of degeneracy in the genetic code and functional constraints on the encoded protein.

As a result, the computational demands are particularly pronounced for those queries that use TBLASTX. TBLASTX compares a nucleotide query sequence translated in all six reading frames against a nucleotide sequence database dynamically translated in all six reading frames. Because of the computational requirements of TBLASTX, this program is restricted to searching only the DBEST, DBSTS, and ALU databases.

■ **Evaluating results from the server**

A return e-mail message by a BLAST program in response to a query e-mail message is organized into three sections: (1) a histogram of the statistical significance of the matches found; (2) one-line descriptions of the database sequences that satisfied the statistical significance threshold (E parameter); and (3) the high-scoring segment pairs containing the matched sequence data. Each section of the reply e-mail message can be suppressed selectively by setting the parameters for the directives HISTOGRAM, DESCRIPTIONS, and ALIGNMENTS, respectively, to a value of 0.

Lowering the CUTOFF score from its default value by even a few units often produces a profound increase in the number of matching database sequences reported. The number reported tends to increase exponentially with decreasing CUTOFF score. However, the lowest scoring alignments are not only statistically insignificant but usually of limited biological relevance as well. Hence, the DESCRIPTIONS and ALIGNMENTS parameters can be used to govern the size of the e-mail message returned by the server.

Even when a high CUTOFF score (low EXPECT value) is used, the actual number of sequences reported may be great, depending on the number of true homologs present in the database and on the prevalence of regions of low compositional complexity in the query and database sequences. The latter characteristic can yield estimates of statistical significance that are not in accordance with the biological interest in such alignments; matches between regions of low compositional complexity should often be discounted heavily. Use of the FILTER directive can greatly improve the potential biological significance of a match if the problem of low compositional complexity in a search is causing an inordinate number of matches.

The GCODE directive is used to select an alternative genetic code for translation by the BLASTX, TBLASTN, and TBLASTX programs (Table 5.4). The universal genetic code is 1. The specified code is used by the server to translate the query sequence in BLASTX and TBLASTX analyses. For TBLASTN analyses, it is used to translate the database sequences. Detailed descriptions of the known alternative genetic codes appear in Appendix D.

Table 5.4. BLAST Genetic Code Parameters

Parameter	Genetic Code Description
1	Standard or Universal
2	Vertebrate Mitochondrial
3	Yeast Mitochondrial
4	Mold, Protozoan, Coelenterate Mitochondrial, and Mycoplasma/Spiroplasma
5	Invertebrate Mitochondrial
6	Ciliate Macronuclear

continued

Table 5.4 continued

Parameter	Genetic Code Description
9	Echinodermate Mitochondrial
10	Alternative Ciliate Macronuclear
11	Eubacterial
12	Alternative Yeast
13	Ascidian Mitochondrial
14	Flatworm Mitochondrial

■ Interpreting sequence identifiers

The sequence identifiers used by the BLAST server depend on the database from which each sequence was obtained. Table 5.5 outlines the syntax of the sequence identifiers for several databases.

Table 5.5. BLAST Sequence Database Identifiers

Database	Identifier and Syntax
GenInfo Integrated Database	gilnumber
GenBank	gblaccessionllocus
GenPept	gplaccessionllocus_cds#
EBI/EMBL	emblaccessionlname
DDBJ	dbjlaccessionlname
PIR	pirlaccessionlentry
SWISS-PROT	splaccessionlname
Protein Data Bank	pdblnamelchain
Kabat's database	gn1lkabatlname
TFD	gn1ltfdlname
Eukaryotic Promoter Database	gn1lepdlname

BioSCAN

The BioSCAN (Biological Sequence Comparative Analysis Node) server allows users to submit a nucleotide or protein query sequence for a similarity search with an algorithm analogous to BLAST (Singh et al. 1993). The search can be done against either SWISS-PROT, PIR, or GenBank databases. This server allows users to send a specially formatted mail message containing a DNA, RNA, or protein query sequence to the BioSCAN computer system at the University of North Carolina, Chapel Hill. A search is then performed against all or a subset (division or section) of a specified database, using a special-purpose hardware accelerator. BioSCAN finds entry segments that are similar to query segments. Given a score matrix, a query sequence, a database of entry sequences, and an expectation of the number of similar sequences, BioSCAN finds all entry sequences that have sufficient gap-free similarity to the query sequence. The results of this search and analysis process are returned in an e-mail message.

Internet addresses

E-mail:	bioscan@cs.unc.edu	(Server)
	bioscan-info@cs.unc.edu	(Personal help)
Web:	http://genome.cs.unc.edu/bioscan.html	(Home page)
Organization:	Department of Computer Science, CB #3175 Sitterson Hall, University of North Carolina, Chapel Hill, NC 27599-3175, U.S.A.	
Telephone:	+1-919-962-1740	
Facsimile:	+1-919-962-1799	

■ Accessing the BioSCAN server

To access the BioSCAN server, send an e-mail message containing a properly formatted request (as described below) to the BioSCAN server address. To receive the current set of instructions on using the BioSCAN e-mail server, send a help message to the BioSCAN server address. Put the word HELP on a line by itself in the body of the mail message. For further information on the BioSCAN project and for detailed help with problems with the server, send a mail message with a detailed question to project members at the Internet address for personal help. Also, if there is a delay in the response to a submission, send the submission e-mail address and a telephone number to the personal help address with a description of the contents of the submission.

■ Server directives and parameters

Table 5.6 contains a summary of the commands and their associated parameters available for use with the BioSCAN server. (Additional information on these directives and their parameters can be found in Chapter 3.) These directives are a subset of those used by the

Table 5.6. BioSCAN Commands and Parameters

Directive	Attributes	Explanation
BEGIN	Mandatory, none	Must appear after all other parameters and immediately before the FastA-formatted query sequence.
DATALIB	Mandatory, text	Available databases are listed in Tables 5.7 and 5.8.
EXPECT	Optional, numeric	The default value is 10. Fractional values are acceptable.
FILTER	Optional, Boolean	This directive is currently ignored for nucleic acid searches. For protein searches the default is YES.
MATRIX	Optional, text	The default matrix for peptide sequences is BLOSUM62; the default for nucleotide sequences is DPAM47. Alternate matrices that may be specified: PAM5 PAM30 PAM70 PAM160 PAM180 PAM250 BLOSUM30 BLOSUM35 BLOSUM40 BLOSUM45 BLOSUM50 BLOSUM55 BLOSUM60 BLOSUM62 BLOSUM65 BLOSUM70 BLOSUM75 BLOSUM80 BLOSUM85 BLOSUM90 BLOSUM95 BLOSUM100 DPAM47X BLOSUMN.
PROGRAM	Mandatory, text	Presently the only arguments supported are BSCAN and RETRIEVE.
QUERY	Mandatory, text	This alternative to the BEGIN directive specifies by locus name or accession number a query sequence which exists in one of the databases.
REPORT	Optional, text	The default parameter is SEGMENT. Available formats for reports are specified by SEGMENT, NONE, SIMPLE, ALIGN, and COMPRESS.
STRAND	Optional, text	This directive is ignored for protein searches. For nucleic acid searches DOUBLE is the default setting. Additional parameters for nucleic acid searches are DOUBLE (BOTH is accepted as a synonym) and SINGLE (TOP, PLUS, and + are accepted as synonyms).

NCBI BLAST server and they generally work in a similar fashion. The following directives or parameters are accepted by the BioSCAN server: PROGRAM, DATALIB, MATRIX, EXPECT, REPORT, FILTER, STRAND, QUERY, and BEGIN. Of these parameters, only the PROGRAM, DATALIB, and QUERY or BEGIN (but not both) are mandatory.

The following directives are accepted by the BioSCAN server for compatibility with the NCBI BLAST server, but are presently ignored: HISTOGRAM, DESCRIPTIONS, ALIGNMENTS, CUTOFF, PATH, and SPLIT. No other directives or parameters besides those described above are selectable through the e-mail server. Only those commands listed as mandatory in Table 5.6 are required to be present in a message submitted to BioSCAN.

■ Formatting a query

Components of an e-mail message to the BioSCAN server must be provided in this order: first, the two mandatory directives, PROGRAM and DATALIB, followed by the desired optional directives and parameters. Finally, the mandatory directive BEGIN or QUERY (but not both) must be inserted.

There are currently two parameters that can be chosen for the PROGRAM directive, RETRIEVE or BSCAN. The BSCAN program does a comparative analysis of the query sequence against the desired database, while the RETRIEVE program retrieves data from the selected database. The databases that can be selected by the second mandatory directive, DATALIB, are summarized in Tables 5.7 and 5.8. Please note that protein sequences must be used when searching protein databases and nucleic acid sequences must be used when searching nucleic acid databases. Deviation from this causes the server to return an error message. To select a desired database, give the name or abbreviation of the desired database as a parameter to the DATALIB directive. The DATALIB directive may be followed by any of the optional directives. Finally, enter the last mandatory directive, BEGIN or QUERY. The BEGIN directive must be followed immediately on the next line of the query message with text of the sequence in FastA format (see Chapter 3 for details). Alternatively, if the QUERY directive is used, it must be followed immediately with a query sequence accession number or locus name.

■ Examples of queries

In Fig. 5.3, the BSCAN program is used for a comparative analysis of a protein sequence query against the SWISS-PROT protein database. Also specified in the optional MATRIX directive is the PAM120 substitution score parameter. Sequences reported will have scores significant at the 1% level (0.01) or better (EXPECT directive). The query protein sequence is given in FastA format and is followed by a blank line.

In Fig. 5.4, the BSCAN program is used for a comparative analysis of a nucleic acid sequence query against the GenBank nucleotide database. The QUERY directive dictates that the sequence that exists in GenBank with locus name PP2AC will be used. Default values are used for MATRIX (DPAM47) and EXPECT (10) since these optional directives are omitted in the text of the message.

Table 5.7. BioSCAN Peptide Sequence Databases

Database Name	Description
SWISS or SW	SWISS-PROT protein database
PIR	PIR-International protein database
GENPEPT or GP	GenPept: a database of peptide sequences translated from coding regions of GenBank sequences

Table 5.8. BioSCAN Nucleotide Sequence Databases

Database Name	Description
GENBANK, or GB	All divisions of GenBank
GBBCT	GenBank Bacterial Sequences
GBEST	GenBank Expressed Sequence Tag Sequences
GBINV	GenBank Invertebrate Sequences
GBMAM	GenBank Other Mammalian Sequences
GBPAT	GenBank Patent Sequences
GBPHG	GenBank Phage Sequences
GBPLN	GenBank Plant Sequences
GBPRI	GenBank Primate Sequences
GBRNA	GenBank Structural RNA Sequences
GBROD	GenBank Rodent Sequences
GBSYN	GenBank Synthetic Sequences
GBUNA	GenBank Unannotated Sequences
GBVRL	GenBank Viral Sequences
GBVRT	GenBank Other Vertebrate Sequences

You can also retrieve database sequences in FastA format for subsequent searches with the BioSCAN server. Sequences can be retrieved either by locus name or by accession number. Specify the parameter RETRIEVE after the PROGRAM directive. Use the QUERY directive as described above to specify a sequence. The value specified with the DATALIB directive is ignored. Only one sequence can be retrieved per transaction. An example of sequence retrieval is shown in Fig. 5.5.

Figure 5.3. Example of a Query Sent to the BioSCAN Server To Analyze a Protein Sequence

```
From:        annep@acs.auc.eun.eg
Date:        Wed 25 Sep 1996 - 09:58.16 - GMT
To:          bioscan@cs.unc.edu
Subject:

PROGRAM BSCAN
DATALIB SW
MATRIX  PAM120
EXPECT  0.01
BEGIN
>Partial sequence of unknown phosphatase catalytic subunit
KLGPPTPLAIPNFLLSYVTRSSDNISCLIIPPLLVQPMQFSNSSCLFSPS YNSTEEIDLG
HVAFSNCTSITNVTGPICAVNGSVFLCGNNMAYTYLPTNWTGLCVLATLL PDIDIIPGDE
```

Figure 5.4. Example of a Query Sent to the BioSCAN Server To Analyze a Nucleotide Sequence

```
From:        annep@acs.auc.eun.eg
Date:        Wed 25 Sep 1996 - 09:58.16 - GMT
To:          bioscan@cs.unc.edu
Subject:

PROGRAM BSCAN
DATALIB GENBANK
QUERY   GENBANK:PP2AC
```

Figure 5.5. Example of Obtaining FastA Format Sequences

```
From:        annep@acs.auc.eun.eg
Date:        Wed 25 Sep 1996 - 09:58.16 - GMT
To:          bioscan@cs.unc.edu
Subject:

PROGRAM RETRIEVE
QUERY SW:PP_2ACAT
```

BLITZ

BLITZ is an e-mail server based on the MPsrch program developed by the Biocomputing Research Unit, University of Edinburgh, Scotland (Sturrock and Collins 1993). MPsrch is designed for one purpose: to determine which protein sequences in a database are the most similar to or contain similar regions to a query sequence. The MPsrch program used by the BLITZ server performs sensitive and extremely fast comparisons of protein sequences against the SWISS-PROT protein sequence database using the Smith and Waterman (SW) best-local-similarity algorithm (Smith and Waterman 1981). The SW algorithm is arguably the best single algorithm to use in the detection of similarities between a query sequence and a database. However, it is computationally intensive (the analysis procedure can be very time consuming). MPsrch is the fastest implementation of the SW algorithm currently in use. A typical search time for a protein query sequence of 400 amino acids is less than a minute. Additional time is required to reconstruct the alignments, depending on the number of alignments requested.

With the BLITZ server, a query sequence is compared against all sequences in the database, and the best results, as judged by the alignment score, are aligned and included in the output. The algorithm looks for the best "local" match as determined by the amino acid similarity matrix (the PAM value) and the cost of inserting gaps (INDEL cost). Only one match per database sequence is recorded and ranked to give the best results. This matching procedure means that with the BLITZ server, it is possible to detect short matching regions such as binding sites in the middle of long sequences.

Internet addresses

E-mail:	blitz@ebi.ac.uk	(Server)
	nethelp@ebi.ac.uk	(Personal help)
	mpsrch_help@biocomp.ed.ac.uk	(Help software)
Web:	http://www.ebi.ac.uk	(EBI home page)
Organization:	EMBL Outstation, the EBI, Hinxton Hall, Hinxton, Cambridge CB10 1RQ, United Kingdom	
Telephone:	+44-1223-494400	
Facsimile:	+44-1223-494468	

■ Accessing the BLITZ server

Send a properly formatted query (as described below) to the server address and the results of the analysis will be returned by e-mail. Inquiries about the MPsrch software or the algorithms used should be sent to either John Collins or Shane Sturrock at the Help software address. Inquiries about the operation of the BLITZ server should be sent to the EBI at the personal help address. The most recent set of instructions on using the server may be obtained by sending a message to the server with the directive HELP on a single line in the body of the message.

■ **Server directives and parameters**

As BLITZ is an automatic server, it understands a limited set of commands; these are summarized in Table 5.9. The server has preset default parameters for all of the directives except for the query sequence itself. BLITZ can only search the latest full release of the SWISS-PROT database. SWISS-PROT is updated as a full release four times annually. To make up for the length of time between the full releases of the database, it is planned that new entries to SWISS-PROT will be available for searching as a separate cumulative update database.

■ **Formatting a query**

When working with the BLITZ server, it is helpful to follow these general rules for formatting a query. First, the body of a query message must contain only one directive per line. BLITZ has two mandatory directives, SEQ and END. All other directives are optional, and if they are not used in a query, then the default settings will be used. As with most of the other servers, BLITZ is case-insensitive, meaning that a query can contain both uppercase and lowercase characters, or a mixture. In addition, the order of the directives in the query is not mandated except for SEQ (which should be the last directive prior to the query sequence itself) and END (which must be the last line of the query after the sequence). Everything after the SEQ line is treated as part of the query sequence. Unlike most other servers, BLITZ will accept blank lines, numbers, or spaces in the query sequence. Finally, only one search per query message is allowed.

Because of the nature of the SW algorithm used in the BLITZ server software, there are two critical directives: PAM and INDEL. The PAM matrices were developed by Dayhoff and coworkers (Dayhoff et al. 1978). They devised a series of weighted amino acid substitution matrices to help detect distant similarities between proteins. These matrices, called PAM matrices (for Percent Accepted Mutation), assign varying weights to different possible pairs of aligned amino acids.

The weights assigned can be positive or negative. With the BLITZ server, a PAM matrix is specified by a value between 1 and 500. The default value for the server is 120, which is an adequate compromise between the constraints of searching speed and detection of similar sequences. If a query submitted to BLITZ gives only a few or no matches, try a different PAM setting. Basically, low PAM values — those near 40 or less — are best for finding short regions of very strong similarity, while high values — those near 250 or greater — are better suited for finding longer, weaker matches. Because of these factors, the sequences that the BLITZ server reports as the most similar will vary significantly depending on the PAM setting. Table 5.10 summarizes the effect of the PAM setting on the INDEL cost.

The INDEL cost is a penalty that is subtracted from the alignment score for every residue that has been inserted or deleted in the best local alignment. There is a lower limit on the allowed INDEL value because, if it is set too low, the alignment will be filled with many short gaps and will be biologically meaningless. This lower limit will depend on the particular PAM setting used. The default INDEL cost will usually work best in most cases. Reducing this cost will encourage gaps; increasing it will discourage them.

The default and lower limit INDEL costs for a range of PAM values are shown in Table 5.10.

Table 5.9. BLITZ Directives and Parameters

Directive	Attributes	Explanation
ALIGN	Optional, numeric	Regulates the number of sequence alignments to display in the reply message. The default is 30 and the maximum is 100.
END	Mandatory, none	Indicates the end of the query sequence.
HELP	Optional, none	Used to request a help document.
INDEL	Optional, numeric	Sets the gap penalty where the parameter is an integer value, typically in the range 5 to 30. If not specified, the default value is dependent on the PAM matrix used. The default INDEL cost for a PAM matrix setting of 120 is 13. The lowest value that is acceptable for this PAM matrix is 7. If an illegal value for INDEL is used, then the server will select a suitable default value.
NAMES	Optional, numeric	Regulates the number of matching sequences to include in a summary list. The lowest setting is 1; the upper limit is not specified.
PAM	Optional, numeric	The PAM matrix is specified where the parameter is a number between 1 and 500. This directive sets the amino acid weight matrix that is used to score nonidentical amino acids in the search. The default value used for analysis is 120.
SEQ	Mandatory, none	Everything that follows this line in the query message, up to the END directive or the end of the query message, is treated as part of the query sequence.
TITLE	Optional, text	Indicates that text on this line is the title of the query. The parameter is a maximum of one line of text. The first word will be used as the name of the query sequence; the rest of the line will be used as a description; i.e., TITLE NEW PHOSPHATASE ANALYSIS USING PAM 130 OF THE PHOSPHATASE will use "NEW PHOPHATASE" as the name and "ANALYSIS USING PAM 130 OF THE PHOSPHATASE" as description. Do not use quotes or double quotes in this line.

Table 5.10. Effect of PAM Settings on INDEL

PAM Setting	Default INDEL Cost	INDEL Cost Lower Limits	
1	53	27	
20	29	15	
40	22	11	
60	19	10	
80	16	8	
100	14	7	
120	13	7	DEFAULT value
150	11	6	
200	9	5	
250	7	4	
300	6	3	
400	5	3	
500	4	2	

■ Examples of queries

A couple of examples of queries to the BLITZ server are shown in the following figures. Note that the server is case-insensitive and the sequences that are submitted to the server do not have to be in FastA format. The example in Fig. 5.6 asks for the 50 best alignments using an INDEL cost of 10 and the PAM 200 matrix. Alternatively, the query shown in Fig. 5.7 will use the default server settings of PAM 120, INDEL 13, and ALIGN 30.

There are several important restrictions associated with queries to and replies from the BLITZ server. First, the maximum number of alignments that can be requested is 100. Second, the PAM matrix value must be between 1 and 500 inclusive. Third, the maximum query sequence length is 10,000 amino acids. Finally, as with most other servers, only one sequence analysis per query message is allowed.

After a query is sent to the BLITZ server, two messages will be sent back in reply. The first is a message that gives the status of the query. The second is the analysis results from the server.

Figure 5.6. Example of a Query Sent to the BLITZ Server

```
From:         annep@acs.auc.eun.eg
Date:         Wed 25 Sep 1996 - 09:58.16 - GMT
To:           blitz@ebi.ac.uk
Subject:

TITLE BLITZ ANALYSIS OF PUTATIVE PHOSPHATASE CATALYTIC SUBUNIT
PAM 200
INDEL 10
ALIGN 50
SEQ

   1 MGAFLDKPKMEKHNAQGQGNGLRYGLSSMQGWRVEMEDAHTAVIGLPSGL

  51 ESWSFFAVYDGHAGSQVAKYCCEHLLDHITNNQDFKGSAGAPSVENVKNG

 101 IRTGFLEIDEHMRVMSEKKHGADRSGSTAVGVLISPQHTYFINCGDSRGL

 151 LCRNRKVHFFTQDHKPSNPLEKERIQNAGGSVMIQRVNGSLAVSRALGDF

 201 DYKCVHGKGPTEQLVSPEPEVHDIERSEEDDQFIILACDGIWDVMGNEEL

 251 CDFVRSRLEVTDDLEKVCNEVVDTCLYKGSRDNMSVILICFPNAPKVSPE

 301 AVKKEAELDKYLECRVEEIIKKQGEGVPDLVHVMRTLASENIPSLPPGGE

 351 LASKRNVIEAVYNRLNPYKNDDTDSTSTDDMW

END
```

Figure 5.7. Example of a Query Sent to the BLITZ Server Using Default Parameters

```
From:         annep@acs.auc.eun.eg
Date:         Wed 25 Sep 1996 - 09:58.16 - GMT
To:           blitz@ebi.ac.uk
Subject:

SEQ
MGAFLDKPKMEKHNAQGQGNGLRYGLSSMQGWRVEMEDAHTAVIGLPSGL
ESWSFFAVYDGHAGSQVAKYCCEHLLDHITNNQDFKGSAGAPSVENVKNG
IRTGFLEIDEHMRVMSEKKHGADRSGSTAVGVLISPQHTYFINCGDSRGL
LCRNRKVHFFTQDHKPSNPLEKERIQNAGGSVMIQRVNGSLAVSRALGDF
DYKCVHGKGPTEQLVSPEPEVHDIERSEEDDQFIILACDGIWDVMGNEEL
CDFVRSRLEVTDDLEKVCNEVVDTCLYKGSRDNMSVILICFPNAPKVSPE
AVKKEAELDKYLECRVEEIIKKQGEGVPDLVHVMRTLASENIPSLPPGGE
LASKRNVIEAVYNRLNPYKNDDTDSTSTDDMW
END
```

DDBJ operates three e-mail servers: FastA, BLAST, and MAlign. The first two servers are mainly derivatives of existing Internet analysis tools such as the NCBI BLAST server and the EBI Mail-FastA server and work in similar fashion. However, there are some major differences in the directives and parameters used by both of these servers, which we discuss below. As with their counterpart servers at NCBI and EBI, the DDBJ FastA and BLAST servers are designed to compare sequence data against available protein and nucleic acid databases. The other DDBJ server, the MAlign server, is designed to perform a multiple sequence alignment of either a series of protein or DNA sequences. It reports out the best alignment of the submitted sequences. As an option, it can also generate a putative ancestral sequence based on the submitted sequences.

Internet addresses

E-mail:	fasta@nig.ac.jp	(FastA server)
	blast@nig.ac.jp	(BLAST server)
	malign@nig.ac.jp	(MAlign server)
	trouble@nig.ac.jp	(Personal help)
	ddbj@ddbj.nig.ac.jp	(Information)
Web:	http://www.ddbj.nig.ac.jp	(DDBJ home page)
Organization:	DNA Data Bank of Japan (DDBJ), Center for Information Biology, National Institute of Genetics, 1111 Yata, Mishima, 411, Japan	
Telephone:	+81-559-81-6853	
Facsimile:	+81-559-81-6849	

Accessing the DDBJ servers

To access any of these servers, send a properly formatted e-mail query to the respective server address listed. The automated help system for these servers functions in much the same fashion as for other servers, but with one major exception: the HELP file that is returned is available in either English or Japanese. To get the latest HELP file in Japanese, send a message to the desired server address containing the word HELP on a single line in the body of the message. To get the instructions in English, send a message to the desired server address containing the word HELP-E on a single line in the body of the message.

FastA server directives and parameters

The DDBJ FastA server supports the following directives: PROGRAM, DATALIB, KTUP, SCORES, ALIGNMENTS, BEGIN, >, and // or END. We discuss these directives and their associated parameters in more detail in Table 5.11. In a query to the server, directives should be written from the beginning of each line and followed by a space and then the parameter. Each directive and its parameter must be written in one line.

Table 5.11. DDBJ FastA Server Directives and Parameters

Directive	Attributes	Explanation
// or END	Mandatory, none	Indicates the end of the sequence. Either // or END should be used.
>	Mandatory, none	Indicates that the sequences begin from the next line. Comments can be written after character on the same line. The > character is necessary even when there is no comment text.
ALIGNMENTS	Optional, numeric	If not specified, the default value is 20.
BEGIN	Mandatory, none	Indicates the beginning of query sequence. After the BEGIN statement, no empty line is allowed until the end of the sequence.
DATALIB	Mandatory, text	Specify the database to search (see Table 5.12).
INVERT	Optional, numeric	This directive is usable only with the PROGRAM directive set to FASTA and a DNA query sequence. If the parameter for INVERT is set to 1, then the query sequence is converted to its complement prior to the analysis. The default is not to invert the sequence. There are no other values for the parameter.
KTUP	Optional, numeric	Specifies the degree of sensitivity of the search. If not specified, KTUP value will be set to 4 when DATALIB specified is DDBJ, GenBank, EMBL, or DNA, and 1 when any other DATALIB is used.
MATRIX	Optional, text	This directive sets the matrix for the analysis of protein sequences. The default is PAM250. Other matrices are PAM120, CODAA, IDNAA, and IDPAA (see Table 5.14).
PROGRAM	Mandatory, text	Sets the analysis program to be used by the server. FASTA is the default setting with TFASTA as the option.
SCORES	Optional, numeric	Sets the number of similar sequences to report. If not specified, the default value is 100.

■ Formatting a FastA query

As with the other servers, the body of a query message must contain only one directive per line. The FastA server has five mandatory directives, PROGRAM, DATALIB, BEGIN, >, and either END or //. The directives should be provided in the following order. First, the two mandatory directives PROGRAM and DATALIB should be listed. These directives should be followed by the desired optional directives and parameters. Next, the BEGIN directive is listed to mark the start of the sequence data. Immediately below the BEGIN directive should be the query sequence data in FastA format (see Chapter 3). Finally, the END or // directive should be provided at the end of the sequence.

The PROGRAM directive has the following parameters: FASTA, for comparing a DNA sequence against a DNA database or a protein sequence against a protein database, and TFASTA, for comparing an amino acid sequence against a DNA database. DATALIB specifies the database to be searched. (The parameters for this directive are shown in Table 5.12.) The BEGIN command indicates the beginning of a query sequence. After the BEGIN directive, no empty lines are permitted until the end of the sequence. The next line after the BEGIN directive contains the > character, which indicates to the server that this line is a comments field. Comments may be written after the > character on the same line. The character > is a mandatory directive even when there is no text to be entered as a comment. On the following lines, the sequence is entered. The final directive of the query is either the END directive or the characters //, but not both, to indicate the end of the sequence.

Table 5.12. DDBJ FastA and BLAST Databases

DNA databases

DNA	DDBJ and DDBJNEW sequence databases
DDBJ	Last release of the DDBJ database
DDBJNEW	New entries to the DDBJ database
EMBL	Last release of the EMBL database
GENBANK	Last release of the GenBank database
EPD	Eukaryotic Promoter Database

Protein databases

PROTEIN	PIR and SWISS-PROT sequence databases
PIR	Last release of the PIR database
SWISS	Last release of the SWISS-PROT database

Three optional directives that are particular to DDBJ's FastA server are INVERT, KTUP, and SCORES. INVERT does just what it says: it inverts a DNA sequence, generating its complementary strand. This complementary strand is then used by the server in the analysis. KTUP specifies the degree of sensitivity of the search. Usually, the KTUP value is set at 3 to 6 for nucleotide sequences and 1 or 2 for amino acid sequences. If the KTUP value is small, the analysis will be more sensitive, but requires more time to complete. If the query is a DNA sequence longer than 200 nucleotides, the KTUP value should be set to 4 or greater. The third directive, SCORES, specifies the lower limit of similarity score to be reported. Only if the similarity score for a sequence in the database is larger than this value will the sequence be reported.

■ Example of a FastA query

As shown in Fig. 5.8, the query message requests the FastA server to perform the analysis with a KTUP value equal to 6 and for 20 similarity scores and 10 alignments to be displayed in the reply message. If the optional directives were not used, then the server would have performed the analysis using a default KTUP of 4 and given a reply containing 100 similarity scores and 20 alignments.

Figure 5.8. Example of a Query to the DDBJ FastA Server

```
From:        annep@acs.auc.eun.eg
Date:        Wed 25 Sep 1996 - 09:58.16 - GMT
To:          fasta@nig.ac.jp
Subject:

PROGRAM FASTA
DATALIB DDBJ
KTUP 6
SCORES 20
ALIGNMENTS 10
BEGIN
>
TGCTTGCATTACCGGATAGATAGCATGATCGATCATGCATGCATGCATGCATGCTAGCT
ACACCGATGCACATGGCTAGCTAGCTAGCTACGATCGATGCATCGAGCATGATCGTTGA
TCACAGTCAGT
//
```

■ BLAST server directives and parameters

In Table 5.13 we show the directives used by the DDBJ BLAST server. The directives supported by the DDBJ version of the BLAST server are different from those supported by NCBI and EBI BLAST servers. The DDBJ version supports the directives ALIGNMENTS, CUTOFF, DATALIB, DESCRIPTION or SCORES, END or //, EXPECT, HISTOGRAM, MATRIX, PROGRAM, SORT, and WORDSIZE.

Table 5.13. DDBJ BLAST Server Directives and Parameters

Directive	Attributes	Explanation
// or END	Mandatory, none	Indicates the end of the sequence. Either // or END should be used.
>	Mandatory, none	Indicates that the sequences begin from the next line. Comments can be written after this character on the same line. The > character is necessary even when there is no comment text.
ALIGNMENTS	Optional, numeric	Specify an integer such as 50. Default value is 20.
BEGIN	Mandatory, none	Indicates the beginning of the query sequence. After the BEGIN statement, no empty line is allowed until the end of the sequence.
CUTOFF	Optional, numeric	The threshold of homology score for matching sequences that are reported. A homologous sequence in a database is reported only when its homology score is larger than the CUTOFF value. By default, the value is calculated from the EXPECT value. If both values are specified, the more severe condition that will result in a lower number of reported sequences is adopted. The values actually used in the search are shown by E (expect) and S (cutoff) in the "Parameters" section near the end of the e-mail reply from the server.
DATALIB	Optional, text	This directive is used to specify the database in which homologous sequences are searched (see Table 5.12). One of the databases listed can be used. The default database is DNA for BLASTN and TBLASTN and PROTEIN for BLASTP and BLASTX.
DESCRIPTION or SCORES	Optional, numeric	Sets the number of similar sequences to report. SCORES and DESCRIPTION are interchangeable. Default value for this command is 20.

continued

Table 5.13 continued

Directive	Attributes	Explanation
EXPECT	Optional, numeric	Default value is 10. If sequences with lower homology scores are desired, increase the value. If only sequences with very high homology scores are desired, decrease the value.
HISTOGRAM	Optional, numeric	This command instructs the server whether or not to report a histogram of homology scores. If set to a value of 0, the server will not report the histogram. The default setting is 1, which means the server generates a histogram.
MATRIX	Optional, text	This directive sets the matrix for the analysis of nucleotide sequences. The default is BLOSUM62. Other matrices are PAM40, PAM120, PAM250, IDENTITY, DAYOFF, and GONNET (see Table 5.14).
PROGRAM	Optional, text	This directive is used to specify the search program. One of the four following programs can be used: BLASTN, BLASTP, BLASTX, or TBLASTN. If this directive is not used in a query, the server will use BLASTN as the default. DNA databases are used with BLASTN and TBLASTN, while protein databases are for BLASTP and BLASTX. All six possible reading frames are searched for homologous sequences when BLASTX and TBLASTN are selected.
SORT	Optional, text	Sets the method for sorting sequence matches. The default parameter is P-VALUE. Other parameters are COUNT, HIGH-SCORE, and TOTAL-SCORE.
WORDSIZE	Optional, numeric	The default value is 12 for BLASTN and 3 for the other BLAST programs.

Most of the mandatory and optional directives used by the DDBJ BLAST server, i.e., ALIGNMENTS, BEGIN, DESCRIPTION and SCORES, END and //, MATRIX, and PROGRAM, function in an identical manner to their counterparts in the DDBJ FastA server. There are several options for the MATRIX directive. These are summarized in Table 5.14. The directives DESCRIPTION and SCORES are interchangeable. They specify how many homologous sequences are reported in a list of homology scores. Three other directives, CUTOFF, EXPECT, and HISTOGRAM, function in an identical manner to the same directives in the NCBI BLAST server.

Table 5.14. Matrix Options for Protein Analysis with FastA and for Nucleotide Sequence Analysis with BLAST

FastA matrices

PAM250	PAM 250 substitution matrix, default setting for the server
PAM120	PAM 120 substitution matrix
CODAA	Genetic code matrix
IDNAA	Protein identity matrix
IDPAA	Weighted protein identity matrix

BLAST matrices

BLOSUM62	BLOSUM 62 scoring matrix, default setting for the server
PAM40	PAM 40 substitution matrix
PAM120	PAM 120 substitution matrix
PAM250	PAM 250 substitution matrix
IDENTITY	Nucleotide identity matrix
DAYOFF	Specialized substitution matrix
GONNET	Modified PAM 250 substitution matrix

Two of the optional directives, WORDSIZE and SORT, are unique to the DDBJ BLAST server. WORDSIZE sets the analysis window for the BLAST program to use when scanning a sequence. The effect of WORDSIZE on the stringency of the analysis is inversely proportional to the value for its parameter. Larger values allow faster but less stringent comparisons, while smaller values give more stringent searches, but at the expense of the speed of analysis. SORT specifies the method for sorting sequence matches. The default setting is P-VALUE, where sequences are sorted by increasing probability. Other options are COUNT, which ranks sequences by the number of high-scoring pairs (HSPs) found in a sequence; HIGH-SCORE, which ranks sequences by the score of the highest-scoring HSP in the sequence; and finally TOTAL-SCORE, which ranks sequences based on the cumulative sum of all HSPs in a sequence.

■ Formatting a query

The structure of a query to the DDBJ BLAST server is nearly identical to that for the DDBJ FastA server. As with the DDBJ FastA server, the directives of the BLAST server should be written from the beginning of each line and followed by a space and then the parameter. The mandatory directives for the BLAST server are the same as for the FastA server. However, the parameters available for use in the PROGRAM directive are:

- **BLASTN:** DNA sequence vs. DNA database

- **BLASTP:** Protein sequence vs. protein database

- **BLASTX:** DNA sequence vs. protein database

- **TBLASTN:** Protein sequence vs. DNA database

The DATALIB directive has the same parameters as the FastA server (these are summarized in Table 5.12). A query should thus start with the PROGRAM directive and then DATALIB. Any optional directives would be entered next, followed by the query sequence in FastA format (see Chapter 3).

■ Example of a BLAST query

The example query message shown in Fig. 5.9 instructs the server to use the program BLASTN to compare a nucleotide sequence against the DDBJ sequence database using the default settings for the optional directives. The reply message from the server would be similar in structure to that shown for the NCBI BLAST server in Fig. 5.2.

Figure 5.9. Example of a Query to the DDBJ BLAST Server

```
From:        annep@acs.auc.eun.eg
Date:        Wed 25 Sep 1996 - 09:58.16 - GMT
To:          blast@nig.ac.jp
Subject:

PROGRAM BLASTN
DATALIB DDBJ
BEGIN
> TEST SEQUENCE
AATTCCGGAATTCCGGAATTCCGGAATTCCGG
AATTCCGGAATTCCGGAATTCCGGAATTCCGG
//
```

■ MAlign server directives and parameters

The directives shown in Table 5.15 are used by the MAlign server. As with other servers, directives should be written from the beginning of each line and followed by a space and then the parameter. Each directive and its parameter must be written in one line. Briefly, the mandatory directives for the MAlign server are BEGIN, >, and END or //. They function in an identical manner to those directives in the DDBJ FastA and BLAST servers.

Several optional directives are unique to MAlign: ANCES, GAPA, GAPB, MOLTYPE, and TREE. ANCES specifies that an ancestral sequence at each node is printed out. This means that at each branch point or node of the phylogenetic tree, the postulated ancestral sequence at that point of sequence divergence would be reported by the server. The GAPA and GAPB directives set the values A and B, which are used to calculate the gap penalty in the equation [gap penalty = A + B (K)], where K is gap length. The MOLTYPE directive indicates the molecular sequence type. The values for the directive parameter are either DNA or PROTEIN. The final optional directive, TREE, instructs the server to create a phylogenetic tree showing the hypothetical evolutionary relationships of the query sequences.

■ Formatting a MAlign query

The basic format of a query to the MAlign server is different from that of the other DDBJ servers. A query message must have the mandatory directive BEGIN to mark the start of the sequences to be aligned. After this directive comes the comment field directive, >, followed by two or more sequences to be aligned. The sequences must be in FastA format (see Chapter 3) and be separated by the character set // on a separate line. After the last

Table 5.15. DDBJ MAlign Server Directives and Parameters

Directive	Attributes	Explanation
// or END	Mandatory, none	Indicates the end of a sequence. Either // or END should be used.
>	Mandatory, none	Indicates that the sequences begin on the next line. Comments can be written after this character on the same line. The > character is necessary even when there is no comment text.
ANCES	Optional, none	Specifies that an ancestral sequence at each node is printed out.
BEGIN	Mandatory, none	This directive must be the last one immediately before the sequence data. Two or more sequences must follow this directive.
GAPA	Optional, numeric	This directive sets the value A used in the determination of the gap penalty.
GAPB	Optional, numeric	This directive sets the value B used in the determination of the gap penalty.
MOLTYPE	Optional, text	The values for the directive parameter are either DNA or PROTEIN. The default value is set as DNA.
TREE	Optional, none	If this directive is used, the server will print a phylogenetic tree.

sequence, either the END directive or the // characters should be placed to mark the end of the sequences. If any of the optional directives are used, they must be placed prior to the BEGIN directive in the body of the query message.

■ Example of a MAlign query

In Fig. 5.10, the query file requests the MAlign server to compare three DNA sequences and print out a phylogenetic tree of the relationship between them. At each node of the tree a predicted ancestral sequence will be shown. The directive MOLTYPE is not necessary because the sequence data being submitted is DNA, the default parameter setting for this optional directive. If the sequences to be aligned were proteins, then the MOLTYPE directive would be required with the parameter setting of PROTEIN.

Additionally, in this query, the default gap penalty setting for the server has been modified by using values of 7 and 3 for the GAPA and GAPB directives, respectively.

Figure 5.10. Example of a Query to the DDBJ MAlign Server

```
From:       annep@acs.auc.eun.eg
Date:       Wed 25 Sep 1996 - 09:58.16 - GMT
To:         malign@nig.ac.jp
Subject:

ANCES
GAPA    7
GAPB    3
MOLTYPE DNA
TREE
BEGIN
>PHOSPHATASE SEQUENCE 1
GCCGAAGTGAAGAACATCAGTACCGGCTACATGGTGACCAACGACTGCACCAATGATAGC
ATTACCTGGCAACTCCAGGCTGCTGTCCTCCACGTCCCCGGGTGCGTCCCGTGCGAGAAA
CATTGGGGCGTCATGTTCGGCTTAGCCTACTTCTCTATGCAGGGAGCGTGGGCAAAAGTC
GTTGTCATTCTTTTGCTGGCCGCCGGGGTGGACGCG
//
>PHOSPHATASE SEQUENCE 2
TACCAAGTGCGCAACTCCACAGGGCTTTATCATGTCACCAATGATTGCCCTAACTCGAGT
ATTGTGTACGAGGCGCACGATGCCATCCTGCATACTCCGGGGTGTGTCCCTTGCGTTCGC
CACTGGGGAGTCCTGGCGGGCATAGCGTATTTCTCCATGGTGGGGAACTGGGCGAAGGTC
CTGGTAGTGCTGTTGCTGTTTGCCGGCGTCGACGCG
//
>PHOSPHATASE SEQUENCE 3
TATGAAGTGCGCAACGTGTCCGGGATATACCATGTCACGAACGACTGCTCCAACTCAAGC
ATTGTGTATGAGGCAGCGGACGTGATCATGCATACTCCCGGGTGCGTGCCCTGTGTTCGG
CACTGGGGAGTCCTGGCGGGCCTTGCCTACTATTCCATGGTAGGGAACTGGGCTAAGGTT
CTGATTGTGGCGCTACTCTTTGCCGGCGTTGACGGG
//
```

GenQuest

GenQuest is an integrated sequence comparison server that makes use of a wide variety of sequence comparison methods and target databases. The purpose of the system is to allow rapid and sensitive comparison of DNA and protein sequences to existing DNA and protein sequence databases.

Internet addresses

E-mail:	Q@ornl.gov	(Server)
	grailmail@ornl.gov	(Personal help)
Web:	http://avalon.epm.ornl.gov/grail-bin/	(GenQuest home page)
Organization:	Informatics Group, Oak Ridge National Laboratory, Oak Ridge, TN 37831-6364, U.S.A.	
Telephone:	+1-615-574-6134	
	+1-615-574-8934	
Facsimile:	+1-615-574-7860	

■ Accessing the GenQuest server

Queries can be submitted to the server at the e-mail address listed. A detailed help document can be received by sending a query to the server with the directive HELP on the subject line or as the first line of the body of the message. If specific problems occur or questions arise that cannot be addressed by the help document, then contact the staff of the GenQuest server at the e-mail address for personal help.

■ Server directives and parameters

Table 5.16 summarizes the directives and parameters that are supported by the GenQuest server. The databases that can be accessed from the GenQuest e-mail server are summarized in Table 5.17.

The server has two mandatory directives, SEQ and TYPE. The remainder of the directives shown in Table 5.16 are optional and have default settings. SEQ marks the start of the query sequence. Any text after this directive until the end of the query message or the END directive is treated as part of the query sequence. TYPE tells the server what kind of sequence data is being submitted to the server. This directive has three different parameters, PROTEIN, DNA6, and DNA. PROTEIN specifies that the input is an amino acid sequence and to search it against a protein database; DNA6 specifies that the query sequence is DNA and to translate it in all six reading frames for searching against protein databases; DNA specifies a DNA sequence that can be searched against DNA target databases, or if a protein database is selected as a target, then the DNA sequence will be trans-

Table 5.16. GenQuest Directives and Parameters

Directive	Attributes	Explanation
ALIGN	Optional, text	ALIGN specifies the number of alignments to be reported in the results of the analysis. The default setting is 10 alignments. This directive is only for use when SW is the parameter for the directive METHOD.
COMMENT	Optional, text	This optional directive allows 1 line of text to be added to the query for record-keeping purposes.
END	Optional, none	The END directive can be used to mark the end of the query sequence. All text between the SEQ and END directives is treated as part of the query sequence.
FILTER	Optional, none	Specifies if filtering of the query sequence is to be done prior to analysis. If this directive is not invoked, then no filtering of the query sequence is done.
MATRIX	Optional, text	The optional MATRIX directive is used to select the substitution matrix that the server will use in the search. Options are all PAM matrices in multiples of 10 such as PAM250 and the BLOSUM62 and BLOSUM80 matrices. The default directive is BLOSUM62. This directive is ignored for DNA-DNA comparison.
METHOD	Optional, text	This directive is used to set the alignment method for the analysis of the query sequence. Options are: SW(Smith-Waterman), FASTA, BLAST, and FLASH. The default is SW.
SCORE	Optional, numeric	This directive specifies how many matches from an analysis will be scored. The default setting is 10. This directive is only for use when SW is the parameter for the directive METHOD.
SEQ	Mandatory, none	The mandatory SEQ directive is used to mark the start of the sequence data in the query. All text after this directive up to the END directive is treated as part of the query sequence.
TARGET	Optional, text	This directive specifies the database to which the sequence will be compared. Databases that are currently available are listed in Table 5.17. The default is GSDB for DNA sequences and SWISSPROT for protein sequences.
TYPE	Mandatory, text	This directive specifies the type of sequence being submitted to the server. Options are PROTEIN, DNA, and DNA6.

Table 5.17. GenQuest Databases

Database	Abbreviation
SWISS-PROT	SWISSPROT
PDB	PDB
Prosite	PROSITE
GSDB	GSDB
BLOCKS Protein Motif	BLOCKS
EST Database	DBEST
Human Repetitive DNA	REPETITIVE
PIR International	PIR

lated starting from the first base, then searched against protein databases. The DNA6 option is computationally intensive and is not recommended for DNA sequences of more than 1,000 to 2,000 bases.

Three optional directives are also useful in constructing a basic query: METHOD, TARGET, and END. METHOD is used to specify what type of sequence analysis algorithm is to be used. Several options are available: Smith-Waterman or SW (Smith and Waterman 1981), FastA (Pearson and Lipman 1988), BLAST (Altschul et al. 1990), and FLASH (Califano et al. 1993). TARGET selects the database to search. Available databases are listed in Table 5.17. Note that the abbreviations listed must be used as the parameter value. Unlike most other servers, GenQuest allows multiple target databases in a single query message to specify comparison of the sequence data against more than one database. END marks the end of the query sequence. This is useful because most Internet e-mail clients add a signature block at the end of an e-mail message. If the END directive is not used, GenQuest might treat the text of the signature block as part of the query sequence.

Several other optional directives can be used. As with the other optional directives, these must be placed prior to the SEQ directive. Among the most useful of these is the directive COMMENT. COMMENT is used to enter text pertaining to a query (or query sequence) for record-keeping purposes. The server will include the text of this line in the reply message. If the COMMENT directive is used in a query, it should precede any other directives.

Another optional directive is FILTER. This directive is useful to prescreen query sequences for repetitive DNA sequences. These repetitive elements are then masked by the server to reduce uninteresting matches against the DNA sequence database.

Two optional directives are applicable only to queries that use the Smith-Waterman algorithm as the analysis method. The ALIGN directive specifies the number of alignments to be reported in the results of the analysis. The default setting is 10 alignments. The other directive that is applicable to queries that use the Smith-Waterman algorithm is SCORE. This directive specifies how many matches from an analysis will be scored. The default setting is 10. For the analysis of proteins, a parameter value in the range from 10 to 200 is optimal. Note that the value of the parameter for SCORE should be greater than or equal to the number of alignments that are reported.

■ Formatting a query

Messages to the GenQuest server begin with a set of directives that specify the options to be used in the search. As previously mentioned, queries to GenQuest have two mandatory directives: TYPE and SEQ. (As the server is not case-sensitive, upper- and lowercase text can be used interchangeably within a query.) A simple query using a combination of mandatory and optional directives has the following organization. The first directive should be COMMENTS, followed on that line by any information for record keeping. On the next line, the directive TYPE should be entered. At this point any optional directives such as TARGET, METHOD, MATRIX, SCORE, or ALIGN can be used. Finally, the directive SEQ, the actual query sequence, and the directive END should be entered at the end of the query.

The query sequence can be either a standard single-letter protein (see Appendix D) or a DNA sequence. Note that letter characters other than ACTG in DNA sequences are treated as C by the server. The length of the sequence lines should be less than 512 characters. Also, the GenQuest server deletes any blank spaces or non-letter characters from the query sequence.

When using the METHOD parameters FASTA or BLAST, the standard defaults for these algorithms are used by the server when conducting the analysis. Descriptions of these settings and how to modify them are available by sending a message to the personal help address with the words HELP FASTA or HELP BLAST in the body of the message. When SW is specified as the analysis method, the default gap penalty setting may be changed using the modifier -G. For example, SW -G 10 sets the gap penalty to 10. The default setting for this optional modifier is 13. The METHOD parameter FLASH is the only one that will permit a search of the PIR database for matching sequences. FLASH can be used in the analysis of the other databases for matching sequences as well. Finally, note that when using the BLOCKS and PROSITE databases, METHOD does not need to be specified since the server uses analysis methods that are specific for those databases.

■ Examples of queries

The example in Fig. 5.11A uses the query protein sequence and searches the SWISS-PROT database using the default Smith-Waterman algorithm. The server uses the default gap penalty setting of 13 and the default BLOSUM62 matrix and scores the top 10 matches. The reply message will show the top 10 alignments. The additional examples shown in Fig. 5.11B and 5.11C illustrate typical queries for various types of searches. The example in Fig. 5.11B searches the given DNA sequence against the default GSDB and the human repetitive sequence database using the Smith-Waterman algorithm with the default settings. The example in Fig. 5.11C shows a more complex query. It translates the query DNA in all six frames, searches the SWISS-PROT and PROSITE databases using the SW algorithm with a gap penalty of 10 and the PAM130 matrix, and then scores the top 100 matches. The reply message from the server will show the top 50 alignments.

When using the SW method for analysis, the server normally does a local alignment of matching sequences. A global alignment of matching sequences may be requested using the -G parameter modifier with ALIGN directive. For example, the directive ALIGN 15 -G returns global alignments of the top 15 matching sequences.

Figure 5.11. Examples of GenQuest Queries

A: Analysis Using the Smith-Waterman Algorithm and the SWISS-PROT Database

```
From:       annep@acs.auc.eun.eg
Date:       Wed 25 Sep 1996 - 09:58.16 - GMT
To:         Q@ornl.gov
Subject:
COMMENT UNKNOWN PROTEIN SEQUENCE TRANSLATION OF CLONE 12-7 TYPE
PROTEIN
SEQ
LKASDEFYTRRGGIPASDECCLGPSTDELIIHHIL
END
```

B: GSDB and Repetive DNA Library Using the Smith-Waterman Algorithm

```
From:       annep@acs.auc.eun.eg
Date:       Wed 25 Sep 1996 - 09:58.16 - GMT
To:         Q@ornl.gov
Subject:
COMMENT UNKNOWN DNA SEQUENCE FROM CLONE 12-7
TYPE DNA
TARGET REPETITIVE
SEQ
ATAGATAAAGGGTGCTGTTTGGCGAAATATTGCTGCTGGCGCCGTAGATATATAG
CTGTGCTGTGATGTCGCTCGTAGATATAGCTAGTCTAGTCGATCG
END
```

C: SWISS-PROT and PROSITE Using Smith-Waterman Algorithm, Gap Penalty Setting of 10

```
From:       annep@acs.auc.eun.eg
Date:       Wed 25 Sep 1996 - 09:58.16 - GMT
To:         Q@ornl.gov
Subject:
COMMENT UNKNOWN DNA SEQUENCE FROM CLONE 12-7
TYPE DNA6
TARGET SWISSPROT
TARGET PROSITE
METHOD SW -g 10
MATRIX PAM130
SCORE 100
ALIGN 50
SEQ
ATCTATCGTCGAGCTGGTGTCTGTGCTAGTCCACAGACAGHCTCGCTATATATGCT
CGTTTTAAAGCTCGTATATATGCTCTCGCTAGTCCGATCGATGCTCGATCGCTAGT
ATCGTATGATTCTTGCTCGGGGAAATATGAGCCTCGTATCTAGTCT
END
```

Mail-FastA

The Mail-FastA server is based on the FastA program developed by Pearson and Lipman (Pearson and Lipman 1988) as implemented in the GCG suite of sequence analysis programs (Devereux et al. 1984). Mail-FastA was designed to perform sensitive and rapid comparisons of DNA or protein sequences to a range of databases and to report which are similar to the query sequence. The two Mail-FastA servers, one located at EBI and the other at EMBnet in Heidelberg, Germany, are identical except for the databases that are available for searching.

 Internet addresses

E-mail:	fasta@ebi.ac.uk	(EBI server)
	mfasta@genius.embnet.dkfz-heidelberg.de	
		(EMBnet server)
	nethelp@ebi.ac.uk	(Personal help)
Gopher:	gopher.ebi.ac.uk	(EBI Gopher server)
Web:	http://www.ebi.ac.uk	(EBI home page)
Organization:	EMBL Outstation, the EBI, Hinxton Hall, Hinxton, Cambridge CB10 1RQ, United Kingdom	
Telephone:	+44-1223-494400	
Facsimile:	+44-1223-494468	

■ **Accessing the Mail-FastA server**

Send any inquiries, questions, or comments to the personal help address listed above. To access the server, send a properly formatted e-mail message to either the EMBnet or EBI servers at the Internet addresses listed above. To request a current help file, send a query to the server containing the directive HELP. The help file will be sent back as a reply message.

■ **Server directives and parameters**

Table 5.18 lists the parameters that are available for the Mail-FastA server. There is only one mandatory directive, SEQ. All text lines after this directive until the end of the query message (or the directive END is reached) are treated by the server as part of the query sequence. The maximum length for a query sequence is 32,000 nucleotides or amino acids. No special format of the query sequence is required other than the use of the standard IUB/IUPAC codes (see Appendix D); however, be sure to remove all comments and other text.

Table 5.18. Mail-FastA Directives and Parameters

Directive	Attributes	Explanation
ALIGN	Optional, numeric	Sets the number of top scoring sequences to be aligned with the query sequence. The maximum is 30 and the default is 20.
END	Optional, none	Indicates the end of the sequence data. If used, END must be the last directive in the query.
HELP	Optional, none	When used in a query, this optional directive returns a help file.
LIB	Optional, text	The LIB directive is used to specify the sequence library to be searched. Refer to Tables 5.19 and 5.20 for a listing of databases. Default libraries for the EBI server are EMALL or SWALL. Default libraries for the EMBnet server are GEALL or SW.
LIST	Optional, numeric	Number of top scoring sequences to be listed in the output file. The maximum value is 100, and the default value is 40.
ONE	Optional, none	Sets the strand of a DNA sequence to analyze. Default for DNA sequences is to analyze both strands. This directive is not usable with protein sequences.
PROT	Optional, none	Denotes to the server that the query sequence is protein. Otherwise the server determines if a sequence is protein or DNA based on the sequence itself.
SEQ	Mandatory, none	The SEQ directive is used to mark the start of the sequence data in the query. All text after this directive up to the END directive is treated as part of the query sequence.
TITLE	Optional, text	Text on the TITLE line of the query will be used by the server on the Subject line of the reply e-mail message.
WORD	Optional, numeric	Sets the value for the window used in analyzing sequences. Parameters may be 1 or 2 for proteins (default is 2) or 3 to 6 for DNA sequences (default is 6).

Several optional directives can be used to modify the default settings of the Mail-FastA server. The most useful of these for basic queries are LIB, PROT, and TITLE. The LIB directive is used to specify the sequence library to be searched. The parameter for this directive can be any one of the sequence databases listed in Table 5.19 for the EBI version of the server or Table 5.20 for the EMB version of the server. For the EBI version

Table 5.19. EBI Version Databases

Nucleotide databases

EMNEW	All new entries to the EMBL database since the last release
EMALL	All EMBL entries: last release and new entries
EEST1	EST division of EMBL database
EFUN	Fungi division of EMBL database
EINV	Invertebrate division of EMBL database
EMAM	Other mammals division of EMBL database (no rodents or primates)
EORG	Cellular organelle division of EMBL database
EPHG	Bacteriophage division of EMBL database
EPLN	Plant division of EMBL database
EPRI	Primates division of EMBL database
EPRO	Prokaryotes division of EMBL database
EROD	Rodents division of EMBL database
ESTS	STS division of EMBL database
ESYN	Synthetic sequences division of EMBL database
EUNA	Unannotated division of EMBL database
EVRL	Viruses division of EMBL database
EVRT	Other vertebrates division of EMBL database (no mammals)

Protein databases

SWALL	All SWISS-PROT entries: last release and new entries
SWNEW release	All new entries to the SWISS-PROT database since the last
SW	Last release of the SWISS-PROT protein database

Table 5.20. EMBnet Version Databases

Nucleotide databases

GEALL	EMBL + GenBank + new EMBL entries
EMNEW	All new entries to the EMBL database since the last release
EMALL	All EMBL entries: last release and new entries
EMEST	EMBL expressed sequence tags
EMFUN	EST division of EMBL database
EMINV	Invertebrate division of EMBL database
EMMAM	Other mammals division of EMBL database (no rodents or primates)
EMORG	Cellular organelle division of EMBL database
EMPHG	Bacteriophage division of EMBL database
EMPLN	Plant division of EMBL database
EMPRI	Primates division of EMBL database
EMPRO	Prokaryotes division of EMBL database
EMROD	Rodents division of EMBL database
EMSTS	STS division of EMBL database
EMSYN	Synthetic sequences division of EMBL database
EMUNA	Unannotated division of EMBL database
EMVRL	Viruses division of EMBL database
EMVRT	Other vertebrates division of EMBL database (no mammals)
GBONLY	Entries only in GenBank, but not in EMBL
GENEMBL	All GenBank and EMBL entries

Protein databases

SW	Last release of the SWISS-PROT protein database
NBRF	Last release of the PIR protein database
PIR1	Annotated/classified division of the PIR database
PIR2	Preliminary division of the PIR database
PIR3	Unverified division of the PIR database
PIRONLY	Entries only in PIR, but not in SWISS-PROT
SWISSPIR	All SWISS-PROT and PIR entries

of the server, the default library is EMALL or SWALL. For the EMB version of the server, the default library is GEALL or SW. The PROT directive is useful to force the server to treat a query sequence as a protein. The Mail-FastA server checks whether a query sequence is protein or DNA; however, it may treat very small protein sequences as DNA. By using the PROT directive, this potential problem can be avoided. The TITLE directive can be used to keep track of replies from the server. When TITLE is used in a query, any text that is entered on the same line will be reported in the Subject line of the reply message from the server, which makes this directive a handy tool for adding a tag to the results of an analysis.

Two other directives, WORD and ONE, also merit some extra comment. When using the directive WORD to modify the window that the program uses when analyzing sequences, the parameter value must be between 3 and 6 for DNA sequences, or 1 and 2 for protein sequences. Note that the smaller the parameter value for the WORD directive, the higher the sensitivity of the analysis, but the smaller the value, the longer the search time. Therefore, using a parameter value that is smaller than the default parameter value is not recommended.

The ONE directive is specific for DNA sequences. When analyzing a query DNA sequence, the Mail-FastA server normally uses both strands in the analysis. The ONE directive specifies that only the strand given in the query is compared against the target database. If this directive is not used, then the complement of the query strand is searched against the database also.

Proper use of the remaining optional directives and their associated parameters can be found in Table 5.18.

■ Formatting a query

As a query to the Mail-FastA server needs only one directive, SEQ, and a query sequence to analyze, this server is among the simplest to use. All of the optional directives, except for END, must be placed prior to the SEQ directive. Otherwise the server may treat the optional directives as part of the sequence or be unable to complete the analysis. A suggested order for using the optional directives in a basic query message is first TITLE, then LIB, LIST, and ALIGN. After the query sequence, use of the END directive is recommended to prevent any signature blocks that might be attached to the query message by the e-mail client program from being included in the analysis.

As with nearly every server, all directives must be on their own line and only one database may be searched per query message.

■ Examples of queries

Shown in Fig. 5.12 are two examples of properly formatted query messages to the Mail-FastA server. In Fig. 5.12A, the query sequence is analyzed against all new entries to the EMBL nucleotide sequence database. The server will send back a reply that lists the best 15 matching sequences and alignments of the top five to the query sequence.

Figure 5.12. Examples of Queries to the Mail-FastA Server

A: Query Using EMNEW with a List of 15 Sequences and an Alignment of 5

```
From:        annep@acs.auc.eun.eg
Date:        Wed 25 Sep 1996 - 09:58.16 - GMT
To:          fasta@ebi.ac.uk
Subject:

TITLE PARTIAL SEQUENCE OF A PUTATIVE PHOSPHATASE GENE
LIB EMNEW
LIST 15
ALIGN 5
ONE
SEQ
201   ACAACTTTGA CTTTGAGAAA AGAGAGGTGG AAATGAGGAA AATGACTTTT

251   CTGTATTAGA TTCCAGTAGA AAGAACTTTC ATCTTTCCCT CGTTTTTTTT

301   GTTTTAAAAC ATCTATCTGG AGGCAGGACA AGTATGGTCG TTAAAAAGAT

351   GCAGGCAGAA GGCATATATT GGCTCAGTCA AAGTGGGGAA CTTTGGTGGC

401   CAAACATACA TTGCTAAGGC TATTCCTATA TCAGCTGGAC ACATATAAAA

451   TGCTGCTAAT GCTTCATTAC AAACTTATAT CCTTTAATTC CAGATGGGGG

501   CAAAGTATGT CCAGGGGTGA GGAACAATTG AAACATTTGG GCTGGAGTAG

551   ATTTTGAAAG TCAGCTCTGT GTGTGTGTGT GTGTGTGCGC GCACGTGTGT
END
```

B: Query using the SWISSPIR Database and Default Program Settings

```
From:        annep@acs.auc.eun.eg
Date:        Wed 25 Sep 1996 - 09:58.16 - GMT
To:          mfasta@genius.embnet.dkfz-heidelberg.de
Subject:

LIB SWISSPIR
SEQ
GHFTEEDKATITSLWGKVNVEDAGGETLGRLLVVYPWTQRFFDSFGNLSS
ASAIMGNPKVKAHGKKVLTSLGDAIKHLDDLKGTFAQLSELHCDKLHVDP
ENFKLLGNVLVTVLAIHFGKEFTPEVQASWQKMVTAVASALSSRYH
```

In Fig. 5.12B, the server is searching the query sequence against all entries in the SWISS-PROT and PIR protein sequence databases. Because none of the optional directives is being used in this analysis, the Mail-FastA server will use the default settings.

After sending a query to the server, two messages will be sent in reply. The first one is sent immediately after the query message is received and initially processed. If the Mail-FastA server had any problem processing the query, it would be reported in this first reply message. Otherwise the server would report that the analysis query was being processed. The second reply message would report the results of the analysis.

Mail-QUICKSEARCH

The Mail-QUICKSEARCH e-mail server grew out of the QUICKSEARCH and QUICK-SHOW programs developed by Devereux and coworkers (Devereux et al. 1988). In Mail-QUICKSEARCH, the server allows the user to perform rapid comparisons of nucleic acid sequences against the EBI and GenBank databases, including the most recent entries. QUICKSEARCH was developed for rapid searching of databases for identical or closely related DNA sequences. Therefore, it is most useful for comparing newly determined unknown DNA sequences to the database to find out whether there are similar sequences. It is not suited for the detection of more distantly related sequences and cannot search for or with protein sequences. (For this purpose use the Mail-FastA e-mail server described on page 85.)

Internet addresses

E-mail:	quick@ebi.ac.uk	(Server)
	nethelp@ebi.ac.uk	(Personal help)
Gopher:	gopher.ebi.ac.uk	(EBI Gopher server)
Web:	http://www.ebi.ac.uk	(EBI home page)
Organization:	EMBL Outstation, the EBI, Hinxton Hall, Hinxton, Cambridge CB10 1RQ, United Kingdom	
Telephone:	+44-1223-494400	
Facsimile:	+44-1223-494468	

■ Accessing the Mail-QUICKSEARCH server

To access the Mail-QUICKSEARCH server, send a properly formatted query message to the server address. Upon completion of the analysis, the server will return the results of the analysis by a reply e-mail message. A help file can be retrieved by sending a message to the server containing only the directive HELP in the body of the message. Detailed questions regarding the operation of the server should be sent to the personal help address.

■ Server directives and parameters

As with other search servers, the Mail-QUICKSEARCH server understands a limited set of directives. The accepted directives and their parameters are summarized in Table 5.21.

Table 5.21. Mail-QUICKSEARCH Directives and Parameters

Directive	Attributes	Explanation
BEST	Optional, none	Determines the algorithm used for the alignment of the query sequence to the database sequences. If this option is used, the Mail-QUICK-SEARCH server will use the classic Smith-Waterman algorithm to align sequences. Default is a Needleman-Wunsch algorithm to find the best overall alignment.
END	Optional, none	Indicates the end of the e-mail message. If this directive is used it must be the last line in the message.
HELP	Optional, none	Returns a help file.
LIB	Optional, text	This directive specifies which sequence database to search in the analysis. LIB can be selected from the list below. The default setting is ALL.
		ALL All EBI/EMBL and GenBank database entries.
		GENEW New EBI/EMBL and GenBank database entries since latest release only.
MATCH	Optional, numeric	Only database entries that show overlaps of more than n% identity to the query sequence will be presented. The default value is 90.
ONE	Optional, none	Only the strand given in a mail message is compared against the database. If this directive is not specified, then the complementary strand is searched as well.
PERFECT	Optional, none	This directive reports only exact matches. It is equivalent to specifying a MATCH value of 100.
WINDOW STRINGENCY	Optional, numeric	The directives WINDOW and STRINGENCY set the sensitivity of a search. The default values are calculated from the length of the query sequence. WINDOW is the sequence length/20 minus 1 with a maximum value of 15, while STRINGENCY is the value of WINDOW/2. Decreasing the window size and decreasing STRINGENCY may find more distantly related sequences at the expense of biological and statistical significance. Increasing the window size and increasing STRINGENCY will permit fewer mismatches and return more closely related sequences. For exact matches, use the directive PERFECT.

continued

Table 5.21 continued

Directive	Attributes	Explanation
TITLE	Optional, text	The QUICKSEARCH program will use the text entered on the TITLE line as the subject line of the return e-mail message that contains the output of the job. Do not use quotes or double quotes.
SEQ	Mandatory, text	Everything following this line up to either the end of the submitted e-mail message or a line starting with the word END will be treated as part of the sequence. Sequence information cannot be put on the same line as the SEQ command or the END command.

■ Formatting a query

The format for a query submitted to the Mail-QUICKSEARCH server is relatively simple, but it must be adhered to closely. First, the e-mail message must contain only one directive per line. There is only one mandatory directive, SEQ. All the other directives are optional, and default values are used whenever they are not specified. The server is not case-sensitive, so both upper- and lowercase letters can be used. The order of the directives is not important, but make sure that SEQ is the last directive, since everything following this line will be treated as sequence data.

The query sequence within an e-mail message may not be longer than 100,000 bases. No special format of the query sequence itself beyond use of the standard IUB/IUPAC symbol codes (see Appendix D) is required. Numbering may be included, but comments and other information must be removed or the server will likely treat them as sequence data. Short query sequences can be sent to the server, but the QUICKSEARCH algorithm was designed to look for sequences longer than 200 bases. If a match longer than 32,000 bases is found, only the first part of the analysis may be reported.

■ Examples of queries

Shown in Fig. 5.13 and 5.14 are example queries sent to the Mail-QUICKSEARCH server. The example in Fig. 5.13 uses the optional directive TITLE. Immediately following the TITLE directive is the parameter analysis of a putative phosphatase catalytic subunit. The next line contains the mandatory directive, SEQ, followed by the nucleotide sequence of the query on the next line. Note that numbers and spaces are acceptable within the query sequence. Finally, the last line of the query contains the END directive. Default values are used for LIB (ALL), MATCH (90), WINDOW (determined by length of sequence), and STRINGENCY (determined from the WINDOW directive).

Figure 5.13. Example of a Query Sent to the QUICKSEARCH Server

```
From:        annep@acs.auc.eun.eg
Date:        Wed 25 Sep 1996 - 09:58.16 - GMT
To:          quick@ebi.ac.uk
Subject:

TITLE QUICKSEARCH ANALYSIS OF PUTATIVE PHOSPHATASE CATALYTIC SUBUNIT
SEQ
      1 cggaacgtgg ttggggaggg ggggtgggg gggactctag acagctgagg cgcgaaagca
     61 tgagtcctcg gctcttcctc ctccttctcc gggacccgct ctctgcctcc ctctccaacg
    121 cccggatgat ctgagccgcg agggcgccga cagccggggg cccggacgca gcccggctcc
    181 tcccctcctc cgccccttcc ccagcctgac ctggcccgcc gctgcagcgg tgaccctcc
    241 cccggctgcc gccgtcgccg ccgcggtgac cccctccccg gctgccgccg ccgccgcctc
    301 ggccgaccag ggacctgccc gcctgcggct gctccggacc tagaggatca agacataatg
    361 ggagcatttt tagacaagcc aaagatggaa aagcataatg cccaggggca gggtaatggg
    421 ttgcgatatg ggctaagcag catgcaaggc tggcgtgttg aaatggagga tgcacatacg
    481 gctgtgatcg gtttgccaag tggacttgaa tcgtggtcat tctttgctgt gtatgatggg
    541 catgctggtt ctcaggttgc caaatactgc tgtgagcatt tgttagatca catcaccaat
    601 aaccaggatt ttaaagggtc tgcaggagca ccttctgtgg aaaatgtaaa gaatggaatc
    661 agaacaggtt ttctggagat tgatgaacac atgagagtta tgtcagagaa gaaacatggt
    721 gcagatagaa gtgggtcaac agctgtaggt gtcttaattt ctccccaaca tacttatttc
    781 attaactgtg gagactcaag aggtttactt tgtaggaaca ggaaagttca tttcttcaca
    841 caagatcaca aaccaagtaa tccgctggag aaagaacgaa ttcagaatgc aggtggctct
    901 gtaatgattc agcgtgtgaa tggctctctg gctgtatcga gggcccttgg ggattttgat
    961 tacaaatgtg tccatggaaa aggtcctact gagcagcttg tctcaccaga gcctgaagtc
   1021 catgatattg aaagatctga agaagatgat cagttcatta tccttgcatg tgatggtatc
   1081 tgggatgtta tgggaaatga agagctctgt gattttgtaa gatccagact tgaagtcact
   1141 gatgaccttg agaaagtttg caatgaagta gtcgacacct gtttgtataa gggaagtcga
   1201 gacaacatga gtgtgattt gatctgtttt ccaaatgcac ccaaagtatc gccagaagca
   1261 gtgaagaagg aggcagagtt ggacaagtac ctggaatgca gagtagaaga aatcataaag
   1321 aagcaggggg aaggcgtccc cgacttagtc catgtgatgc gcacattagc gagtgagaac
   1381 atccccagcc tcccaccagg gggtgaattg gcaagcaaga ggaatgttat tgaagccgtt
   1441 tacaatagac tgaatcctta caaaaatgac gacactgact ctacatcaac agatgatatg
   1501 tggtaaaact gctcatctag ccatggagtt taccttcacc tccaaaggag agtacagctc
   1561 aactttgttg aaactttaa catccatcct caactttaag gaaggggata tgacatgggt
   1621 gagaatgatt acatcagaga acttcagcag tacaacagct agcccagaac tgattttttt
   1681 ttttttttt tttgtaaatt tgagacttat gtaagcgtga tttcaaacca taattcgtgt
   1741 tgtaaatcag actccagcaa tttttgttgt atgattttgt tttttgtaa agtgtaattg
   1801 tccttgtaca aaatgctcat atttaattat gaactgcttt aaatcactat caaagttaca
   1861 agaaatgttt ggcttattgt gtgatgcaac agatatatag cccttttcaag tcatgttgtg
   1921 tttggacttg gggttggaac agggagagca gcagccatgt cagctacacg ctcaaatgtg
   1981 cagatgatta tggaaaataa cctcaaaatc ttacaaagct gaacatccaa ggagttattg
   2041 aaaactatct taaatgttct tggtagggga gttggcattg ttgataaagc cagtcccttc
   2101 atttaactgt ctttcaggat gttccttcgt tgtttccatg agtattgcag gtaataatac
   2161 agtgtgttcc ataagaatct caatcttggg gctaaatgcc ttgtttcttt gcacctcttt
   2221 tcaagtcctt acatttaatt actaattgat aagcagcagc ttcctacata tagtaggaaa
   2281 ctgccacatt tttgctatca tgattggctg ggcctgctgc tgttcctagt aagatattct
   2341 gaattc
END
```

Figure 5.14. Example of a Query Sent to the QUICKSEARCH Server with Optional Directives

```
From:        annep@acs.auc.eun.eg
Date:        Wed 25 Sep 1996 - 09:58.16 - GMT
To:          quick@ebi.ac.uk
Subject:

TITLE QUICKSEARCH ANALYSIS OF PUTATIVE PHOSPHATASE CATALYTIC SUBUNIT
WINDOW 30
STRINGENCY 10
MATCH 95
BEST
SEQ
     1 cggaacgtgg ttggggaggg gggggtgggg gggactctag acagctgagg cgcgaaagca
    61 tgagtcctcg gctcttcctc ctccttctcc gggacccgct ctctgcctcc ctctccaacg
   121 cccggatgat ctgagccgcg agggcgccga cagccggggg cccggacgca gcccggctcc
   181 tcccctcctc cgccccttcc ccagcctgac ctggcccgcc gctgcagcgg tgacccctcc
   241 cccggctgcc gccgtcgccg ccgcggtgac cccctccccg gctgccgccg ccgccgcctc
   301 ggccgaccag ggacctgccc gcctgcggct gctccggacc tagaggatca agacataatg
   361 ggagcatttt tagacaagcc aaagatggaa aagcataatg cccaggggca gggtaatggg
   421 ttgcgatatg ggctaagcag catgcaaggc tggcgtgttg aaatggagga tgcacatacg
   481 gctgtgatcg gtttgccaag tggacttgaa tcgtggtcat tctttgctgt gtatgatggg
   541 catgctggtt ctcaggttgc caaatactgc tgtgagcatt tgttagatca catcaccaat
   601 aaccaggatt ttaaagggtc tgcaggagca ccttctgtgg aaaatgtaaa gaatggaatc
   661 agaacaggtt ttctggagat tgatgaacac atgagagtta tgtcagagaa gaaacatggt
   721 gcagatagaa gtgggtcaac agctgtaggt gtcttaattt ctccccaaca tacttatttc
   781 attaactgtg gagactcaag aggtttactt tgtaggaaca ggaaagttca tttcttcaca
   841 caagatcaca aaccaagtaa tccgctggag aaagaacgaa ttcagaatgc aggtggctct
   901 gtaatgattc agcgtgtgaa tggctctctg gctgtatcga gggcccttgg ggattttgat
END
```

The example in Fig. 5.14 also uses the TITLE directive and the same parameter as shown in Fig. 5.13. However, the WINDOW directive is set at 30, and the STRINGENCY directive is set at 10. Additionally, only sequences with overlaps of 95% or greater will be reported (MATCH 95) and the Smith-Waterman algorithm will be used to search the databases (BEST). The SEQ is followed on the next line by the query nucleotide sequence, and the END directive is used to signal the end of the sequence.

After a query is sent to the server, as with other EBI servers such as Mail-FastA and BLITZ, two e-mail messages will be sent back in reply. The first one is sent immediately after the query e-mail message is processed. If the Mail-QUICKSEARCH server had any problem with the submitted query, this message would report this information. Otherwise the server will report that the query has been successfully submitted to the QUICKSEARCH batch queue and that the results will be mailed back after completion of the analysis. The second message received from the server will be the result of the actual analysis.

Retrieval

Several avenues exist on the Internet for the retrieval of nucleotide and protein sequence data. We will discuss five different server systems in this section: the GENIUSnet/NetServ sequence servers (of European origin); and Retrieve, the STS and EST mail servers, and the Query server (all from NCBI in the United States). The GENIUSnet/NetServ sequence servers are relatively simple but fast systems. GENIUSnet/NetServ retrieve sequence files based on the accession numbers of the entries. The matching database records are returned by e-mail message to the requester.

Retrieve is arguably the most widely used and popular of the current sequence retrieval servers in widespread use. It is more sophisticated than the GENIUSnet/ NetServ sequence servers, allowing for more complex searches; however, with the increase in sophistication comes an increase in complexity. Sequence records can be retrieved by keyword, author, text, or field identifier in addition to retrieval by accession number.

Two other sequence retrieval servers are based on Retrieve: the STS and EST sequence servers. Both of these retrieval servers use the Retrieve engine to access the burgeoning data sets of the STS and EST databases that are available for many organisms, including humans. Like Retrieve, they can perform searches based on a variety of different criteria, including chromosomal location, organism, or sequence identifiers.

The latest entry in the sequence record retrieval field is Query. If the GENIUSnet/NetServ system is viewed as the most limited but simplest of the retrieval systems, and Retrieve is viewed as the current standard retrieval system because of its searching capabilities coupled with a relatively straightforward set of directives, then Query must be viewed as a powerful heir-apparent. Query is based on the Entrez search engine developed at NCBI. This relational system is designed to query an integrated database of sequence files and literature records. Further, because these files and records are neighbored or linked, Query permits the user to retrieve related records and data in a fashion much like that of the Web. With Query, a search for a specific sequence can lead not only to that sequence, but to those most closely related based on BLAST analysis and related literature references from MEDLINE.

With this power and flexibility comes a significant increase in the complexity of operation. Query has a steep learning curve. Once this initial hurdle is overcome, the relational structure of the system makes it possible to perform true searches or queries of the databases, rather than straight sequence downloads as with the GENIUSnet/NetServ system or limited searches of discrete databases as with the Retrieve server.

Each of these five retrieval systems has its place in the research community. Whether making use of them all as the needs of the project or query dictate or standardizing on one as the laboratory sequence data retrieval system, the ability to download the latest information using these servers is a major advance in the evolution of molecular biology databases.

GENIUSnet/NetServ

The GENIUSnet/NetServ sequence servers can be used to retrieve nucleotide or protein sequence files based on the locus name or accession number only; these servers cannot retrieve sequence files based on other parameters such as text, author, or keyword.

Internet addresses

E-mail:	netserv@ebi.ac.uk	(NetServ server)
	netserv@genius.embnet.dkfz-heidelberg.de	
		(GENIUSnet server)
Web:	http://www.ebi.ac.uk	(EBI home page)
Organization:	EMBL Outstation, the EBI, Hinxton Hall, Hinxton, Cambridge CB10 1RQ, United Kingdom	
Telephone:	+44-1223-494400	
Facsimile:	+44-1223-494468	

■ Accessing the server

The servers can be accessed by sending retrieval requests to either of the two Internet e-mail addresses listed above. Both servers operate in a like fashion. To get some brief introductory information on using the servers, send an e-mail message to either server with the directive HELP in the body of the message.

■ Server directives and parameters

A single directive is used by the server, GET, with two modifiers, PROT and NUC, that are used to restrict the search to either a protein database or a nucleotide database, respectively. Sequence files can be retrieved either from two nucleotide or two protein sequence databases. For nucleotide sequences, matching records from the latest full public release of the EMBL database and the cumulative updates to the EMBL database will be retrieved. For protein sequences, matching records in the PIR International and SWISS-PROT databases will be retrieved. Remember, entries can be retrieved by locus name or accession number only. Additionally, there is no limit to the number of parameters that follow the GET directive.

■ Examples of queries

Send a query to either of the servers that contains one directive per line. The general syntax for the query is:

```
GET database:accnumber
```

where the database directive modifier is either NUC for searching the nucleotide databases or PROT for searching the protein databases as in the directive:

```
GET NUC:Q01897
```

In the query shown in Fig. 5.15, the sequence PP2C_HUM is being requested from
the SWISS-PROT and PIR protein databases; the sequence with the accession number
X09714 is being requested from the EMBL nucleotide sequence databases. Subject lines
are ignored by the server.

Figure 5.15. Retrieval of Sequence Records Using the GENIUSnet/NetServ Sequence Servers

```
From:        annep@acs.auc.eun.eg
Date:        Wed 25 Sep 1996 - 09:58.16 - GMT
To:          netserv@genius.embnet.dkfz-heidelberg.de
Subject:

GET PROT:PP2C_HUM
GET NUC:X09714
```

It should be noted by the users of these retrieval servers that some sequence files available from the databases are much larger than 100 kilobytes in size, which can present
problems for the transmission of data from the servers back to the requester. Although
this file size should present no problem for current Internet mail systems, some older
mail systems are unable to process files of this size and refuse to accept them. To accommodate users of these older mail systems, the GENIUSnet/NetServ sequence servers
automatically split files into packets of approximately 100 kilobytes in size. The Subject
lines of the reply e-mail messages received from these two servers contain information
that tells the number and order of message packets that were sent in response to a
retrieval query. With this information, the original sequence files can be reconstructed in
a word processing program by pasting together the individual parts.

Query

The Query e-mail server allows users to retrieve records from the sequence and MED-LINE databases at the National Center for Biotechnology Information (NCBI), National Library of Medicine, NIH, in Bethesda, Md. Query uses the Entrez query engine to retrieve records of interest. Under the Entrez query system, information is arranged in a relational fashion by domains rather than by source databases. Further information on Entrez can be obtained by contacting NCBI by e-mail at net-info@ncbi.nlm.nih.gov or by browsing Web Entrez at the URL http://www.ncbi.nlm.nih.gov/. This relational model of data access enables the Query server to retrieve a record or records of interest from a target domain and related or "neighboring" records from the same domain or from other domains.

 Internet addresses

E-mail:	query@ncbi.nlm.nih.gov	(Server)
	info@ncbi.nlm.nih.gov	(Information)
Web:	http://www.ncbi.nlm.nih.gov	(NCBI home page)
Organization:	National Center for Biotechnology Information, National Library of Medicine, Building 38A, Room 8S-803, 8600 Rockville Pike, Bethesda, MD 20894, U.S.A.	
Telephone:	+1-301-496-2475	
Facsimile:	+1-301-480-9241	

■ Accessing the server

Search requests can be sent to the Query server at the e-mail address listed above. A help document can be obtained by sending a message to the Query server that contains the directive HELP in the body of the message. Detailed questions, comments, or suggestions on the operation of the Query server should be send to the information address listed above.

■ Server directives and parameters

The directives supported by the Query server are DB, UID, TERM, DISPMAX, DOPT, and HTML. Table 5.28 briefly describes the directives and their associated parameters.

■ Formatting a query

The domains that make up the relational database used by the Query server may be searched by either unique identifiers (UID) or by text terms. To conduct either type of search, send an e-mail message to the server containing a single query structured as follows. The first line of the query should list the domain (DB) to be searched. The following list gives domains currently supported by the Query server. (The letter abbreviation in parentheses is the parameter which should be given with the DB directive.)

- **Nucleotide sequences (n):** GenBank, EMBL, DDBJ, dbEST, dbSTS

- **Protein sequences (p):** GenPept (translated coding regions from DNA), PIR, SWISS-PROT, PRF, PDB

- **MEDLINE (m):** Literature records from the molecular biology subject area

- **3-D structure (t):** MMDB database, derived from PDB

- **Nucleotide and protein (s):** GenBank, EMBL, DDBJ, dbEST, dbSTS, GenPept (translated coding regions from DNA), PIR, SWISS-PROT, PRF, PDB

Table 5.28. Query Server Directives and Parameters

Directive	Attributes	Explanation
DB	Mandatory, text	Specifies the domain to be searched. Options are N, P, M, or T. Refer to the text for details.
UID	Mandatory, text	Specifies a unique identifier that is carried by the domain specified in the DB parameter. Either UID or TERM must be specified in a query.
TERM	Mandatory, text	Specifies a term to be used for the search. Either UID or TERM must be specified in a query.
DISPMAX	Optional, numeric	Sets the number of reports to be sent back to the the user. The most recent report is displayed first. The default value is 200 reports.
DOPT	Optional, text	Specifies how to display the record or citation. It consists of zero or more neighbor or link characters followed by a single report type character. Refer to Table 5.29 for default parameters and Table 5.30 for all permissible parameter values.
HTML	Optional, none	Specifies that the records be shown in HTML format.

The second line should list the type of search, either UID or TERM, followed by the search string. A UID used in a search can be any of the following:

- MEDLINE UID (Unique Identifier) of the desired entry

- Sequence ID (GI) of the entry

- GenBank Accession Number of the entry

- A FastA (see below) specification for the entry

More than one UID or combination of UIDs can be used in a query to the server. Multiple UIDs must be separated by commas, but without spaces between them. If more than one entry is specified, use the DOPT parameter D in order to view summary information for each entry before seeing detailed information on any individual entry. Further details on the parameters for the DOPT directive can be found in Tables 5.29 and 5.30.

Table 5.29. Query Server Default Parameters for the DOPT Directive

Default Parameter	Explanation
G	UID searches for designated records in the nucleotide or protein domains
R	UID searches for designated records in the MEDLINE domain
D	UID searches for neighbors or links
D	TERM searches

Table 5.30. Query Server DOPT Parameter Options

MEDLINE entries
Report types:

R	Standard report format: citation, title, abstract, indexing terms
B	Abstract format: Citation, title, abstract only
L	MEDLARS format
A	ASN.1 format
D	Entrez document summary format

Neighbors/Links:

M	MEDLINE neighbors
P	Protein links
N	Nucleotide links
T	Structure links

continued

Table 5.30 continued

Protein sequence entries
Report types:

G   GenPept format

R   Report format

F   FastA format

A   ASN.1 format

D   Entrez document summary format

Neighbors/links:

M   MEDLINE links

P   Protein neighbors

N   Nucleotide links

T   Structure links

Nucleotide sequence entries
Report types:

G   GenBank format

R   Report format

F   FastA format

A   ASN.1 format

D   Entrez document summary format

Neighbors/links:

M   MEDLINE links

P   Protein links

N   Nucleotide neighbors

T   Structure links

Structure entries
Report types:

S   Structure summary

D   Entrez document summary format

Neighbors/links:

M   MEDLINE links

P   Protein links

N   Nucleotide links

T   Structure neighbors

The UID for a sequence (protein or nucleotide) entry can be specified using the FastA format, if desired. FastA-formatted UIDs are of the general form:

```
DATABASE_NAME | ID1 | ID2
```

where ID1 and ID2 are identifier fields appropriate to the database. Normally only one of the fields is used, even if both are filled in. If an identifier field is not used, the specified number of vertical bar separators must still be used, as in DATABASE_NAME | | ID2. Supported FastA specifications are shown in Table 5.31.

Table 5.31. Query Server UID FastA Specifications

Name	Format To Use
DDBJ	DBJ|ACCESSION|LOCUS
EMBL	EMB|ACCESSION|LOCUS
GENBANK	GB|ACCESSION|LOCUS
GI	GI|INTEGER
GIBBMT	BBM|INTEGER
GIBBSQ	BBS|INTEGER
PATENT	PAT|COUNTRY|PATENT NUMBER (string)|SEQ NUMBER (integer)
PDB	PDB|ENTRY NAME (string)|CHAIN ID (single character)
PIR	PIR|ACCESSION|NAME
PRF	PRF|ACCESSION|NAME
SWISSPROT	SP|ACCESSION|NAME

For a TERM search, terms can be followed by optional field specifiers. These are summarized in Table 5.32. If no field specifier is used in a TERM search, then all fields will be searched.

When executing TERM searches, a text string may be entered and a field to search specified. This helps to refine a search, rather than linking the TERM search to certain UIDs. For TERM searches containing multiple terms, Boolean operators can be used. For example, in a WORD field, if the terms are separated by spaces rather than Boolean operators, a default AND will be used between each term. In addition, text used in a TERM search can be truncated, or cases can be ranged to refine a search. (Examples of truncation and ranging of searches are given in the sections that follow.)

Another useful option to refine searches is the field option S, which is used in conjunction with the WORD field name to retrieve special records. Special records contain the search term or terms in the title or main definition line of the record. To retrieve special records, add the field option S to the WORD field name, e.g., [WORD,S]. (Further examples of the use of this field option are shown in the sections below.)

The third line of the query message should give the display option directive, DOPT, so that records from each domain can be viewed in various formats. Display option is a code for how to show the citation. It consists of zero or more neighbor or link characters followed by a single character that designates the report type. The parameter options for the DOPT directive are given in Table 5.30.

Table 5.32. Query Server Field Specifiers

MEDLINE

AUTH	Author name
DATE	Publication year
ECNO	E.C. number
GENE	Gene name
JOUR	Journal name
KYWD	Substance
MESH	Mesh term
WORD	Text word

Protein sequence

ACCN	Accession number
AUTH	Author name
DATE	Publication year
DATM	Last modification date
ECNO	E.C. number
GENE	Gene name
JOUR	Journal name
KYWD	Keyword
ORGN	Organism
PROP	Property
PROT	Protein name
SLEN	Sequence length
SUBS	Substance
WORD	Text word

If DOPT is included in a search, then we recommend the report type R for a single or a few MEDLINE reports and G for a single or a few protein or nucleic acid sequence reports. For multiple documents in any database, or for neighbor and link searches, we recommend that D be used to first retrieve a title list of records. A complete record of interest can be retrieved from the list by searching for the query's UIDs and having their records displayed in the desired format.

The fourth line of the query message is reserved for any other optional search parameters such as HTML or DISPMAX.

■ Examples of queries

Because of the search engine and the structure of the data sets that can be searched by the Query server, complex queries can be submitted to the server. The following figures show examples in which a variety of different search strategies are demonstrated, illustrating the power of this server.

- UID searches. Figure 5.25 gives examples of queries based on UID searches. In Fig. 5.25A, the query instructs the server to retrieve the nucleotide record whose accession number is X67189 and then display it in the default GenBank format. The query in Fig. 5.25B instructs the server to retrieve the nucleotide records whose accession numbers are X67189 and U41233 and then display them in FastA format. Figure 5.25C is a UID query of the MEDLINE database whereby the record with MEDLINE UID 98065732 is being retrieved and will be displayed in the MEDLINE standard report format. Finally, the example in Fig. 5.25D will retrieve the MEDLINE links of the protein sequence with the SWISS-PROT accession number P56149. The information will be displayed in the standard report format and also in the HTML format, which is suitable for loading into a Web browser. In all of the UIDs searched above, the most recent 200 entries will be dis played (default DISPMAX).

Figure 5.25. UID Searches

A

```
From:        annep@acs.auc.eun.eg
Date:        Wed 25 Sep 1996 - 09:58.16 - GMT
To:          query@ncbi.nlm.nih.gov
Subject:
DB N
UID X67189
```

B

```
From:        annep@acs.auc.eun.eg
Date:        Wed 25 Sep 1996 - 09:58.16 - GMT
To:          query@ncbi.nlm.nih.gov
Subject:
DB N
UID X67189,U41233
DOPT F
```

continued

Figure 5.25 continued

```
C
        From:       annep@acs.auc.eun.eg
        Date:       Wed 25 Sep 1996 - 09:58.16 - GMT
        To:         query@ncbi.nlm.nih.gov
        Subject:
        DB M
        UID 98065732
        DOPT R


D
        From:       annep@acs.auc.eun.eg
        Date:       Wed 25 Sep 1996 - 09:58.16 - GMT
        To:         query@ncbi.nlm.nih.gov
        Subject:
        DB P
        UID SP|P56149|
        DOPT M
        HTML
```

■ **Term searches**. In Fig. 5.26 are shown examples of queries based on searches for
terms. The query in Fig. 5.26A instructs the server to search for protein entries
with the term ESCHERICHIA in all fields, then display the most recent 200 entries
in the Entrez Document Summary format. Figure 5.26B displays a query request-
ing the server to display the most recent 200 document summaries for MEDLINE
articles containing the text word FOS. In Fig. 5.26C, the search is for the term FOS
as a text word; the most recent 100 entries will be displayed as an Entrez
Document Summary. In Fig. 5.26D, the search is for MEDLINE articles with titles
that contain the word FOS; the server will then locate protein sequence links for
these articles and finally display the ten most recent of these protein sequences in
GenPept format. In another example of a term search, Fig. 5.26E, the server will
search for nucleotide entries by the author KLEMM DK and display the 15 most
recent entries in the default Entrez Document Summary format, but the entries
will be written in HTML format for viewing through a Web browser. In Fig. 5.26F,
the server will search for the nucleotide entry whose accession number is U45719
and display it in GenBank format. Finally, in the last example (Fig. 5.26G) the server will search for nucleotide entries by the author KLEMM DK and display the
most recent 15 in the default Entrez Document Summary format.

■ **Boolean searches**. Boolean searches can also be used to extend or refine a
search in much the same manner as for the Retrieve server. Query can understand
complex Boolean expressions in place of UIDs or single TERM-and-FIELD con-
structions. The search expression is created from one or more terms, separated
by operators. The field specifiers shown in Table 5.32 can follow terms if desired.

Figure 5.26. Term Searches

A

```
        From:       annep@acs.auc.eun.eg
        Date:       Wed 25 Sep 1996 - 09:58.16 - GMT
        To:         query@ncbi.nlm.nih.gov
        Subject:
        DB P
        TERM ESCHERICHIA
```

B

```
        From:       annep@acs.auc.eun.eg
        Date:       Wed 25 Sep 1996 - 09:58.16 - GMT
        To:         query@ncbi.nlm.nih.gov
        Subject:
        DB M
        TERM FOS[WORD]
```

C

```
        From:       annep@acs.auc.eun.eg
        Date:       Wed 25 Sep 1996 - 09:58.16 - GMT
        To:         query@ncbi.nlm.nih.gov
        Subject:
        DB M
        TERM FOS[WORD]
        DISPMAX 100
```

D

```
        From:       annep@acs.auc.eun.eg
        Date:       Wed 25 Sep 1996 - 09:58.16 - GMT
        To:         query@ncbi.nlm.nih.gov
        Subject:
        DB M
        term FOS[WORD,S]
        DISPMAX 10
        DOPT PG
```

E

```
        From:       annep@acs.auc.eun.eg
        Date:       Wed 25 Sep 1996 - 09:58.16 - GMT
        To:         query@ncbi.nlm.nih.gov
        Subject:
        DB N
        TERM KLEMM DK [AUTH]
        DISPMAX 15
        HTML
```

F

```
        From:       annep@acs.auc.eun.eg
        Date:       Wed 25 Sep 1996 - 09:58.16 - GMT
        To:         query@ncbi.nlm.nih.gov
        Subject:
        DB N
        TERM U45719 [ACCN]
        DOPT G
```

G

```
        From:       annep@acs.auc.eun.eg
        Date:       Wed 25 Sep 1996 - 09:58.16 - GMT
        To:         query@ncbi.nlm.nih.gov
        Subject:
        DB N
        TERM KLEMM DK [AUTH]
        DISPMAX 15
```

Symbols that are used by the Query server to represent Boolean operators are
& for AND, | for OR, and - for NOT. Since the Boolean operator - (NOT) can also
be used as a hyphen in a word, it must be separated from the terms around it by
one or more spaces. Boolean expressions are normally processed left to right. To
process part of a Boolean expression out of order, enclose it in parentheses.

In Fig. 5.27A, the server will first retrieve all nucleotide entries containing the
text word FOS and the journal name SCIENCE, then find nucleotide sequences
that neighbor these entries and display the most recent 200 as document summaries. Figure 5.27B shows another application of Boolean searches. In this example, the server will display all the nucleotide sequence records for the organism
Escherichia coli that were added to the nucleotide databases on 16 February 1996
and display them in GenBank format.

Figure 5.27. Boolean Searches

A

```
From:      annep@acs.auc.eun.eg
Date:      Wed 25 Sep 1996 - 09:58.16 - GMT
To:        query@ncbi.nlm.nih.gov
Subject:
DB N
TERM FOS[WORD]&SCIENCE[JOUR]
DOPT N
```

B

```
From:      annep@acs.auc.eun.eg
Date:      Wed 25 Sep 1996 - 09:58.16 - GMT
To:        query@ncbi.nlm.nih.gov
Subject:
DB N
TERM ESCHERICHIA COLI[ORGN] & 1996/02/16[DATM]
DOPT G
```

■ **Truncation searches.** Truncation searches are useful when the spelling of the
full term is unknown or when it is necessary to expand a search. All of the terms
that begin with a given string can be searched on by appending three dots (...) to
the end of the term. An asterisk (*) can also be used for truncation. For instance,
RHIZOP* will retrieve all terms that begin with the characters RHIZOP.

In Fig. 5.28A, the server is directed to search for nucleotide entries by the
author KLEMM, whose initials are D or D and any other character, and display the
most recent 20 in the default Entrez Document Summary format. In Fig. 5.28B, the
server will obtain the protein sequence records that contain the characters
CAMPY and display the most recent 200 in document summary format. In the last
example of a truncation search, Fig. 5.28C, the server will locate the MEDLINE
articles which have MESH terms that begin with the characters MYCOBACT and
display the most recent 50 in the document summary format.

Figure 5.28. Truncation Searches

A

```
From:       annep@acs.auc.eun.eg
Date:       Wed 25 Sep 1996 - 09:58.16 - GMT
To:         query@ncbi.nlm.nih.gov
Subject:
DB N
TERM KLEMM D* [AUTH]
DISPMAX 20
```

B

```
From:       annep@acs.auc.eun.eg
Date:       Wed 25 Sep 1996 - 09:58.16 - GMT
To:         query@ncbi.nlm.nih.gov
Subject:
DB P
TERM CAMPY...[WORD]
```

C

```
From:       annep@acs.auc.eun.eg
Date:       Wed 25 Sep 1996 - 09:58.16 - GMT
To:         query@ncbi.nlm.nih.gov
DB M
TERM MYCOBACT*[MESH]
DISPMAX 50
```

- **Range searches.** Range searches are a powerful tool to retrieve all records that occur in a sequential pattern, such as those records entered over a specific period of time or those records that belong to a series or accession number. Ranging can be used for numerical fields such as accession number, date, and sequence length. The symbol for ranging is the colon (:). The search should be constructed in a manner similar to a Boolean search.

  In Fig. 5.29A, the server will obtain sequence records with accession numbers ranging from U23475 through U23590 and display them in FastA format. In Fig. 5.29B, the server will obtain sequence records that are between 50,000 and 200,000 base pairs in length and display their document summaries. In the last example of a range search, Fig. 5.29C, the server will obtain all nucleotide records from *Escherichia coli* that were added to the nucleotide sequence database between 1 February and 31 March 1996, and display a maximum of 350 of these records in document summary format.

Figure 5.29. Range Searches

A

```
From:      annep@acs.auc.eun.eg
Date:      Wed 25 Sep 1996 - 09:58.16 - GMT
To:        query@ncbi.nlm.nih.gov
Subject:
DB N
TERM U23475 : U23590 [ACCN]
DOPT F
```

B

```
From:      annep@acs.auc.eun.eg
Date:      Wed 25 Sep 1996 - 09:58.16 - GMT
To:        query@ncbi.nlm.nih.gov
Subject:
DB N
TERM 050000 : 100000 [SLEN]
DOPT D
```

C

```
From:      annep@acs.auc.eun.eg
Date:      Wed 25 Sep 1996 - 09:58.16 - GMT
To:        query@ncbi.nlm.nih.gov
Subject:
DB N
TERM ESCHERICHIA COLI[ORGN] & (1996/02/01:1996/03/31 [DATM])
DISPMAX 350
```

■ **Display options.** The query in Fig. 5.30A requests that the server display the nucleotide sequence record with accession number X12387 in the GenBank format. In Fig. 5.30B, the server will display the record for the nucleotide sequence with the accession number X12387 in the FastA format. In Fig. 5.30C, the server will display the FastA format of protein sequence records linked to the nucleotide record X12387.

The last two examples in Fig. 5.30 demonstrate how to retrieve special records. In the example query shown in Fig. 5.30D, the server will display the most recent 200 document summaries for all MEDLINE articles that contain the word PRION in their title. In the final example, Fig. 5.30E, the server will display the GenBank format of protein sequence records linked to any MEDLINE records that have the word PRION in their title.

Figure 5.30. Display Options

A

```
From:       annep@acs.auc.eun.eg
Date:       Wed 25 Sep 1996 - 09:58.16 - GMT
To:         query@ncbi.nlm.nih.gov
Subject:
DB N
UID X12387
```

B

```
From:       annep@acs.auc.eun.eg
Date:       Wed 25 Sep 1996 - 09:58.16 - GMT
To:         query@ncbi.nlm.nih.gov
Subject:
DB N
UID X12387
DOPT F
```

C

```
From:       annep@acs.auc.eun.eg
Date:       Wed 25 Sep 1996 - 09:58.16 - GMT
To:         query@ncbi.nlm.nih.gov
Subject:
DB N
UID X12387
DOPT PF
```

D

```
From:       annep@acs.auc.eun.eg
Date:       Wed 25 Sep 1996 - 09:58.16 - GMT
To:         query@ncbi.nlm.nih.gov
Subject:
DB M
TERM PRION [WORD,S]
```

E

```
From:       annep@acs.auc.eun.eg
Date:       Wed 25 Sep 1996 - 09:58.16 - GMT
To:         query@ncbi.nlm.nih.gov
Subject:
DB M
TERM PRION [WORD,S]
DOPT PG
```

Retrieve

The Retrieve e-mail server is designed to search for and retrieve protein and nucleic acid sequence data records from the sequence databases at NCBI. Searches can be done using different parameters such as accession number, locus, author name, or text words. Several sequence databases are available for searching, including GenBank, EMBL, SWISS-PROT, PIR, and OMIM.

 Internet addresses

E-mail:	retrieve@ncbi.nlm.nih.gov (NCBI server)
	retrieve@ebi.ac.uk (EBI server)
	retrieve-help@ncbi.nlm.nih.gov (Personal help)
Web:	http://www.ncbi.nlm.nih.gov (NCBI home page)
Organization:	National Center for Biotechnology Information, National Library of Medicine, Building 38A, Room 8S-803, 8600 Rockville Pike, Bethesda, MD 20894, U.S.A.
Telephone:	+1-301-496-2475
Facsimile:	+1-301-480-9241

■ **Accessing the server**

To use Retrieve, send an e-mail message to one of the server e-mail addresses listed above or access the server through the NCBI home page on the Web. The latest information on changes to the server and how to use it can be obtained by sending an e-mail message to the server. The body of the message should have a single line with the word HELP. A subject line is not necessary. Instructions on searching and retrieving records from the Expressed Sequence Tag database (dbEST) or the Sequence Tag Site database (dbSTS) maintained at NCBI can be obtained by sending an e-mail message to the Retrieve server as well (refer to the section on dbEST and dbSTS in Chapter 5 for further information). The body of the message to the server must contain the following text:

```
DATALIB DBEST (OR DBSTS)
HELP
```

For questions that are not answered by the help files returned by the server, the NCBI staff maintains an e-mail address for personal help which is also listed above. To use this service, e-mail a description of the problem to the staff at NCBI, including an example of the query sent to the server and any output received back.

■ **Search directives and parameters**

An e-mail message to the Retrieve server, like most of the other servers described in this chapter, is composed of a set of directives, each on a separate line of the message, followed by a query on one or more lines. Table 5.22 lists and briefly describes the server directives. The format of the message is strictly defined. First is a mandatory line with the directive DATALIB and a parameter composed of a database abbreviation. This tells the server which database to search. (Table 5.23 lists the databases which can be searched by the Retrieve server.) Next, several optional directives can be specified. After the optional directives comes another mandatory line with the BEGIN directive, followed by the query on subsequent lines.

Table 5.22. Retrieve Server Directives and Parameters

Directive	Attributes	Explanation
DATALIB	Mandatory, numeric	Specifies the database to be searched. Only one database can be searched per message. Available databases are listed in Table 5.23.
MAXDOCS	Optional, numeric	Sets the maximum number of sequence documents sent back. The default value is 20 and the upper limit is 2,400.
MAXLINES	Optional, numeric	Sets the maximum number of lines of output in the reply e-mail message. The default value is 1,000 and the upper limit is 50,000.
TITLES	Optional, none	If included in a query message, only the title of the sequence records will be sent in the reply e-mail message.
STARTDOC	Optional, numeric	Sets the starting record number; the default value is 1. Can be used in successive mail messages to retrieve blocks of records when a query to the server generates data that exceeds the limits of the MAXDOCS or MAXLINES directives.
PATH	Optional, text	Gives a return e-mail address with the PATH directive if there is a problem in having the server send reply messages. An example for use of this directive is PATH ANNEP@ACS.AUC.EUN.EG
BEGIN	Mandatory, none	This directive must be the last directive before the query. There is no parameter associated with this directive.

Table 5.23. Retrieve Databases

Nucleotide sequence databases

GB or GENBANK	GenBank DNA sequence: database and current update
GBU or GBUPDATE	GenBank update: current updates of GenBank only
GBONLY	Last public release of the full GenBank database
EMBL or EMB	EMBL DNA sequence database: database and current update
EMBLUPDATE or EMBLU	EMBL update: current updates of EMBL only
EMBLONLY	Last public release of the full EMBL database

Protein databases

SP or SWISS or SWISSPROT	SWISS-PROT protein database
SPU or SWISSPROTUPDATE	SWISS-PROT updates
PIR	PIR protein database
GP or GENPEPT	GenPept, a translation of GenBank
GPU or GPUPDATE	GenPept update: current updates of GenPept only

Specialized databases

OMIM	Online Mendelian Inheritance in Man
VECTOR	Vector sequence subset of GenBank
VECBASE	Vecbase (1987 version of the database)
KABATNUC	Kabat's database of sequences of immunological interest: nucleotide
KABATPRO	Kabat's database of sequences of immunological interest: protein
EPD	Eukaryotic Promoter Database
PDB	Protein Data Bank
TFD	Transcription Factors Database

■ **Formatting a query**

Through the DATALIB directive there are several sequence and related databases that can be queried using the Retrieve server. Table 5.23 lists the databases that can be accessed by the Retrieve server and briefly details what type of information the database contains.

Some of the databases listed in Table 5.23 have several fields within each database record. These databases can be searched for specific terms within those fields, which can both speed the search and refine it to retrieve a more focused group of database records. Listed in Table 5.24 are these databases and their associated fields. Abbreviations for the database fields are given in brackets. Use the field abbreviations to define field restrictions for searches using the Retrieve server. The use of field restrictions in searching is described in more detail in the examples shown below. Please note that not all of the databases listed in Table 5.24 have fields that can be searched.

Table 5.24 Retrieve Sequence Databases and Fields

EMBL and EMBLUPDATE	**OMIM**
DEFINITION [DEF]	MIM Number [NO]
ID [LOC]	Title [TI]
ACCESSION [ACC]	Mini-Min [MN]
KEYWORDS [KEY]	Text [TX]
DATES [DAT]	Allelic Variants [AV]
SOURCE [SRC]	See Also References [SA]
CROSS-REF [DXR]	References [RF]
REFERENCE [REF]	Clinical Synopses [CS]
COMMENT [COM]	Creation Date [CD]
FEATURES [FEA]	Edit History [ED]
GENBANK and GBUPDATE	**PDB (Brookhaven)**
DEFINITION [DEF]	DEFINITION [DEF]
LOCUS [LOC]	HEADER [HDR]
ACCESSION NO. [ACC]	ACCESSION [ACC]
KEYWORDS [KEY]	DATE [DAT]
SEGMENT [SEG]	SOURCE [SRC]
SOURCE [SRC]	AUTHOR [AUT]
REFERENCE [REF]	SUPERSEDE [SPR]
COMMENT [COM]	REFERENCE [REF]
FEATURES [FEA]	COMMENTS [COM]
ORIGIN [ORG]	FOOTNOTE [FTN]
	HETEROGENS [HET]

continued

Table 5.24 continued

PIR (NBRF)	SWISS-PROT
DEFINITION [DEF]	DEFINITION [DEF]
ALT-NAME [ALT]	ID [LOC]
SUMMARY [SUM]	ACCESSION [ACC]
DATE [DAT]	KEYWORDS [KEY]
SUPERFAMILY [SUP]	DATES [DAT]
ACCESSION NO [ACC]	GENE NAME [GEN]
HOST [HST]	SOURCE [SRC]
KEYWORDS [KEY]	ORGANISM CLASSIFICATION [CLS]
SOURCE [SRC]	ORGANELLE [ORG]
GENETICS [GEN]	REFERENCE [REF]
INCLUDES [INC]	COMMENT [COM]
REFERENCE [REF]	FEATURES [FEA]
COMMENT [COM]	CROSS REFERENCE [DCR]
FEATURES [FEA]	SEQUENCE DATA [BAS]
SEQUENCE	

■ **Examples of queries**

Shown in Fig. 5.16 through 5.22 are several example messages to the Retrieve server. Like most other servers, the Retrieve server is case-insensitive. These examples show the search directives and parameters in uppercase letters.

In Fig. 5.16, a typical message is being sent to the server. As we mentioned in Chapter 3, the first four lines of the message form the header. The Subject line can be left blank, or a phrase or word may be included to identify the message. The Retrieve server will repeat the information in the Subject line in the return e-mail message. For example, if the text "phosphatase search" was entered into the subject line of the query submitted to the server, then the subject line of the reply would contain, "Results-RETRIEVE Server: phosphatase search."

Figure 5.16. Example of a Retrieve Server Search Limited by Field

```
From:      annep@acs.auc.eun.eg
Date:      Wed 25 Sep 1996 - 09:58.16 - GMT
To:        retrieve@ncbi.nlm.nih.gov
Subject:   phosphatase search

DATALIB GENBANK
BEGIN
U01749 A03211
```

The body of the e-mail message to the server starts with the mandatory search parameter DATALIB. Immediately following is the name of the database that is to be searched, in this case, GENBANK. The next line contains the mandatory search parameter BEGIN. Any text after this line is treated as part of the query. In this query, the server will retrieve the complete GenBank records for the GenBank accession numbers U01749 and A03211. It is important not to have any blank lines after the BEGIN directive or before any portion of the query.

After the BEGIN line comes the query part of the message. The query can be made up of multiple lines, each containing terms from different fields within the database records. For example, if a query contains lines of accession numbers, with each accession number on its own line, then the Retrieve server will send back all sequence records containing any of those accession numbers. In this manner, each line of the query is treated as a separate request, much like using the word OR between terms on the same line.

In addition, this default term OR applies to terms listed on the same line. If the accession numbers were all placed on the same line as in Fig. 5.16, then the server would again send back all sequence records that contained any of the accession numbers. The default logical term OR can be overridden by using the logical terms AND or NOT. To expand a search term to related words or to permit misspelling, wild card characters can be used within the query terms as shown in Table 5.25.

Table 5.25. Retrieve Wild Card Characters

Character	Description
#	match any single character
$	match zero or one character
*	match zero or more characters

In two of the databases, SWISS-PROT and GenPept, underscores are used to identify locus names for sequence records. To retrieve a record with an underscore such as CAT_PPSUB from the SWISS-PROT database or PP2A_B from the GenPept database, the locus name being used as a query must be enclosed in double quotes as shown in Fig. 5.17.

Double quotes are needed to retrieve records by locus name for these two databases because the retrieve server treats the underscore as the word delimiter, OR. So, it would treat PP2A_B as a query for PP2A or B and CAT_PPSUB as CAT or PPSUB. In addition, the use of double quotation marks is useful to restrict a search. A search performed using more than one term in double quotes will only retrieve records that have those terms occurring together within the same single field of a record. To retrieve several entries from SWISS-PROT, the OR operator must be specified as shown in Fig. 5.18.

Figure 5.17. Example of a Retrieve Server Search Using Double Quotes To Retrieve a Record by Locus

A

```
From:          annep@acs.auc.eun.eg
Date:          Wed 25 Sep 1996 - 09:58.16 - GMT
To:            retrieve@ncbi.nlm.nih.gov
Subject:       phosphatase search

DATALIB SWISSPROT
BEGIN
"CAT_PPSUB"
```

B

```
From:          annep@acs.auc.eun.eg
Date:          Wed 25 Sep 1996 - 09:58.16 - GMT
To:            retrieve@ncbi.nl.nih.gov
Subject:       phosphatase search

DATALIB GENPEPT
BEGIN
"PP2A_B"
```

Figure 5.18. Example of the Retrieve Server Use of the OR Operator To Retrieve Several Entries

```
From:          annep@acs.auc.eun.eg
Date:          Wed 25 Sep 1996 - 09:58.16 - GMT
To:            retrieve@ncbi.nlm.nih.gov
Subject:       phosphatase search

DATALIB SWISSPROT
BEGIN
"CAT_PP2B" OR
"CAT_PP2C" OR
"CAT_PPX"
```

To refine searches for specific types or classes of sequence records, the Retrieve server also offers the ability to write Boolean or logical expression-based queries. As we mentioned, the server has a default logical setting of OR for any searches submitted to it. To override this default logical setting, the logical operators AND, OR, and NOT can be used singly or in combination. While the logical operator OR is still the default setting, it can be part of the query that is submitted to the server. The logical operator NOT can be used to retrieve records that match one query term, but not another query term. These three logical operators can be used both within single lines or between lines to construct Boolean queries. The Retrieve server can process queries of up to 240 characters on a single line, provided that the ENTER or RETURN key is not used. Note that when constructing a query with more than one logical operator, parentheses can be useful to group terms and operators. When parentheses are used in a query, the server will carry out the part of the query in parentheses first, then attempt to complete the remaining parts of the query.

Figure 5.19 shows a search request whereby 30 records on protein phosphatase X in humans or yeast would be retrieved.

Figure 5.19. Example of the Retrieve Server Use of Logical Operators To Refine a Search with Retrieve

```
From:        annep@acs.auc.eun.eg
Date:        Wed 25 Sep 1996 - 09:58.16 - GMT
To:          retrieve@ncbi.nlm.nih.gov
Subject:     phosphatase search

DATALIB GENBANK
MAXDOCS 30
BEGIN
PROTEIN AND PHOSPHATASE AND X AND HUMAN OR YEAST
```

If too many records are retrieved, try using double quotes around the query terms. The use of double quotes is an effective tool to narrow a search because it mandates that the terms in the double quotes be located within a single field of the database record, such as the COMMENT or TITLE fields for GenBank. Note that if more than one set of double quotes is used in a query, then a logical operator must be used between each of the double quote sets.

■ Limiting searches to specific fields

Another common problem when using Retrieve concerns queries based on accession numbers. Often a query specifying accession numbers results not only in the desired database record or records being returned, but many seemingly unrelated and highly extraneous records as well. This is because accession numbers can appear in several other fields of a database record. For example, in GenBank, accession numbers are found in the ACCESSION field and the COMMENT field. This means that if the accession number used in a query appears in the COMMENT field of any other database records, then those records will be returned in response to the query (Fig. 5.20).

To refine searches, queries to Retrieve can be limited to specific database fields of records. To do this, after the parameter used for the BEGIN directive, add the abbreviation of the field in brackets. A listing of databases and their fields is given in Table 5.24. This search strategy will limit the search of the database to a user-defined term or expression to the specified field. In the example shown in Fig. 5.20, a search for the accession numbers J02459 and J01636 is limited to the ACCESSION field of GenBank. Similarly, in the example shown in Fig. 5.21, a search for the term "phosphatase" is limited to the KEYWORDS field of GenBank.

Figure 5.20. Example of a Retrieve Server Search Limited to the Accession Number Field

```
From:        annep@acs.auc.eun.eg
Date:        Wed 25 Sep 1996 - 09:58.16 - GMT
To:          retrieve@ncbi.nlm.nih.gov
Subject:     phosphatase search

DATALIB GENBANK
BEGIN
J02459 [ACC]
J01636 [ACC]
```

Figure 5.21. Example of a Retrieve Server Search Limited to the Keyword Field

```
From:        annep@acs.auc.eun.eg
Date:        Wed 25 Sep 1996 - 09:58.16 - GMT
To:          retrieve@ncbi.nlm.nih.gov
Subject:     phosphatase search

DATALIB GENBANK
BEGIN
PHOSPHATASE [KEY]
```

As we mentioned, double quotes are used around query terms to narrow searches. This strategy can be used to further narrow a search by restricting the search to a specific field of the database. Figure 5.22 shows how to retrieve records from the update to GenBank database that contain a group of words, "protein ser/thr phosphatase," in the keyword field. As before, if more than one set of double quotes are used in a query, then a logical operator must be used between each of the double quote sets.

Figure 5.22. Further Example of a Retrieve Server Search Limited to the Keyword Field

```
From:        annep@acs.auc.eun.eg
Date:        Wed 25 Sep 1996 - 09:58.16 - GMT
To:          retrieve@ncbi.nlm.nih.gov
Subject:     phosphatase search

DATALIB GBUPDATE
BEGIN
"PROTEIN SER/THR PHOSPHATASE" [KEY]
```

STS and EST

The Sequence Tagged Site (STS) and the Expressed Sequence Tag (EST) report servers are designed for the retrieval of detailed reports on entries in the NCBI databases or in the nucleotide sequence databases dbSTS and dbEST. Retrieval of entries is done by specifying either the unique NCBI identifier (ID) for each record of interest or certain other fields of the record, or by utilizing a chromosome/map string.

Internet addresses

E-mail:	retrieve@ncbi.nlm.nih.gov	(NCBI server)
	retrieve@ebi.ac.uk	(EBI server)
	retrieve-help@ncbi.nlm.nih.gov	(Personal help)
Web:	http://www.ncbi.nlm.nih.gov	(NCBI home page)
Organization:	National Center for Biotechnology Information, National Library of Medicine, Building 38A, Room 8S-803, 8600 Rockville Pike, Bethesda, MD 20894, U.S.A.	
Telephone:	+1-301-496-2475	
Facsimile:	+1-301-480-9241	

Accessing the server

To access either one of these servers, send a properly formatted e-mail message to the Retrieve server at the address listed above. To receive the latest set of instructions on using either server, send a message as shown in Fig. 5.23 to the Retrieve server.

Figure 5.23. Requesting a Help File on the dbSTS Server

```
From:      annep@acs.auc.eun.eg
Date:      Wed 25 Sep 1996 - 09:58.16 - GMT
To:        retrieve@ncbi.nlm.nih.gov
Subject:

DATALIB DBSTS
HELP
```

■ **Search parameters and directives**

Queries to either of these servers are composed of directives and associated parameters that are used to identify the EST and STS records to retrieve. There are four directives that are shared by these two servers: DATALIB, DATATYPE, RPT, and ORG. Of these four, DATALIB and RPT are mandatory. DATALIB is used to specify which report server to access. The parameters for this directive are DBEST for the EST Report Server or DBSTS for the STS Report Server. The RPT directive functions in an analogous fashion to the BEGIN or SEQ directives of the search servers discussed earlier in this chapter; it marks the start of the report identifiers that will be used to retrieve the files. Unlike the BEGIN or SEQ directives, the RPT directive and its report identifiers appear on the same line of the query message.

If the DATATYPE directive is not used in a query, the EST Report Server will use the parameter DBEST as the default setting, while the STS Report Server will use NCBI. If the DATATYPE directive is used in a query, then it must be followed by one of the parameters shown in Table 5.26.

Table 5.26. STS and EST Server DATATYPE Parameters

dbSTS DATATYPE parameters

NCBI (the ID assigned by NCBI) (*default setting*)

SOURCE (the name given by the contributor)

ATCC_H (the ATCC inhost number)

ATCC_D (the ATCC pure DNA number)

GENBANK (the GenBank accession number)

CLONE (the clone ID)

CHROM (the chromosome)

MAPSTR (the map address string)

dbEST DATATYPE parameters

DBEST (the ID assigned by dbEST) (*default setting*)

GI (the GenBank identification number)

GENBANK (the GenBank accession number)

NAME (the name given by the contributor)

SOURCE_H (the clone source inhost number)

SOURCE_D (the clone source pure DNA number)

CLONE (the clone ID)

CHROM (the chromosome)

MAPSTR (the map address string)

■ **Formatting a query**

The first line of the body of the query must specify the data set to be searched, as follows:

```
DATALIB DBSTS
```

If the directive DATATYPE is used in a query, it must appear on the line immediately after the DATALIB directive. If either of the DATATYPE parameters CHROM or MAPSTR is used, the directive ORG and its parameter must be used to specify the organism to be searched. This directive and its parameter must appear on the next line after the DATATYPE directive. The organisms that can be specified as parameters are shown in Table 5.27. The next directive that must be in the query is RPT. The IDs of the records to be retrieved may appear on one or more lines of the query message, each line commencing with the directive RPT. If multiple IDs appear on a line, they must be separated by a blank character.

Table 5.27. Organisms in the dbSTS and dbEST Databases

Organisms in the dbSTS database

Homo sapiens

Drosophila melanogaster

Oryza sativa

Organisms in the dbEST database

Homo sapiens

Arabidopsis thaliana

Caenorhabditis elegans

Oryza sativa

Plasmodium falciparum

Zea mays

Mus musculus

Mus domesticus

Saccharomyces cerevisiae

Capra hircus

Pyrococcus furiosus

Macropus eugenii

Gallus domesticus

Gallus gallus

Sus scrofa

While both the STS server and the EST server function in a nearly identical fashion, there are some important differences. When using the STS server, it is not necessary to include the full organism name as a parameter with the ORG directive as long as the organism string is long enough to be unambiguous. For example, ORG HOMO SAPIENS and ORG HOMO will both search the *Homo sapiens* section of the database. Further, the STS server is completely case-insensitive. In contrast, the EST server is partially case-sensitive. The organism name for the ORG directive and the data used with the RPT directive are both case-sensitive. For example, GenBank accessions must appear as T00871, not t003871, and the organisms such as *Zea mays* cannot appear as ZEA MAYS. Lastly, both servers require that each line of a query be less than 80 characters in length. Longer lines will be truncated.

■ Examples of queries

Figure 5.24 shows four example queries. The example query in Fig. 5.24A requests reports on five sequences. The default NCBI IDs for the reports are used. The second example query, Fig. 5.24B, also requests reports, but uses GenBank accession numbers instead. In the third example query, Fig. 5.24C, all human entries that are mapped to chromosome 7 are requested. The last example, Fig. 5.24D, uses the MAPSTR parameter as the retrieval method to request all reports on entries that map the p arm of chromosome 18 in the organism *Homo sapiens*.

Figure 5.24. Examples of STS and EST Server Queries Using Other Datatypes

A: Query Using NCBI IDs

```
From:        annep@acs.auc.eun.eg
Date:        Wed 25 Sep 1996 - 09:58.16 - GMT
To:          retrieve@ncbi.nlm.nih.gov
Subject:

DATALIB DBSTS
RPT 5560 5431 12774 17221 15173 367
```

B: Query Using GenBank Accession Numbers

```
From:        annep@acs.auc.eun.eg
Date:        Wed 25 Sep 1996 - 09:58.16 - GMT
To:          retrieve@ncbi.nlm.nih.gov

DATALIB DBEST
DATATYPE GENBANK
RPT G00001 Z16571 L10012 L14180 G00002 G00103
```

continued

Figure 5.24 continued

C: Query Using Chromosome Location

```
From:        annep@acs.auc.eun.eg
Date:        Wed 25 Sep 1996 - 09:58.16 - GMT
To:          retrieve@ncbi.nlm.nih.gov
Subject:


DATALIB DBSTS
DATATYPE CHROM
ORG HOMO SAPIENS
RPT 7
```

D: Query Using a Map String

```
From:        annep@acs.auc.eun.eg
Date:        Wed 25 Sep 1996 - 09:58.16 - GMT
To:          retrieve@ncbi.nlm.nih.gov
Subject:


DATALIB DBEST
DATATYPE MAPSTR
ORG Homo sapiens
RPT 18P
```

Selected readings

For additional information on individual servers it is highly recommended that the help documentation from the respective server of interest be consulted. Help documentation can usually be obtained by sending an e-mail request to the server with the word HELP on a single line in the body of the message. The server will return the help document in a reply message. If the server is also offered on a Web site, another way to obtain a help file is to browse the site. Usually the help documentation is maintained on line and can be accessed via the Web; in many cases it is maintained as a single text-based document that can be downloaded or viewed through a browser.

Journal Articles

Altschul, S. F. 1993. A protein alignment scoring system sensitive at all evolutionary distances. *J. Mol. Evol.* **36**:290-300.

Altschul, S. F., W. Gish, W. Miller, E. W. Myers, and D. J. Lipman. 1990. Basic local alignment search tool. *J. Mol. Biol.* **219**:403-410.

Altschul, S. F., and D. J. Lipman. 1990. Protein database searches for multiple alignments. *Proc. Natl. Acad. Sci. USA* **87**:5509-5513.

Appel, R. D., A. Bairoch, and D. F. Hochstrasser. 1994. A new generation of information retrieval tools for biologists: the example of the ExPASy WWW server. *Trends Biochem. Sci.* **19**:258-260.

Califano, A., and I. Rigoutsos. 1993. FLASH: a fast look-up algorithm for string homology. *ISMB* **1**:56-64.

Collins, J. F., and A. F. W. Coulson. 1990. Significance of protein sequence similarities. *Methods Enzymol.* **183**:474-486.

Dayhoff, M. O. 1976. The origin and evolution of protein superfamilies. *Fed. Proc.* **35**:2132-2138.

Dayhoff, M. O., W. C. Barker, and L. T. Hunt. 1983. Establishing homologies in protein sequences. *Methods Enzymol.* **91**:524-545.

Devereux, J., P. Haeberli, and O. A. Smithies. 1984. Comprehensive set of sequence analysis programs for the VAX. *Nucleic Acids Res.* **12**:387-395.

Doolittle, R. F. 1981. Similar amino acid sequences: chance or common ancestry? *Science* **214**:149-159.

Doolittle, R. F. 1989. Similar amino acid sequences revisited. *Trends Biochem. Sci.* **14**:244-245.

Doolittle, R. F. 1990. Searching through sequence databases. *Methods Enzymol.* **183**:99-110.

Doolittle, R. F. 1992. Stein and Moore Award address. Reconstructing history with amino acid sequences. *Protein Sci.* **1**:191-200.

Doolittle, R. F., and D. F. Feng. 1990. Nearest neighbor procedure for relating progressively aligned amino acid sequences. *Methods Enzymol.* **183**:659-669.

Feng, D. F., and R. F. Doolittle. 1990. Progressive alignment and phylogenetic tree construction of protein sequences. *Methods Enzymol.* **183**:375-387.

Feng, D. F., M. S. Johnson, and R. F. Doolittle. 1984. Aligning amino acid sequences: comparison of commonly used methods. *J. Mol. Evol.* **21**:112-125.

Green, P., D. Lipman, L. Hillier, R. Waterston, D. States, and J. M. Claverie. 1993. Ancient conserved regions in new gene sequences and the protein databases. *Science* **259**:1711-1716.

Gribskov, M., A. D. McLachlan, and D. Eisenberg. 1987. Profile analysis: detection of distantly related proteins. *Proc. Natl. Acad. Sci. USA* **84**:4355-4358.

Henikoff, S., and J. G. Henikoff. 1992. Amino acid substitution matrices from protein blocks. *Proc. Natl. Acad. Sci. USA* **89**:10915-10919.

Henikoff, S., and J. G. Henikoff. 1994. Position-based sequence weight. *J. Mol. Biol.* **243**:574-578.

Henikoff, S., J. C. Wallace, and J. P. Brown. 1990. Finding protein similarities with nucleotide sequence databases. *Methods Enzymol.* **183**:111-132.

Lipman, D. J., S. F. Altschul, and J. D. Kececioglu. 1989. A tool for multiple sequence alignment. *Proc. Natl. Acad. Sci. USA* **86**:4412-4415.

Lipman, D. J., and W. R. Pearson. 1985. Rapid and sensitive protein similarity searches. *Science* **227**:1435-1441.

Lipman, D. J., and W. R. Pearson. 1986. Multiple sequence alignment by consensus. *Nucleic Acids Res.* **14**:9095-9102.

Lipman, D. J., W. J. Wilbur, T. F. Smith, and M. S. Waterman. 1984. On the statistical significance of nucleic acid similarities. *Nucleic Acids Res.* **12**:215-226.

Needleman, S. B., and C. D. Wunsch. 1970. A general method applicable to the search for similarities in the amino acid sequence of two proteins. *J. Mol. Biol.* **48**:443-453.

Pearson, W. R. 1990. Rapid and sensitive sequence comparison with FASTP and FASTA. *Methods Enzymol.* **183**:63-98.

Pearson, W. R., and D. J. Lipman. 1985. Rapid and sensitive protein similarity searches. *Science* **227**:1435-1441.

Pearson, W. R., and D. J. Lipman. 1988. Improved tools for biological sequence comparison. *Proc. Natl. Acad. Sci. USA* **85**:2444-2448.

Pearson, W. R., and W. Miller. 1991. Searching protein sequence libraries: comparison of the sensitivity and selectivity of the Smith-Waterman and FASTA algorithms. *Genomics* **11**:635-650.

Smith, T. F., and M. S. Waterman. 1981. Identification of common molecular subsequences. *J. Mol. Biol.* **147**:195-197.

Smith, T. F., M. S. Waterman, and C. Burks. 1984. Efficient sequence alignment algorithms. *J. Theor. Biol.* **108**:333-337.

Sonnhammer, E. L. L., and D. Kahn. 1994. Modular arrangement of proteins as inferred from analysis of homology. *Protein Sci.* **3**:482-492.

Tatusov, R. L., S. F. Altschul, and E.V. Koonin. 1994. Detection of conserved segments in proteins: iterative scanning of sequence databases with alignment blocks. *Proc. Natl. Acad. Sci. USA* **91**:12091-12095.

Uberbacher, E. C., and R. J. Mural. 1991. Locating protein-coding regions in human DNA sequences by a multiple sensor-neural network approach. *Proc. Natl. Acad. Sci. USA* **88**:11261-11265.

Wilbur, W. J., and D. J. Lipman. 1983. Rapid similarity searches of nucleic acid and protein data banks. *Proc. Natl. Acad. Sci. USA* **80**:726-730.

Books and Monographs

Dayhoff, M. O., R. M. Schwartz, and B. C. Orcutt. 1978. A model of evolutionary change in proteins: matrices for detecting distant relationships, p. 345-358. *In* M. O. Dayhoff (ed.), *Atlas of Protein Sequences and Structure*, vol. 5, Suppl. 3. National Biomedical Research Foundation, Washington, D.C.

Devereux, J. 1988. A rapid method for identifying sequences in large nucleotide sequence databases. Ph.D. thesis. Reprints available from GCG, Madison, Wis.

Gribskov, M., and J. Devereux (ed.). 1991. *Sequence Analysis Primer.* Stockton Press, New York.

Singh, R. K., S. G. Tell, C. T. White, D. Hoffman, V. L. Chi, and B. W. Erickson. 1993. A scalable systolic multiprocessor system for analysis of biological sequences, p. 168-182. *In Research on Integrated Systems: Proceedings of the 1993 Symposium.* MIT Press, Cambridge, Mass.

Sturrock, S. S., and J. F. Collins. 1993. *MPsrch Manual, version 1.3.* Biocomputing Research Unit, University of Edinburgh, Edinburgh.

Uberbacher, E. C., J. R. Einstein, X. Guan, and R. J. Mural. 1992. Gene recognition and assembly in the GRAIL system: progress and challenges, p. 465-476. *In* H. A. Lim, J. W. Fickett, C. R. Cantor, and R. J. Robbins (ed.), *Proceedings of the Second International Conference on Bioinformatics, Supercomputing, and Complex Genome Analysis.* World Science Publishing Co., Inc., River Edge, N.J.

Internet Publications

Hayden, D. 1994. *Guide to Molecular Biology Databases.* School of Library and Information Studies, University of Alberta, Calgary, Alberta, Canada.

Shah, M. B., X. Guan, J. R. Einstein, S. Matis, Y. Xu, R. J. Mural, and E. C. Uberbacher. 1994. *User's Guide to GRAIL and GenQuest (Sequence Analysis, Gene Assembly And Sequence Comparison Systems) E-mail Servers and XGRAIL (Version 1.2) and XGENQUEST (Version 1.1) Client-Server Systems.* Bioinformatics Group, Oak Ridge National Laboratory, Oak Ridge, Tenn.

6 Structure and Function

■ Servers designed to determine gene structure, detect and identify protein motifs and domains, and predict secondary structure

Blocks	CBRG	Domain	GeneID
GeneMark	MotifFinder	NetGene	nnPredict
PredictProtein	ProDom	PSORT	SBASE

In this chapter we describe servers that detect coding regions, motifs, and domains and predict secondary structure. These servers operate like the search and retrieval servers of Chapter 5, but those tools were not designed to search for motifs and structure-function relationships; these 12 servers were designed specifically with this purpose in mind. Like the search and retrieval servers, the systems we present in this chapter are complementary tools. The search and retrieval servers were designed to answer one question; the servers we describe here cover many distinct aspects of sequence analysis.

Several of these servers look for sequence patterns, profiles, and motifs within proteins. Blocks is a powerful server for searching the Blocks and PROSITE structure-function motif databases. In addition, Blocks offers a retrieval system to obtain detailed information on entries in the Blocks and PROSITE databases. MotifFinder is another example of this class of server; it uses a proprietary pattern database and the well-known PROSITE database to analyze query sequences. Three other servers, Domain, ProDom, and SBASE, perform similar tasks with proprietary database sets.

Three of the servers, GeneID, GeneMark, and NetGene, are tools that analyze sequences for coding, structural, and regulatory regions; in other words, they annotate nucleotide sequence data with respect to organization at the genomic level. They can predict coding regions, splice sites, introns and exons, promoter domains and other regulatory regions. With the surge in projects to sequence whole genomes, each of these three tools is finding expanding roles in the analysis and annotation of sequence data.

The PSORT server is an interesting tool that can offer clues to the transport of proteins within cells. It was designed to predict the sorting and the compartmentalization of proteins within prokaryotic and eukaryotic cells. As such, it can offer clues to cellular organization that can be verified in the laboratory.

Two other servers, nnPredict and PredictProtein, do just what their names imply: they predict the secondary structure of a protein. While both servers use a neural network-based system to analyze the sequence, they take different approaches to the art of secondary structure prediction. PredictProtein aligns the query sequences against entries in the SWISS-PROT database to create a multiple sequence alignment. The composite of this alignment is fed into the neural network system to create the secondary structure model. In contrast, nnPredict determines the propensity for each amino acid residue of a protein sequence to have a certain secondary structure based on a series of predetermined weights for each residue.

One of the servers we describe in this chapter performs a multitude of functions. This system, the CBRG server, offered by the Computational Biochemistry Research Group in Zurich, is designed to look at the evolutionary aspects and structure-function relationships of sequence. The CBRG server has some of the functional aspects of the servers we described in Chapter 5 in that it can analyze an unknown protein or nucleotide sequence for similarities to entries in the SWISS-PROT database. However, it also offers tools to look at phylogenetic and evolutionary relationships as well as to identify proteins based on estimated mass and proteolytic enzyme digestion pattern.

In keeping with the organization of Chapter 5, these servers are listed in alphabetical order, starting with the Blocks server and ending with SBASE. As before, the information that is given for each system is adapted both from the help files of the particular server and from actual use of the system. Since each of these servers functions within a narrow niche, use them to refine an analysis. For a protein sequence, one approach could be to first use Blocks, Domain, and SBASE in parallel to look for motifs and conserved regions that might suggest functional regions, then use PSORT for predicting localization, and finally nnPredict and PredictProtein for the prediction of structure. For genomic nucleotide sequences, the GeneID, GeneMark, and NetGene servers can be used in a parallel fashion as computer workbenches to annotate individual sequences or determine structural organization of complex genomic sequences. As we have said before, each investigator will probably settle on a few of the servers for most analysis work. The key is to experiment with each one to learn which tool is best suited to the analysis tasks performed for a given laboratory project.

Blocks

The Blocks server aids in the detection and verification of protein sequence homology (Pietrokovski et al. 1996). It compares a protein or a DNA sequence to a database of protein blocks. The Blocks server is based on the PATMAT searching tool (Wallace and Henikoff 1992) which accepts protein or DNA sequences. The database of blocks used by the server was constructed by successive application of the PROTOMAT system (Henikoff and Henikoff 1991) to individual entries in the PROSITE catalog of protein groups (Bairoch et al. 1996) and cross-referenced to the SWISS-PROT protein sequence databank (Bairoch and Apweiler 1996). The Blocks database is updated after each revision of the PROSITE database.

The rationale behind searching a database of blocks is that information from multiple-aligned sequences is present in a concentrated form, reducing background and increasing sensitivity to distant relationships. Protein blocks are short, multiple-aligned, ungapped segments derived from the most highly conserved regions of proteins. Typically, a group of related or similar proteins has more than one region or domain in common, and this relationship can be represented as a series of sequence blocks separated by unaligned regions.

In addition, the server system allows users to define their own unique sets of protein blocks based on a family of query sequences they submit to the server. This function, called Block Maker, is a recent addition to the server (Henikoff et al. 1995). Block Maker is designed to generate protein blocks in a group of similar or related query protein sequences. These user-defined blocks can be used to search sequence databases or to aid in the study of protein family relationships.

Internet addresses

E-mail:	blocks@howard.fhcrc.org	(Blocks server)
	blockmaker@howard.fhcrc.org	(Block Maker server)
	henikoff@howard.fhcrc.org	(Personal help)
Web:	http://blocks.fhcrc.org	(Home page)
Organization:	Fred Hutchinson Cancer Research Center, 1124 Columbia Street, Seattle, WA 98104, U.S.A.	

■ Accessing the Blocks and Block Maker servers

To access the Blocks server, send a query message to the server address. To receive an informational help file on using the Blocks server or the Block Maker server, send a message to the server address containing the word HELP in the body of the message. To access the Block Maker server, send a query message to the server address. For further information on the Blocks project or to report problems with either server, send a message to the personal help e-mail address.

■ Formatting a query to the Blocks server

No mandatory or optional directives are needed to analyze data using the Blocks server. A Subject line can be included in the header of the query message but is ignored by the server, except when it requests information on entries in either the Blocks or PROSITE databases. The only thing that needs to be entered is the sequence (either nucleic acid or amino acid). These should be prepared in FastA format, although other sequence formats such as GenePro, GenBank, EBI, SWISS-PROT, GCG, or PIR are acceptable (see Chapter 3). DNA sequences are translated in all six frames for searching.

When the query sequence is in FastA format or one of the other formats mentioned above, the server decides if the sequence is protein or DNA, based on a heuristic method that recognizes the standard IUB/IUPAC codes. Non-alphabet characters including numbers and symbols are allowed but ignored by the server.

To simplify the usage of sequence data that is used in queries to a variety of servers, the FastA format is strongly suggested. This format is accepted by nearly all servers or can be readily converted into an acceptable format for a query.

Following up a potential match is often aided by examining the Blocks or PROSITE database entries for that match. Furthermore, since each group in Blocks corresponds to a group in PROSITE, the PROSITE annotations give further information and provide useful references. To obtain this information for a single database entry, send a blank message to the Blocks server containing the Subject heading directive GET, followed by the Blocks identification number. In contrast to other server directives, which are placed in the body of the query message, this directive and its parameter must be placed in the subject line of the mail header of a query message. Leave the body of the message blank.

■ Formatting a query to the Block Maker server

As with the Blocks server, a query to the Block Maker server has no mandatory or optional directives. The only thing that needs to be entered in a query message is the protein sequences. DNA sequences are not permitted. Sequences should be prepared in the FastA format, although other sequence formats such as GenePro, GenBank, EBI, SWISS-PROT, GCG, or PIR are acceptable. The server can accept a minimum of two and a maximum of 250 protein sequences for analysis. Blank lines are ignored, but sequence formats cannot be mixed in a query message. In addition, the first ten characters after the greater-than character, >, must be unique for each query sequence.

■ Examples of queries

Figures 6.1 and 6.2 display two examples of hypothetical queries to the Blocks server. In Fig. 6.1, the protein sequence HYPOTHETICAL PROTEIN PHOSPHATASE is sent to the server. Note that there are no directives in the body of the query message and that the query sequence is in FastA format. The second example, Fig. 6.2, demonstrates the format of a query message to the server to retrieve the sequence or block designated BL00281.

Figure 6.1. Analysis Query to the Blocks Server

```
From:       annep@acs.auc.eun.eg
Date:       Wed 25 Sep 1996 - 09:58.16 - GMT
To:         blocks@howard.fhcrc.org
Subject:
>YEAST PROTEIN - HYPOTHETICAL PROTEIN PHOSPHATASE
MKAVVIEDGKAVVKEGVPIPELEEGFVLIKTLAVAGNPTDWAHIDYKVGPQGSILGCDAA
GQIVKLGPAVDPKDFSIGDYIYGFIHGSSVRFPSNGAFAEYSAISTVVAYKSPNELKFLG
EDVLPAGPVRSLEGAATIPVSLTTAGLVLTYNLGLNLKWEPSTPQRNGPILLWGGATAVG
QSLIQLANKLNGFTKIIVVASRKHEKLLKEYGADQLFDYHDIDVVEQIKHKYNNISYLVD
CVANQNTLQQVYKCAADKQDATVVELTNLTEENVKKENRRQNVTIDRTRLYSIGGHEVPF
GGITFPADPEARRAATEFVKFINPKISDGQIHHIPARVYKNGLYDVPRILEDIKIGKNSG
EKLVAVLN
```

Figure 6.2. Retrieval Query to the Blocks Server

```
From:       annep@acs.auc.eun.eg
Date:       Wed 25 Sep 1996 - 09:58.16 - GMT
To:         blocks@howard.fhcrc.org
Subject:    GET BL00281
```

Figure 6.3 shows an example of a query to the Block Maker server. In this example, three protein sequences in FastA format are being used to generate a set of user-defined protein blocks. The server will analyze these query sequences using two separate runs of the PROTOMAT system. The first analysis run will use the MOTIF algorithm and the second analysis run will use a modified Gibbs sampler algorithm. Each of these analyses will create a set of protein blocks that will be returned to the user by e-mail. Note that while the server attempts to align all query sequences submitted in a message, occasionally some query sequences may be excluded if they are too divergent from the other members of the query group of sequences for the server to detect any similarity. This exclusion would prevent the construction of any meaningful blocks.

Figure 6.3. Analysis Query to the Block Maker Server

```
From:      annep@acs.auc.eun.eg
Date:      Wed 25 Sep 1996 - 09:58.16 - GMT
To:        blockmaker@howard.fhcrc.org
Subject: Phosphatases, human ser/thr cat subunits


>PP1_CAT HUMAN PROTEIN SER/THR PHOSPHATASE
MSDSEKLNLDSIIGRLLEGSRVLTPHCAPVQGSRPGKNVQLTENEIRGLCLKSREIFLSQ
PILLELEAPLKICGDIHGQYYDLLRLFEYGGFPPESNYLFLGDYVDRGKQSLETICLLLA
YKIKYPENFFLLRGNHECASINRIYGFYDECKRRYNIKLWKTFTDCFNCLPIAAIVDEKI
FCCHGGLSPDLQSMEQIRRIMRPTDVPDQGLLCDLLWSDPDKDVQGWGENDRGVSFTFGA
EVVAKFLHKHDLDLICRAHQVVEDGYEFFAKRQLVTLFSAPNYCGEFDNAGAMMSVDETL
MCSFQILKPADKNKGKYGQFSGLNPGGRPITPPRNSAKAKK


>PP2A_CAT HUMAN PROTEIN SER/THR PHOSPHATASE
MDEKVFTKELDQWIEQLNECKQLSESQVKSLCEKAKEILTKESNVQEVRCPVTVCGDVHG
QFHDLMELFRIGGKSPDTNYLFMGDYVDRGYYSVETVTLLVALKVRYRERITILRGNHES
RQITQVYGFYDECLRKYGNANVWKYFTDLFDYLPLTALVDGQIFCLHGGLSPSIDTLDHI
RALDRLQEVPHEGPMCDLLWSDPDDRGGWGISPRGAGYTFGQDISETFNHANGLTLVSRA
HQLVMEGYNWCHDRNVVTIFSAPNYCYRCGNQAAIMELDDTLKYSFLQFDPAPRRGEPHV
TRRTPDYFL


>PP2B_CAT HUMAN PROTEIN SER/THR PHOSPHATASE
MSEPKAIDPKLSTTDRVVKAVPFPPSHRLTAKEVFDNDGKPRVDILKAHLMKEGRLEESV
ALRIITEGASILRQEKNLLDIDAPVTVCGDIHGQFFDLMKLFEVGGSPANTRYLFLGDYV
DRGYFSIECVLYLWALKILYPKTLFLLRGNHECRHLTEYFTFKQECKIKYSERVYDACMD
AFDCLPLAALMNQQFLCVHGGLSPEINTLDDIRKLDRFKEPPAYGPMCDILWSDPLEDFG
NEKTQEHFTHNTVRGCSYFYSYPAVCEFLQHNNLLSILRAHEAQDAGYRMYRKSQTTGFP
SLITIFSAPNYLDVYNNKAAVLKYENNVMNIRQFNCSPHPYWLPNFMDVFTWSLPFVGEK
VTEMLVNVLNICSDDELGSEEDGFDGATAAARKEVIRNKIRAIGKMARVFSVLREESESV
LTLKGLTPTGMLPSGVLSGGKQTLQSATVEAIEADEAIKGFSPQHKITSFEEAKGLDRIN
ERMPPRRDAMPSDANLNSINKALTSETNGTDSNGSNSSNIQ


>PPX_CAT HUMAN PROTEIN SER/THR PHOSPHATASE
MAEISDLDRQIEQLRRCELIKESEVKALCAKAREILVEESNVQRVDSPVTVCGDIHGQFY
DLKELFRVGGDVPERNYLFMGDFVDRGFYSVETFLLLLALKVRYPDRITLIRGNHESRQI
TQVYGFYDECLRKYGSVTVWRYCTEIFDYLSLSAIIDGKIFCVHGGLSPSIQTLDQIRTI
DRKQEVPHDGPMCDLLWSDPEDTTGWGVSPRGAGYLFGSDVVAQFNAANDIDMICRAHQL
VMEGYKWHFNETVLTVWSAPNYCYRCGNVAAILELDEHLQKDFIIFEAAPQETRGIPSKK
PVADYFL
```

The Computational Biochemistry Research Group (CBRG) server offers four analysis functions: PepPepSearch, NuclPepSearch, AllAll, and MassSearch. The PepPepSearch analysis function will search a submitted amino acid sequence against the entire SWISS-PROT peptide database using the Smith-Waterman version of dynamic programming. It is designed for use in comparing the sequence of an unknown protein against a database of identified proteins. The NuclPepSearch analysis function will search a submitted nucleotide sequence against the entire SWISS-PROT peptide database using the algorithm by Knecht and Gonnet (Knecht and Gonnet 1992). It is designed for use in comparing an unknown nucleic acid sequence against a database of identified proteins. The AllAll analysis function will do an all-against-all analysis of two or more query sequences against each other and use the results of this analysis to do many additional commands, which are described below. Finally, the MassSearch analysis function will search the SWISS-PROT database for sequences that, when digested by a given protease, will match the given set of weights for a query protein digested by the same protease. The rationale behind this analysis function is simple. In some cases, recognition of a protein of unknown sequence can be done by fragmenting the protein with a protease of known specificity and using the determined molecular weights of the fragments as a trace. This method is not effective for finding the composition of an unknown protein, but it is effective in locating an unknown sample if its sequence or one like it is in a protein database. (These analysis functions are discussed in more detail below.) All the queries submitted to the servers are processed with a proprietary analysis system called DARWIN (Data Analysis and Retrieval With Indexed Nucleotide/peptide sequences) and publicly available databases.

Internet addresses

E-mail:	cbrg@inf.ethz.ch	(Server)
	korosten@inf.ethz.ch	(Personal help)
Web:	http://cbrg.inf.ethz.ch	(Home page)
Organization:	Institute of Scientific Computing, Computational Biochemistry Research Group, Swiss Federal Institute of Technology (ETH), 8092 Zurich, Switzerland	
Telephone:	+41-01-632-7479	

■ Accessing the CBRG server

To access the CBRG server, send a query message that contains a properly formatted request (as described below) to the server address. To receive the current set of instructions on using the server, send a help message to the same Internet address. Simply put the word HELP on a line by itself in the body of the e-mail message. For further information or for detailed answers to specific questions, send an e-mail message to the personal help e-mail address.

If a request for help or other information is sent to the server, then a general informational help file is returned. If a help request is modified by a specific topic for the argument, such as in HELP PEPPEPSEARCH, then the server returns a detailed informational help file for that server function. If the help request is modified by the argument ALL, a help file for each function is obtained. A request for HELP NEWS will receive news about the automatic server.

■ Formatting a PepPepSearch query

The mail header is the first part of an e-mail message to be read by the server (see Chapter 3). The subject line of the message is ignored, but is echoed back in the return e-mail message from the server. For this reason it is useful to enter a brief phrase or statement to identify the contents or purpose of the message sent to the server. As stated above, there are no directives for a CBRG sequence analysis; therefore, only the desired function to be run and a query sequence need to be stated. The first line of the body of an e-mail message should contain the directive for the desired analysis function, in this case, PEPPEPSEARCH. The lines following the analysis function statement contain the amino acid sequence to be analyzed by the server. The sequence can have spaces or tabs or be on multiple lines as convenient, but must contain no other extraneous characters. Please note that the submitted sequence data must be given either in all uppercase or all lowercase letters. CBRG accepts the standard IUB/IUPAC codes for the 20 amino acids (see Appendix D) plus X, a symbol for an unknown amino acid. Further, use of the codes B and Z is not supported, and these must be changed to X by the sender. Finally, each e-mail message can only contain one analysis function.

■ Formatting a NuclPepSearch query

A query sent to the CBRG server for an analysis using NuclPepSearch will have the same format as a PepPepSearch query except NUCLPEPSEARCH will be used in the first line of the query instead of PEPPEPSEARCH. The next line of the query contains the nucleotide sequence. As with a PepPepSearch query, the sequence can have spaces or tabs or be on multiple lines as convenient, but no other extraneous characters are allowed. The sequence must be given in either all uppercase or all lowercase letters. The IUB/IUPAC codes (see Appendix D) for the five standard bases are accepted (A, C, G, T, U), plus X as the code for an unknown base. Other codes are not supported and must be changed to X before submitting the sequence data to the server.

■ Examples of PepPepSearch and NuclPepSearch queries

An example PepPepSearch query is shown in Fig. 6.4A. The PEPPEPSEARCH function is indicated on the first line. This is immediately followed by the amino acid query sequence in standard notation. In Fig. 6.4B, an example of a NuclPepSearch query is shown. The NUCLPEPSEARCH function is indicated on the first line and immediately followed by the nucleotide query sequence in standard IUBPAC notation.

Figure 6.4. Examples of PepPep and NuclPep Queries to the CBRG Server

A

```
From:       annep@acs.auc.eun.eg
Date:       02 Dec 1996 22:14:07-GMT
To:         cbrg@inf.ethz.ch
Subject:
PEPPEPSEARCH
SVIIHTACIIDVFGVTHRESIMNVNVKGTQLLLEACVHSKKLAEKAVLAANGWNLKNGGTLYTCALRPM
LRALQDPKKAPSIRGQFYYISDDTPHQSYDNLNYTLSKEFGLRLDSRWSFPLSLMYWIGFLLEIVSFLL
RHIVTLSNSVFTFSYKKAQRDLAYKPLYSWEEAKQKTVEWVGSLVD
```

B

```
From:       annep@acs.auc.eun.eg
Date:       02 Dec 1996 22:14:07-GMT
To:         cbrg@inf.ethz.ch
Subject:
NUCLPEPSEARCH
AAACACCCTTGCAGCTATGGCAGAAATGGGTCAACGCATCATGATTGTAGGTTGCGACCCTAAAGCTGA
CTCCACCCGTCTGATGCTTCACTCCAAAGCTCAAACCACCGTACTACACTTAGCTGCTGAACGCGGTGC
GAACTCCACGAAGTAATGTTGACCGGTTTCCGTGGCGTTAAGTGCGTAGAATCTGGTGGTCCAGAACCC
GAAATGATGGCGATGTATGCTGCTAACAACATCGCTCGCGGTATTTTGAAATATGCTCACTCCGGTGGT
GTTTGATCTGTAACAGCCGTAAGGTTGACCGTGAAGACGAGTTAATCATGAACTTGGCT
```

■ Directives and parameters of an AllAll search

The AllAll analysis function will do an all-against-all analysis of two or more query protein sequences against each other and use the results of this analysis to do any of the directives shown in Table 6.1. The directives for this type of analysis are all optional and have no associated parameters. What this means is that if these directives are used in a query message, then the directive will be executed as part of the analysis.

■ Formatting an ALLALL search

The first line of the query statement contains the ALLALL statement and indicates the analysis function to be run. The lines following the analysis function statement contain the sequences in one of the following formats (each query sequence is complete).

- The query sequence in its entirety.

- A SWISS-PROT accession number with a trailing semicolon refers to an entry in SWISS-PROT, optionally followed by a range of amino acid positions within that entry. For example, P01000; refers to the sequence with accession number

Table 6.1. CBRG Server Directives for an AllAll Query

Directive	Explanation
PAMDATA	PAM distance and variance data for the query sequences
PHYLOTREE, ROOTED TREE	PostScript graphic file of the phylogenetic trees of the query sequences
SPLITDATA, SPLITGRAPH	Isolation indices and graphs
2DPLACEMENT	Approximate two-dimensional placement of the sequences
MULALIGNMENT	Multiple alignment of the sequences
PROBANCESTRAL	Probabilistic ancestral sequence of the sequences
KWINDEX, PROBINDEX, SCALEINDEX	Variation indices
SIAPREDICTION	Surface/Inside/Active site prediction
PARSEPREDICTION	Prediction of breaks in secondary structure

P01000, while P02000; 15-165 refers to positions 15 to 165 of the sequence with accession number P02000.

- A SWISS-PROT identification, including the underscore, refers to an entry in SWISS-PROT, optionally followed by a range of amino acid positions. For example, 4PPS_HUM refers to the sequence with identification 4PPS_HUM, while 4PPS_HUM 33-98 refers to positions 33 to 98 of the sequence with identification 4PPS_HUM.

- A complete entry in SGML format. SGML entries must be enclosed between <E> and </E> tags, with a sequence between <SEQ> and </SEQ> tags, and optional other tagged information (of which <AC>, <ID>, and <DE> will be printed in the cross reference). For example:

```
<E><DE>Experimental protein phosphatase</DE>
<SEQ>AYHDNEWGVP ETDSKKLFEM ICLEGQQAGL SWITVLKKRE
NYRACFHQFD PVKVAAMQEE VERLVQDAGI IRHRGKIQAI IGNARAYLQM
EQTEFGSELK SFPEVVGKTV DQAREYFTLH</SEQ></E>
```

These entry formats may be freely mixed in a query message, but all sequences must be separated by commas and the last sequence must be terminated by a period. As the server is case-sensitive, the sequences must be given either in all uppercase or all lowercase letters. The server accepts standard IUB/IUPAC codes for the 20 amino acids (see Appendix D) plus X, the code for an unknown amino acid. The use of B and Z is not supported, and these codes must be changed to X prior to submitting the sequence to the server.

On the line immediately following the sequences, any number of directives may be requested, with at least one directive required in a submitted query. For example, if the directive PAMDATA is requested, the estimated PAM distances and variances between each pair of sequences are determined.

If PHYLOTREE is requested, a phylogenetic tree of the sequences is constructed, with the result sent back in PostScript format. The tree created is based on the estimated PAM distances between each pair of sequences. For each pair of sequences, the variance of the distance is also computed, which is used to weight the distance in the final tree. In the final tree, the length of each branch is proportional to the evolutionary distance between the nodes. The resulting tree is unrooted, meaning that there is not enough information to decide which common ancestor is the most ancient. A small circle indicates the weighted centroid of the tree. These trees are approximations to maximum likelihood trees.

If the directive ROOTEDTREE is requested, the same tree as constructed by the directive PHYLOTREE is determined with the result sent back in PostScript format. The vertical distance between the nodes is proportional to the evolutionary distance between the nodes. In this directive, the weighted centroid of the tree is assumed to be the root.

If the directive 2DPLACEMENT is requested, the sequences are placed on a plane such that the distance between each is as close as possible to the estimated PAM distance between the sequences. As with PHYLOTREE and ROOTED TREE, the result is given in PostScript format. While forcing the sequences to lie in a two-dimensional (2D) plane necessarily reduces the precision of the representation, this is another mechanism to visualize the relation between sequences.

If SPLITDATA is requested, the split decomposition of the estimated distances is computed with a list of isolation indices and splits being determined. For example, assuming four sequences a, b, c, d, the split data:

```
70.4   {d}
32.9   {b}
29.9   {a}
12.8   {c}
 1.8   {a, d}
 1.7   {c, d}
```

indicates that sequence d is isolated by 70.4 PAM from sequences a, b, and c; sequence b is isolated by 32.9 PAM from sequences a, c, and d; and so on for the remaining relationships. The splittable percentage indicates how much of the estimated PAM distances, on the average, is recovered from the isolation indices.

If SPLITGRAPH is requested, the split decomposition is drawn as a graph similar to a tree. The length of an edge separating two subgraphs is proportional to the isolation index of the split corresponding to the two subgraphs.

If MULALIGNMENT is requested, a multiple alignment of the sequences based on the phylogenetic tree of the sequences is computed. If PROBANCESTRAL is requested, a probabilistic ancestral sequence in the weighted centroid of the phylogenetic tree is printed.

If KWINDEX is requested, the variation index is computed, printed, and then plotted as a histogram. The result is returned as a PostScript file. If PROBINDEX is requested, the probability of the mutations implied by the phylogenetic tree is computed for each position of the multiple alignment, and the negative logarithm to base 10 of this probability is printed and plotted as a histogram in PostScript.

If SCALEINDEX is requested, a scale factor for each position of the multiple alignment is computed such that the probability of the mutations implied by the phylogenetic tree with PAM distances scaled by this factor is maximized. The logarithm to base 10 of this scale factor is printed and plotted as a histogram in PostScript.

If SIAPREDICTION is requested, a prediction according to an unpublished algorithm based on the structure prediction method of Benner and coworkers (Benner et al. 1994; Gerloff et al. 1993) is performed. Each position of the multiple alignment is assigned as being on the surface (strong: S, or weak: s) or the interior (strong: I, or weak: i) of the fold. The procedure assumes that all members of the multiple alignment have the same tertiary structure. Positions which contain a conserved functional amino acid are marked with an A or a. Positions where no prediction is possible are marked with a period. The surface and interior assignments can be used for the prediction of coil regions, helices, and strands. A region in a multiple alignment can be assigned as a beta strand or alpha helix depending on the type of periodicity; a region that contains mainly surface assignments can be assigned as a surface loop or coil; and regions that contain mainly interior assignments can be assigned as internal strands or helices.

If PARSEPREDICTION is requested, a prediction of breaks in the secondary structure of a protein family according to an unpublished algorithm is used with SIAPREDICTION. All positions in a multiple alignment are checked for whether they contain typical motifs (such as a large number of proline residues or deletions) that are between standard secondary structure elements (like helices or strands). The strength of identified parse positions is reported with up to five stars. Together with the SIAPREDICTION routine, the PARSEPREDICTION routine allows an easy search for secondary structure motifs in a protein family based on the pattern of variation and conservation of amino acids.

■ Example of an AllAll query

Figure 6.5A shows the construction of an AllAll query message to the server. While it appears to be a complicated message, the actual structure is straightforward. Note that, similar to other types of analysis queries to the CBRG server, the first line contains the statement that sets the type of analysis to be performed by the server. After that first line come the sequences to be analyzed, followed by the specific analysis directives.

■ Directives and parameters of a MassSearch

The statement MASSSEARCH indicates to the server which analysis function will be run. If a nucleotide database is to be searched, then the MASSSEARCH statement is replaced with DNAMASSSEARCH. Other directives include the name of the enzyme used, the weight of the peptide digest fragments used in the analysis, and the approximate mass of the undigested protein (APPROXMASS). Table 6.2 shows the recognized enzymes and their cleavage rules.

Figure 6.5. Examples of AllAll, MassSearch, and DNAMassSearch Queries to the CBRG Server

A

```
From:        annep@acs.auc.eun.eg
Date:        02 Dec 1996 22:14:07-GMT
To:          cbrg@inf.ethz.ch
Subject:
ALLALL
PIVPLLFGMWQLAREKASNTLLQCVKYYYVFLARNTVAGRRPLSMKYSDKVCSPRKGTKT,
AEPIVPLLFGLWQLAREKASNTLLQCVKYVFLARNTVAGRRPLKMKYSDKVCSPRKGAKT,
TEPIVPLLMWQLAIEKSSNTLLQCVKKVFLARKTVAGRRPLSMKFSDKVCNPRKGTKT,
AEVIVPLLFGVWRLKREERTYTLLQCVKYVFLARNTVAGNRPLSKKFSEKVCSPRK,
AEPIVPLLFGMWRLKRKANNKLLRCVKYTLLARNTSDGREPVACRYSEKICSPRTGTKT,
AKQVVLLIFGSWQLARERLANEMRKAVAYTFLNFDMGRQPLSMHYSDKVCSPRMSTET.
PAMDATA
PHYLOTREE
ROOTEDTREE
2DPLACEMENT
SPLITDATA
SPLITGRAPH
MULALIGNMENT
PROBANCESTRAL
KWINDEX
PROBINDEX
SCALEINDEX
SIAPREDICTION
PARSEPREDICTION
```

B

```
From:        annep@acs.auc.eun.eg
Date:        02 Dec 1996 22:14:07-GMT
To:          cbrg@inf.ethz.ch
Subject:
MASSSEARCH
TRYPSIN: 1264.8, 1520.2, 955.9, 2487.0, 1094.1
ASPN: 1624.4, 2961.4, 718.8, 716.9, 1890.0
```

C

```
From:        annep@acs.auc.eun.eg
Date:        02 Dec 1996 22:14:07-GMT
To:          cbrg@inf.ethz.ch
Subject:
DNAMASSSEARCH
APPROXMASS: 50000
TRYPSIN: M=83.092, 1264.8, 1520.2, 955.9, 2487.0, 1094.1
ASPN: Deuterated, 1624.4, 2961.4, 718.8, 716.9, 1890.0
```

■ **Formatting a MassSearch**

The first line of the query contains the MASSSEARCH statement. The next line(s) contains the peptide digest information in the format:

```
ENZYME: MR, MR, MR
```

where ENZYME is the name of the protease used to digest the protein in question and MR is the molecular weight of the peptide digest fragments. More than one protease can be used in the analysis; they are entered on separate lines.

Table 6.2. CBRG Server Proteolytic Enzyme Properties for a MassSearch

Enzyme	Cuts Between	Exceptions
Armillaria	Xaa-Cys, Xaa-Lys	
ArmillariaMellea	Xaa-Lys	
Chymotrypsin	Trp-Xaa, Phe-Xaa, Tyr-Xaa, Met-Xaa, Leu-Xaa	Trp-Pro, Phe-Pro, Tyr-Pro, Met-Pro, Leu-Pro
Clostripain	Arg-Xaa	
CNBr	Met-Xaa	
AspN	Xaa-Asp	
LysC	Lys-Xaa	
Hydroxylamine	Asn-Gly	
MildAcidHydrolysis	Asp-Pro	
PancreaticElastase	Ala-Xaa, Gly-Xaa, Ser-Xaa, Val-Xaa	
Thermolysin	Xaa-Leu, Xaa-Ile, Xaa-Met, Xaa-Phe, Xaa-Trp, Xaa-Val	
TrypsinCysModified	Arg-Xaa, Lys-Xaa, Cys-Xaa	Arg-Pro, Lys-Pro, Cys-Pro
Trypsin	Arg-Xaa, Lys-Xaa	Lys-Pro
V8AmmoniumAcetate	Glu-Xaa	Glu-Pro
V8PhosphateBuffer	Asp-Xaa, Glu-Xaa	Asp-Pro, Glu-Pro

■ **Examples of MassSearch queries**

The example in Fig. 6.5B shows the text of the body of an e-mail message to the CBRG server using the MassSearch analysis function. The analysis function, MASSSEARCH, indicates the operation to be performed by the server. The subsequent lines contain the name of the digesting enzyme followed by the molecular weights of the resulting fragments. The molecular weight data can be separated by spaces, commas, or tabs or placed on separate lines as convenient, but no other extraneous characters should be present. An e-mail message to the server may contain more than one digestion; the server will interpret the digestions as being on the same protein. As a result, the name of a digesting enyzme can appear only once in an e-mail message sent to the server.

Searching the DNA database for matches is also possible. The example in Fig. 6.5C instructs the server to search the given fragments against the EMBL nucleotide sequence database. The first line contains the name of the analysis function. The second line gives an approximation of the molecular weight of the entire protein being searched. This is an optional value, obtainable from 2D gels, which increases the precision of the search. (In other words, if you know the molecular weight, use it.) The other lines are similar to the previous case, with two additions. The first value for the TRYPSIN line indicates that the weight of methionine (M) is to be considered 83.092 (as if it were transformed into homoserine). This is frequently the case when proteins are preprocessed with reagents that may change some amino acids, hence changing the weights of the fragments. For cases involving digestion with CNBR or TRYPSINCYSMODIFIED, because it is known that cysteine is transformed into homoserine or homoserinelactone by CNBR and to aminoethylcysteine when treated with TRYPSINCYSMODIFIED, this weight correction is done automatically by the server and no change needs to be indicated in the query message. The line for ASPN has the keyword DEUTERATED, which means that the protein has been treated with heavy water and that all the exchangeable hydrogen atoms are now deuterium atoms, which causes an increase in their molecular weight. The approximate mass, special weights for amino acids, or deuterium treatment can also be specified for the protein searches.

The problem of identifying a sampled protein can be reduced to digesting the protein with an enzyme, finding the molecular weights of each of the pieces, and then comparing this set of weights to what would be obtained from the digestion of each protein in the database. The process can be repeated with several different enzymes to increase its selectivity. The analysis function MassSearch locates the best candidates in the SWISS-PROT protein database that fit the given molecular weights, while the analysis function DNAMassSearch locates the best candidates in the EMBL DNA sequence database. This type of sequence analysis is useful in the following circumstances:

- To identify proteins when the amount of protein available is very small, for example a protein purified by 2D gels

- To determine whether an unknown protein is already known in the database before spending a significant effort in sequencing

- To identify more than one protein that cannot be separated by other means (a method successfully used to identify two proteins that were digested together)

Increased precision in the searching is obtained when more than one digestion is available. In general, it is much better to perform two digestions with different enzymes. The precision of the server analysis increases with the number of digestions available.

Domain

The Domain server is designed to compare a submitted protein sequence against the latest release of the SWISS-PROT protein sequence database (Bairoch and Apweiler 1996) and return information on the most stable protein domain homologies (Hegyi and Pongor 1993). The database search is performed using the BLAST algorithm. Based on this analysis, the server selects the most frequent domain homologies. At this point, a specially designed algorithm, FTHOM, processes each alignment from the BLAST analysis to create a domain listing. In this processing, the beginning and end of the alignment are determined and the aligned region is compared with the feature table of the database entry. This step generates a relative score for each domain based on the similarity of the sequence and the description of the entry from the SWISS-PROT database. The server reports back the relative scores obtained by each domain, the list of best domain homologies, and the results of the BLAST search on which the analysis was based.

 Internet addresses

E-mail:	domain@hubi.abc.hu	(Server)
	hegyi@hubi.abc.hu	(Personal help)
Web:	http://www.abc.hu/blast.html	(Home page)
Organization:	Domain, Institute of Biochemistry and Protein Research, Agricultural Biotechnology Center, Szent-Gyorgyi Albert u. 4, Godollo, Hungary	
Telephone:	+36-28-320-095	
Facsimile:	+36-28-330-338	

▪ Accessing the server

To submit a query sequence for analysis, send a properly formatted message to the server address. To receive a general help file on the use of the server, send a message to the server address containing the directive HELP in the body of the message. To receive a help file on the BLAST component of the server, send a message to the server address containing the directive HELP BLAST in the body of the message.

▪ Server directives and parameters

There is only one mandatory directive for a query to the Domain server, BEGIN. This directive must be followed immediately by a sequence in FastA format (as described previously in Chapter 3).

A summary of optional directives is given in Table 6.3. Use ALIGNMENTS to specify whether the BLAST alignments of domains against the query sequence should be shown in the results from the server. The ANNOTATIONS directive asks the server to report back the annotations contained in the SWISS-PROT database for matching sequences used in the domain analysis. The EXPECT directive is nearly identical in function to that used by other servers such as BLAST, except that the default parameter value is set higher, at 25. The MATRIX directive is also similar to that used by other servers. The default matrix is PAM120 with several other options available.

Table 6.3. Domain Server Directives and Parameters

Directive	Attributes	Explanation
ALIGNMENTS	Optional, Boolean	Default parameter setting is YES. Reply from the server will show the alignments of the query sequence against similar sequences from the SWISS-PROT database.
ANNOTATIONS	Optional, Boolean	Default parameter setting is NO. If the setting is changed to YES, the server will show the annotation files for the similar sequences identified by the analysis.
BEGIN	Mandatory, none	Must be the last directive prior to the query sequence. Immediately following this directive is first a comments line, then the sequence in FastA format.
EXPECT PARAMETER	Optional, numeric	The default parameter value is 25.
HELP	Optional, text	Returns a help file from the server. The directive HELP alone returns a file on the general operation of the server. The directive HELP BLAST returns a file on the BLAST component of the server.
MATRIX	Optional, text	The default matrix is PAM120.
OUTPUT_PAGES	Optional, numeric	The default number of pages of results that the server will send back is 10. More pages can be requested in increments of 10.
SCORE PARAMETER	Optional, numeric	Default parameter value is 35.
-SORT_BY_PVALUE	Optional, none	These four directives govern the sorting of output by the server. The default is the directive -SORT_BY_PVALUE. Only one of these directives may be used in a query.
-SORT_BY_COUNT	Optional, none	
-SORT_HIGHSCORE	Optional, none	
-SORT_BY_ TOTALSCORE	Optional, none	

The OUTPUT directive governs the size of the results reported back by the server. In the default mode, the Domain server sends back a message containing the best domain homologies followed by up to ten pages of BLAST-derived alignments of the query sequence with the SWISS-PROT database. Longer outputs can be requested by increasing the parameter value for this directive. The server will report these results in messages of ten-page increments.

Use the directive SCORE to set the number of matching sequences when generating the ranking of the domains. The default parameter value is 35, which means that the server will use the best 35 matches in generating the results.

■ Formatting a query

The BEGIN directive marks the beginning of the query sequence. It is followed on the next line by the query sequence in FastA format (as previously described in Chapter 3). The sequence must be in the standard IUB/IUBPAC code. It is important that there be no blank lines in the query sequence. Further, since the END directive used by many other servers to mark the end of the sequence data is not supported, a blank line at the end of the query sequence may be useful to prevent the server from interpreting the signature block of the e-mail message as part of the query sequence. All optional directives must be stated on separate lines before the BEGIN directive. If these directives are not included in the query, then the server will use the default settings.

Note that in the current version of the Domain server, the analysis functions optimally with query sequences of less than 200 amino acids. Longer sequences may provide lengthy listings of potential domain matches, making interpretation of the results difficult.

■ Examples of queries

The two examples shown in Fig. 6.6 demonstrate hypothetical query messages to the Domain server. In the first example, Fig. 6.6A, an analysis is being requested using the default server settings. Domain will analyze the query sequence against the SWISS-PROT database using an EXPECT value of 25, the PAM120 matrix, and a SCORE value of 35. The server will send back a report with up to 10 pages of results, composed of the relative scores obtained by each matching domain, the list of best domain homologies, and the results of the BLAST search on which the analysis was based.

In the second example, Fig. 6.6B, the Domain server will analyze the query sequence using an EXPECT value of 10, the default PAM120 matrix, and a SCORE value of 50. The server will send back a report composed of up to 20 pages of results, split into ten-page sections. The results will be like those for the first example, except that the annotations in the SWISS-PROT database for the matching entries will also be reported.

Figure 6.6. Examples of Domain Queries

A

```
From:        annep@acs.auc.eun.eg
Date:        Wed 25 Sep 1996 - 09:58.16 - GMT
To:          domain@hubi.abc.hu
Subject:
BEGIN
>YEAST PROTEIN - HYPOTHETICAL PROTEIN PHOSPHATASE
MKAVVIEDGKAVVKEGVPIPELEEGFVLIKTLAVAGNPTDWAHIDYKVGPQGSILGCDAA
GQIVKLGPAVDPKDFSIGDYIYGFIHGSSVRFPSNGAFAEYSAISTVVAYKSPNELKFLG
EDVLPAGPVRSLEGAATIPVSLTTAGLVLTYNLGLNLKWEPSTPQRNGPILLWGGATAVG
QSLIQLANKLNGFTKIIVVASRKHEKLLKEYGADQLFDYHDIDVVEQIKHKYNNISYLVD
CVANQNTLQQVYKCAADKQDATVVELTNLTEENVKKENRRQNVTIDRTRLYSIGGHEVPF
GGITFPADPEARRAATEFVKFINPKISDGQIHHIPARVYKNGLYDVPRILEDIKIGKNSG
EKLVAVLN
```

B

```
From:        annep@acs.auc.eun.eg
Date:        Wed 25 Sep 1996 - 09:58.16 - GMT
To:          domain@hubi.abc.hu
Subject:
ANNOTATIONS YES
EXPECT 10
OUTPUT 20
SCORE 50
BEGIN
>YEAST PROTEIN - HYPOTHETICAL PROTEIN PHOSPHATASE
MKAVVIEDGKAVVKEGVPIPELEEGFVLIKTLAVAGNPTDWAHIDYKVGPQGSILGCDAA
GQIVKLGPAVDPKDFSIGDYIYGFIHGSSVRFPSNGAFAEYSAISTVVAYKSPNELKFLG
EDVLPAGPVRSLEGAATIPVSLTTAGLVLTYNLGLNLKWEPSTPQRNGPILLWGGATAVG
QSLIQLANKLNGFTKIIVVASRKHEKLLKEYGADQLFDYHDIDVVEQIKHKYNNISYLVD
CVANQNTLQQVYKCAADKQDATVVELTNLTEENVKKENRRQNVTIDRTRLYSIGGHEVPF
GGITFPADPEARRAATEFVKFINPKISDGQIHHIPARVYKNGLYDVPRILEDIKIGKNSG
EKLVAVLN
```

GeneID

The GeneID e-mail server is an artificial intelligence system for analyzing vertebrate genomic DNA and the prediction of exons and gene structure (Guigo et al. 1992). Users have the option of having their genomic DNA sequence analyzed by the NetGene e-mail server simultaneously (Brunak et al. 1991). GeneID also has the capability to analyze the coding regions of cDNA sequences. For the analysis of genomic DNA sequences, the server will first analyze a submitted sequence for potential splice sites, start codons, and stop codons. Then the server will try to assemble these features into 5' exons, internal exons, and 3' exons. Exons will be evaluated according to a number of characteristics related to optimal coding region and splicing, with only the most likely exons being retained. The GeneID server then compares predicted exons to a protein database using the BLAST algorithm. Predicted exons with matches in the database will be reported with the locus name of the match, and the server will use this information in the construction of gene models for the submitted sequence data, resulting in a significant increase in prediction accuracy. This information will be used by the server to assemble gene models. Only the 15 highest ranking 5' and 3' exon classes and the 35 highest ranking internal exon classes will be used. These models will be ranked with an open reading frame according to the quality of the exons. The 20 highest ranking models will be included in the return e-mail message from the server. The returned models will also be annotated for exons and sites that the server predicts are present. The GeneID server offers a separate analysis command for cDNA sequences. If this option is invoked, the server will identify potential start codons and potential stop codons, and will evaluate all open reading frames in between these features for coding potential. The evaluation is done with a neural network approach similar to the analysis of genomic sequences. Thus, the logic of evaluating open reading frames in cDNA is very similar to the logic of evaluating exons in genomic DNA.

Internet addresses

E-mail:	geneid@darwin.bu.edu	(Server)
	graf@darwin.bu.edu	(Personal help, GeneID)
	engel@virus.fki.dth.dk	(Personal help, NetGene)
Organization:	Molecular Biology Computer Research Resource (MBCRR), Dana-Farber Cancer Institute, Boston, MA 02146, U.S.A.	

■ Accessing the GeneID server

To submit a query sequence for analysis, send a query message to the server address. To receive a general help file from the server, send a message containing the directive HELP in the first line of the message body.

■ **Server directives and parameters for a genomic sequence query**

For the analysis of genomic sequences, there is only one mandatory directive: GENOMIC SEQUENCE. A listing of all the optional parameters for this analysis is shown in Table 6.4. The directive GENOMIC SEQUENCE informs the server that the query sequence will be from genomic DNA; the server will perform an analysis based on that criterion.

Table 6.4. GeneID Server Directives and Parameters for a Genomic Sequence Query.

Directive	Explanation
-ALT_SPLICING X Y	Specify region where splicing is blocked, starting at position X and ending at position Y. GeneID will attempt to locate alternative splice sites outside this region.
-FIRST_EXON X Y	Specify coordinates of known first exon, starting at position X and ending at position Y. GeneID will use this information in building gene models. Position X can be either the start codon or the beginning of the transcript. Noncoding first exons may prevent gene model building.
-GENEBLAST	Compare top predicted gene model to protein databases using the BLAST algorithm.
-KNOWN_CODING X Y	Specify region of known coding, starting at position X and ending at position Y. GeneID will attempt to use this information in building gene models.
-LAST_EXON X Y	Specify coordinates of known last exon, starting at position X and ending at position Y. GeneID will use this information in building gene models. Position Y can be either the stop codon or the end of the transcript. Noncoding last exons may prevent gene model building.
-NETGENE	Perform NetGene splice site analysis. The results of this analysis will be mailed back in a separate message.
-NOEXONBLAST	Use of this parameter option will cause the server to skip the BLAST search of matches against the open reading frames. This permits a faster analysis of the sequence, but at the expense of the accuracy of the analysis. This is the default setting if the sequence is greater than 8,000 base pairs.
-SMALL_OUTPUT	With this option specified, GeneID will skip updates, output explanation, and atomic sites and show only predicted exons and gene models. This option is the default setting for the server if the query sequence submitted is greater than 50 kilobase pairs.

Several of the optional directives deserve further note. The -SMALL_OUTPUT directive is recommended for query sequences greater than 20 kilobase pairs. This will cause the server to skip all updates, output with explanation, and atomic sites. The output will only be predicted exons and gene models.

The use of the -NOEXONBLAST directive causes the server to skip the BLAST search of the predicted exons against the public protein databases. This greatly speeds up the analysis of the sequence. One significant limitation of the server as presently configured is that sequences longer than 8,000 base pairs can only be analyzed for database matches by submitting them to the server in sections less than 8,000 base pairs.

Use of the -GENEBLAST directive tells the server to use the BLAST algorithm to analyze a translation of the top ranking gene model against the public protein databases. The rationale for doing this search in addition to an analysis of exons by BLAST is that the product of the spliced exons might match a sequence that small independent exons might not, and thus give an indication as to function. The results will be returned in a separate reply message.

The directives -KNOWN_CODING X Y, -FIRST_EXON X Y, -LAST EXON X Y, and -ALT_SPLICING X Y are for indicating known information about the genomic sequence to the server. The GeneID server will attempt to use this information when building gene models. Additionally, if the -FIRST_EXON X Y directive or the -LAST_EXON X Y directive is specified and the first or last exon contains noncoding sequences, then the -GENEBLAST directive and gene model building may not work properly.

■ Formatting a genomic sequence query

The query must begin with the directive GENOMIC SEQUENCE. If it is not specified, the server will not send back a response to the query message. All the optional directives have to be specified on the same line as the GENOMIC SEQUENCE directive, immediately adjacent to the keyword as shown below.

```
GENOMIC SEQUENCE -GENEBLAST
```

The sequence to be analyzed must immediately follow this line, and the sequence data must be in FastA format. On the comments line, the actual sequence name is limited to 20 characters. The actual sequence follows on the successive lines. The sequence data must be composed of the standard IUB/IUPAC codes (Appendix D). Additionally, the maximum line length for the query sequence is 80 characters per line. Finally, the server will not accept sequences smaller than 100 base pairs or larger than 200 kilobase pairs.

The syntax for the optional parameters has to be followed exactly; otherwise errors will result and the server will either perform an incomplete analysis or not be able to return any results. A full list of the optional directives for the analysis of genomic sequences is given in Table 6.4. Please note that the dash character and the first character of the option are sufficient to specify all directives except -NETGENE.

Use of the NETGENE directive causes the server to predict splice sites and determine the likelihood of the site prediction (Brunak et al. 1991). The NETGENE directive functions by predicting both coding regions and the acceptor splicing signals that predict

---

Figure 6.7. Example of a Genomic DNA Query

```
From:      annep@acs.auc.eun.eg
Date:      Wed 25 Sep 1996 - 09:58.16 - GMT
To:        geneid@darwin.bu.edu
Subject:
GENOMIC SEQUENCE
>UNKNOWN GENOMIC SEQUENCE FROM CLONE 67-45
TTGGCCACTCCCTCTCGGCCTCAGTGCGCGACCAAAGGTCGCCCGACGCCCGGGCTTTGCCCTCGC
TCACTGAGGCCGGGCGACCAAAGCAGGGAGTGGCCAACTCCACAGATCAGCTCGGCTAGACCACAC
AGCTGACAGCTCAGTCGTACGCTACGCGCTACGSTCGCAGCTGTCAGTCACATCAGCTCGGCTAGA
CCACACAGCTGACAGCTCAGTCGTACGCTACGCGCTACGSTCGCAGCTGTCAGTCACACATCCATC
GTCAGTGCGACCAAAGCGGGCGGCTGCTACAGTCGACTAGCCCGTACAGCTGBACCGTACCTGATC
AGTCAGTCGATCACTAGGCCATCCATCGTCAGTGCGACCAAAGCGGGCGGCTGCTACAGTCGACTA
GCCCGTACAGCTGBACCGTACCTGATCAGTCAGTCACCAAAGCGGGCGGCGAGCGAGCGAGCGCGC
ATGCCCGGGCGGCCTCAGTGAGCGAGCGAGCGCGCAGAGAGGGAGTGGCCTCAGCTCGGCTAGACC
CACAGCTGACAGCTCAGTCGTACGCTACGCGCTACGSTCGCAGCTGTCAGTCACACATCCATCGTC
GTGCGACCAAAGCGGGCGGCTGCTACAGTCGACTAGCCCGTACAGCTGBACCGTACCTGATCAGTC
TCGATCACTAGGCAACTCCACAGATCAGCTCGGCTAGACCACACAGCTGACAGCTCAGTCGTACGC
ACGCGCTACGSTCGCAGCTGTCAGTCACACATCCATCGTCAGTGCGACCAAAGCGGGCGGCTGCTA
AGTCGACTAGCCCGTACAGCTGBACCGTACCTGATCAGTCAGTCGATCACTAGGCGACCAAAGCGG
GCGGCCCAACTCCATCACTATGCGCGCTCGCTCGCTCACTGCCAACTCCATCACTAGAGGCCGGGG
ATCACGGGCGGCCTCAGTGAGCGAGCGAGCGCGCAGAGAGGGAGTG
```

both small and large exons. Additionally, the length of the query sequence must be between 451 base pairs and 100 kilobase pairs. As with the -GENEBLAST option, the results of the NETGENE analysis will be returned in a separate e-mail message.

■ Example of a query

Figure 6.7 shows an example of a query to the GeneID server. In this query, only the mandatory directive GENOMIC SEQUENCE is used, followed by the sequence data in FastA format.

■ Tips for use in genomic sequence analysis

The GeneID output will contain potential start codons, stop codons, splice sites, exons, exon classes, and gene models. If the option -SMALL_OUTPUT is specified, only potential exons, exon classes, and gene models are output. If the option -GENEBLAST is specified, an additional file containing the results of a protein database search using the top gene model will also be reported. If any gene models are predicted in analysis, the output may contain additional information. For example, if two potential gene models are non-overlapping, they may represent two separate genes. Please note that the GeneID server will only attempt to predict gene models in sequences of less than 50 kilobases.

In its analysis, the GeneID server will try to identify first, internal, and last exons in each of the sequences that are submitted, and then assemble these into models of one gene in each sequence. To avoid missing any exons, the server will look for any possible exon; as a result, the number of exons will initially be overpredicted — only a few of these exons will be true. The true exons are likely to be found in the gene models because those exons should form a continuous open reading frame. Thus, gene models are the key to locating a probable coding region.

If a query sequence turns out to contain more than one gene, the analysis can be unpredictable. GeneID could predict one gene containing exons that actually belong to more than one gene, or it could predict only the gene with the optimal characteristics. By the same token, the server may correctly identify all of the genes in the sequence.

If a query sequence contains only part of a gene, the server will try to identify an entire gene in this sequence. This can also lead to some unpredictability in the analysis. The predicted first exon may, in reality, be part of an internal exon, or the predicted last exon may be part of an internal exon. If that happens, either of the exons will tend to be short in length, less than 15 base pairs, and have a low score, usually less than 0.5.

■ Formatting a cDNA query

As with genomic sequence analysis, a query submitted to the GeneID for a cDNA analysis has one mandatory directive: CDNA. A listing of the optional parameters for the cDNA analysis is shown in Table 6.5. The format for a cDNA query is the same as a Genomic Sequence query except the directive GENOMIC SEQUENCE is replaced with the directive CDNA. The query sequence must be in FastA format and immediately follow the CDNA directive.

Table 6.5. GeneID Server Directives and Parameters for a cDNA Query

Directive	Explanation
-3	This parameter option causes the server to assume the submitted cDNA sequence to be a 3′ section of the cDNA. The server will not predict start codons and will look instead for open reading frames traversing the 5′ end of the sequence. If this parameter option is not specified, the server will assume the cDNA sequence submitted is either a full-length cDNA or a 5′ part of a cDNA.
-NOEXONBLAST	Use of this parameter option will cause the server to skip the BLAST search of matches against the open reading frames. This permits a faster analysis of the sequence, but at the expense of the accuracy of the analysis.

GeneMark

The GeneMark gene identification program was originally developed at the School of Biology, Georgia Institute of Technology (Borodovsky and McIninch 1993; Borodovsky et al. 1994). There are two GeneMark servers, one located at the Georgia Institute of Technology, Atlanta, and the other at EBI in Hinxton, U.K.

GeneMark was designed to identify protein-coding regions within nucleotide sequences from prokaryotic and eukaryotic species. It began as a system for predicting protein coding regions in *Escherichia coli* and closely related species. In its current version, GeneMark can find prokaryotic genes, analyze cDNA sequences, and discriminate both eukaryotic and non-eukaryotic coding regions from stretches of noncoding sequence. The lower limit of detection is a coding region of approximately 100 nucleotides or 30 amino acids. In addition, GeneMark is useful for detecting frameshifts and rare start codons.

The algorithm used by the server is a statistical method that requires a training set for developing the parameters of the models of protein-coding and noncoding sequences; it is based on a special type of Markov chain model of coding and noncoding nucleotide sequences.

Internet addresses

E-mail:
genemark@ford.gatech.edu   (Server, U.S.A.)

genemark@embl-ebi.ac.uk   (Server, EBI)

mb56@prism.gatech.edu   (Personal help, U.S.A.)

gt1619a@prism.gatech.edu   (Personal help, U.S.A.)

nethelp@embl-ebi.ac.uk   (Personal help, EBI)

Web:
http://amber.biology.gatech.edu/~genemark

(Information and news)

Organization:
School of Applied Biology and Office of Information Technology, Georgia Institute of Technology, Atlanta, GA 30332-0230, U.S.A.

■ Accessing the server

To use GeneMark, send an e-mail query containing a properly formatted request to the server address. To receive the current set of instructions on using the server, send a help message to the server address. Put the word HELP in the Subject line of the e-mail message header. For further information on the GeneMark server and/or problems with the server, send an e-mail message with a detailed question to project members at the personal help address.

The GeneMark server automatically maintains an e-mail log of all users. To become a registered user of GeneMark, send a message to the server with the directive REGIS-

TRATION in the subject line of the message. Include the user name, Internet address, organization, and research interests in the body of the message.

■ Server directives and parameters

Unlike most other servers, GeneMark uses two classes of directives in a query to the server. These are listed in Tables 6.6 and 6.7 and described in more detail below.

One class of directives is entered in the body of the query message (as in other servers). The statement DATA is the only mandatory directive of this class of directives required for a GeneMark query. In addition to the DATA directive, there are several optional parameters that can be specified in the body of the query message (these are summarized in Table 6.6).

The other class of directives (summarized in Table 6.7) are entered in the Subject line of the e-mail message header. There are four subject line directives: GENEMARK, HELP, LIST ORG, and REGISTRATION. So, a query to GeneMark must contain, in addition to the mandatory directive DATA, a Subject line directive that sets the operation of the server.

■ Formatting a query

As with most servers, GeneMark is not case-sensitive. But, because of the two classes of directives used by GeneMark, the structure of a query message to the server is a bit more complicated than that of other servers. Components of a query to the server must be provided in a carefully structured order. First, within the e-mail header, a Subject line directive must appear in the Subject line of the message. Second, within the body of the query message, any of the optional parameters may be placed. Finally, the mandatory directive DATA must be given as the last command immediately prior to the query sequence.

The four subject line directives are shown in Table 6.7. One of these subject line directives must be specified when sending an e-mail message to the server. For example, when sending a query sequence for analysis to the server, GENEMARK must be specified. If only a help file is needed, then the directive HELP should be in the subject line. One of the Subject line directives, LIST ORG, is only supported by the EBI/EMBL GeneMark server. This directive instructs the server to return a list of supported species. The last of the Subject line directives, REGISTRATION, is used to become a registered user of the server.

Many of the optional directives deserve further note and will be described in more detail below. Table 6.6 summarizes all of the directives that can be used in the body of a query message to the server. Some of these directives, SPECIES, THRESHOLD, WINDOW, and STEP, are used to change the parameters of the analysis. Others, such as PSGRAPH, ORFLIST, PROTEIN, and NUCSEQ, specify the type of output to be sent. Two directives, DATA and END, are used to help format the query sequence data in the message. Three other directives, TITLE, NAME, and ADDRESS, govern how and where the server will send back reply messages upon the completion of an analysis.

Table 6.6. GeneMark Server Directives and Parameters

Directive	Attributes	Explanation
# or ;	Optional, text	The # and ; directives must be at the beginning of the line to indicate that that line is a comment. This line of your message will be ignored.
ADDRESS	Optional, text	Allows the user to specify an alternate e-mail address for GeneMark to reply to. The default value is the address derived from the e-mail message header.
DATA	Mandatory, none	This is the only mandatory directive for a query sent to GeneMark. It is interpreted by GeneMark as an indication that all text that follows is DNA sequence to be analyzed.
END	Optional, none	Marks the end of a sequence.
NAME	Optional, text	Allows user to specify his or her name. The name is tacked onto replies from GeneMark.
NUCSEQ	Optional, text	NOTE: This option has been temporarily switched off. This option will operate similarly to the directive PROTEIN. It operates with nucleotide sequences instead of protein translations. The nucleotide fragments are sent to the same servers for similarity searches.
ORFLIST	Optional, text	Instructs the server to return a list of areas in the DNA sequence where GeneMark identified a coding region. The three acceptable parameters are ORF, REGION, and EXON. Default is ORF.
PROTEIN	Optional, text	Instructs the server to translate fragments of the DNA sequence which GeneMark has identified as a coding region with respect to the predicted frame. Acceptable parameters are REGION, ORF, EXON (which can be abbreviated -R, -O, or -X, respectively), and <homology server>; the homology server can be BLAST, BLITZ, or FastA.
PSGRAPH	Optional, text	Instructs the GeneMark server to send graphical output in the form of a PostScript file which may be printed on any PostScript compatible laser printer or viewed with programs such as GhostScript. The graph depicts coding potentials (probabilities) in six possible reading frames. The PSGRAPH directive may have an optional argument: REGION(-R), ALT, or LANDSCAPE. There is no graphics output if PSGRAPH is not specified. NOTE: PostScript responses can be quite large; every 2K sequence results in 12K of graphic file.

continued

Table 6.6 continued

Directive	Attributes	Explanation
SPECIES	Optional, text	Included to specify the GeneMark analysis procedure parameters derived for the organism the sequence belongs to (or for some other organism of your choice). See Table 6.8 for available parameters. There may be several parameters available for a single species, or only one matrix that represents several. Default value is ECOLI.
STEP	Optional, numeric	Allows the user to specify the step, in nucleotides, used by the algorithm to advance the scanning analysis window. Default step value is 12. It should not, generally, be necessary to modify this value. NOTE: Changing this value will alter the resolution of the graphical output. Decreasing this value increases the graph resolution and increases the size of the PostScript output.
THRESHOLD	Optional, numeric	Allows the user to specify a value between 0 and 1 which will be used to judge whether a given region should be predicted as coding. By default, this threshold is 0.5 (50%), meaning that any region yielding a sustained coding probability above 50% will be judged coding and be included in the list of predicted coding regions.
TITLE	Optional, text	Specifies the title of the sequence. This string appears in all responses from GeneMark and is important in identifying which e-mail response corresponds to which sequence. No title appears if this directive is not specified.
VIA	Optional, text	The VIA option is introduced as a shortcut that is equivalent to several commands. Currently, VIA BLASTP is equivalent to using the options PSGRAPH REGION, PROTEIN REGION BLAST, and ORFLIST REGION ORF.
WINDOW	Optional, numeric	Specifies the length of the analysis window (in nucleotides) that GeneMark uses in its algorithm. The default is 96 base pairs. In the most cases it shouldn't be necessary to modify this value.
ZOOM	Optional, numeric	After this directive, a parameter value between 0.2 and 5.0 must be given. The function of this option is to cause GeneMark to scale the graphical output. By default, GeneMark plots 176 data points per frame on the graphs or ZOOM parameter of 1.

Table 6.7. GeneMark Server Subject Line Directives

Subject Line Directive	Explanation
GENEMARK	Notifies the server that this query message contains sequence data to be analyzed by GeneMark
HELP	Instructs the server to send back a help file
LIST ORG	Instructs the server to send back a list of supported organisms. Note that this directive is currently only supported by the EBI GeneMark server.
REGISTRATION	Notifies the server that this query contains information to register a user of GeneMark

DATA is the only mandatory directive. It indicates to the server that the query sequence follows on the next line of the message. Only letters are recognized; numbers, spaces, or punctuation are allowed, but will be filtered out before the sequence is analyzed. Note that any text containing letters left between the DATA directive and the actual query sequence, such as in the comment line of a FastA formatted sequence, will result in an incorrect analysis.

If the directive PSGRAPH REGION is used in a query, the server will draw gray bars to indicate regions with high coding potential. The parameter ALT provides a more legible format for the nucleotide sequence positions, while the parameter LANDSCAPE provides landscape orientation of graphic images. The parameters for the directive PROTEIN may be combined in a query message. The parameter ORF instructs GeneMark to send back translated protein sequences for all open reading frames identified as protein coding regions in a query sequence. The parameter REGION invokes a similar action, except that instead of translating open reading frame fragments from start to stop codon, GeneMark will translate regions from stop to stop codon. The parameter EXON will perform the same function, but with predicted exon regions. The first and last amino acid in the translated exon may not be present if the exon does not contain a multiple number of codons. The parameter "homology server" can be used in addition to the other parameters to instruct GeneMark to send the protein sequences to BLAST, BLITZ, or FastA servers for a sequence similarity search. As we previously mentioned, these parameter settings can be combined or used separately. The results of similarity searches will be directed from the BLAST, BLITZ, or FastA server to the initial submitter of the query message.

The directive ORFLIST, when coupled with the parameter ORF, instructs GeneMark to reply with a list of open reading frames where a protein coding region is predicted. Such a list includes an assessment of a probability of the protein coding for the region. The parameter REGION allows the user to get a list of the regions (from stop codon to stop codon) where high coding potential is predicted. The parameter EXON instructs GeneMark to reply with a list of exons that are predicted to be protein coding regions. Since this version of the program does not contain precise predictors for acceptor and donor splicing sites, do not expect the positions of exon boundaries to be reliable. However, in regions where rather long exons (defined as those greater than 100 base pairs) are predicted, the accuracy of these predictions rises to 91%.

Small parameter values for the WINDOW directive will produce a higher rate of false signals, but may also identify smaller coding regions. In contrast, larger parameter values may prevent GeneMark from recognizing smaller coding regions, but will clarify the coding signal of larger regions.

The END directive is usually unnecessary. However, if you send a series of sequences to GeneMark in one query message (rather than sending several separate files), use this directive to separate query sequences. GeneMark will treat everything after the END directive as if it were a new query message except for one very important aspect: whatever options were in the previous query message will also be applied to the following sequence, unless new options are given. Also, note that for additional query sequences to be recognized in such a query, the DATA directive must be put prior to the sequence data.

By specifying a larger parameter value for the ZOOM directive, fewer data points are plotted, which makes the graph easier to read but it takes a larger number of pages. Conversely, a parameter value of less than 1.0 will plot more data points per page, using less space and indicating more nucleotides on a page.

The directive SPECIES and its associated parameters, shown in Table 6.8, are used to tell the server what species of organism the query sequence originated from. Careful use of this directive can help improve the accuracy of the analysis performed by the server. For *E. coli* sequences, both ECOLI and ECOPHAGE parameters should be applied for reliable detection of all genes. In this case, the sequence should be sent for analysis to the server two times.

Table 6.8. GeneMark Server Parameters for the SPECIES Directive

Parameter	Species
ARAB	*Arabidopsis thaliana*
SUBT	*Bacillus subtilis*
PHI80	Bacteriophage ø80
T4	Bacteriophage T4
BJAP	*Bradyrhizobium japonicum*
CELEGANS	*Caenorhabditis elegans*
FRUITFLY	*Drosophila melanogaster*
EBV	Epstein-Barr virus
EHV2	Equine herpesvirus II
ECOLI	*Escherichia coli* "native" genes [DEFAULT]
ECOHIEXP	*Escherichia coli* highly expressed genes
ECOPHAGE	*Escherichia coli* horizontally transferred genes
CHICKEN	*Gallus gallus*
GIARDIA	*Giardia lamblia*
FLU	*Haemophilus influenzae*

continued

Table 6.8 continued

Parameter	Species
PYLORI	*Helicobacter pylori*
SAIMIRI	Herpesvirus saimiri
HUMAN	*Homo sapiens*
HCMV	Human cytomegalovirus
KPNEU	*Klebsiella pneumoniae*
LACTO	*Lactococcus* species
TOMATO	*Lycopersicon esculentum* %GC-separated
TOMATO_LOW	*Lycopersicon esculentum* %GC=0-46
TOMATO_HI	*Lycopersicon esculentum* %GC=46-100
TOMATO_G	*Lycopersicon esculentum* non-classed
URCHIN	*Lytechinus pictus*
MOUSE	*Mus musculus*
LEPRAE	*Mycobacterium leprae*
TUBERC	*Mycobacterium tuberculosis*
CAPRI	*Mycoplasma capricolum* (TGA = Trp)
GENIT	*Mycoplasma genitalium* (TGA = Trp)
RICE	*Oryzae sativa*
RAT	*Rattus norvegicus*
RHODO	*Rhodobacter capsulatus*
YEAST	*Saccharomyces cerevisiae*
SALM	*Salmonella typhimurium*
SPOMBE	*Schizosaccharomyces pombe*
SAUREUS	*Staphylococcus aureus*
SOLF	*Sulfolobus solfataricus*
THEIL	*Theileria parva*
FROG	*Xenopus laevis*

■ **Example of a query**

An example query to the GeneMark server is shown in Fig. 6.8. The subject line directive is set to GENEMARK to indicate that this message contains an analysis query. This query uses several of the optional directives in the body of the message. Note that multiple sequences are being sent to the server in the same query message.

Figure 6.8. Example of a GeneMark Query

```
From:       annep@acs.auc.eun.eg
Date:       Wed 25 Sep 1996 - 09:58.16 - GMT
To:         genemark@ford.gatech.edu
Subject: GENEMARK
# FROM ABI RUN OF 04 JULY 1996
# FILE STORED IN DIRECTORY ANNE'S DATASET ON LAB MAC
TITLE UNKNOWN SEQUENCE FROM CLONE 14-14
SPECIES HUMAN
WINDOW 100
STEP 10
PROTEIN REGION BLAST BLITZ FASTA
PSGRAPH REGION
ORFLIST REGION
DATA
GGGACGATCGATCGATGGCGCTCGGATGCGATCGCTGGCTAGGCNNNCGATCNCGATCNNCGCTTGCT-
GCTAGCGCCTCCCCGGGGATGGGTTTTCTGCCGGGGGCTGATCGACCTGATTCGATGGCGCTCGGATGC-
GATCGCTGGCTAGGCNNNCGATCNCGATCNNCGCTTGCTGCTAGCGCCTCCCCGGGGTGCTGCCCCTCC-
CATGGCCCCGGCGCCTCCCCGGGGATGGCTXGGATCGACGCTXGGTTTAAAACCG
TITLE UNKNOWN SEQUENCE FROM CLONE 14-15
DATA
CTGATTCGATGGCGCTCGGATGCGATCGCTGGCTAGGCNNNCGATCNCGATCNNCGCTTGCTGC-
TAGCGCCATCGATCGATGGCGCTCGGATGCGATCGCTGGCTAGGCNNNCGATCNCGATCNNCGCTTGCT-
GCTAGCGCCTCCCCGGGGATGGCTXGGGGTTTTNNCTCTGCTGAAAAAAATTTTCTGCTGCCTCCC-
CGGGGTGCTGCCCCTCCCATGGCCCCGGCGCCTCCCCGGGGATGGCTXGGATCGACGCTXGGTT-
TAAAACCCTATCGNGAXGATGCCACGGATCGACCTGATTCGATGGCGCTCGGATGCGATCGCTGGCTAG-
GCNNNCGATCNCGATCNNCGCTTGCTGCTAGCGCCATCGATCGATGGCGCTCGGATGCGATCGCTGGC-
TAGGCNNNCGATCNCGATCNNCGCTTGCTGCTAGCGCCTCCCCGGGGATGGCTXGGGGTTTTNNCTG
```

MotifFinder

The server MotifFinder is designed to analyze a query protein sequence for patterns, profiles, and sequence motifs. The program compares the query sequence against either the PROSITE database or a proprietary database called MotifDic (Ogiwara 1992). MotifFinder was designed to help understand the structure and function of a protein sequence. First, MotifFinder searches for conserved patterns that appear in the query sequence. If any patterns are found, MotifFinder then searches the SWISS-PROT and PIR protein sequence databases along with the PDB structural database for those entries that contain the matching conserved patterns. The results of the analysis are given as graphic representations in the computer image description language PostScript. To view the location of the conserved patterns in the three-dimensional (3D) structure of the protein and their relation to known biological function sites annotated in the sequence databases, use a PostScript viewing program or print out the files on a PostScript-compatible printer.

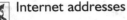 **Internet addresses**

E-mail:	motiffinder@genome.ad.jp	(Server)
	motif-manager@genome.ad.jp	(Personal help)
Web:	http://www.genome.ad.jp/SIT/MOTIF.html	(Home page)
Organization:	Supercomputer Laboratory, the Institute for Chemical Research, Uji, Kyoto 611, Japan	
Telephone:	+81-774-38-3267	
Facsimile:	+81-774-32-8235	

■ Accessing the server

To access the server, send an e-mail message containing the formatted query sequence to the server address. To receive instructions on using the MotifFinder server, send a mail message to the server address containing the directive HELP on a single line in the body of the message.

■ Server directives and parameters

MotifFinder has only a single directive, DATALIB, which is optional. This directive is used to specify the motif database library that will be used by the server to analyze the query sequence. The options are PROSITE, described in Chapter 4, and a proprietary library called MotifDic. The sequence to be analyzed must be in FastA format (see Chapter 3).

■ Formatting a query

Because directives for the server are limited, formatting a query to MotifFinder is simple. For the analysis of a sequence using the default settings, the body of the query message must contain the sequence in FastA format. To change the motif database library from the default parameter, PROSITE, the directive DATALIB should be used. This directive and its parameter must be entered prior to the query sequence in the message to the server.

■ Examples of queries

Two examples of query messages to MotifFinder are shown in Fig. 6.9. In the first query (Fig. 6.9A) the default settings for the server are used. This query will use the default PROSITE database to analyze the query sequence for putative motifs. In the second query example, the optional directive DATALIB is used to change the motif database library. For this query, the server will use the MotifDic database in the analysis.

Figure 6.9. Example Queries to MotifFinder

A

```
From:        annep@acs.auc.eun.eg
Date:        Fri 04 Oct 1996 - 16:14.27 - GMT
To:          motiffinder@genome.ad.jp
Subject:
>UNKNOWN PROTEIN PHOSPHATASE CATALYTIC SUBUNIT
MNVAINGFGRIGRLVLRASAKNPLINIVAINDPFVSTTYMEYMLEYDTVHGKFDGSLSHD
ETHIFVNGKPIRVFNEMNPENIKWGEEQVQYVVESTGAFTTLEKASTHLKNGVEKVVISA
PSSDAPMFVMGVNHELYEKNMHVVSNASCTTNCLAPLAKVVNDKFGIKEGLMTTVHAVTA
TQKTVDGPSKKDWRGGRGACFNIIPSSTGAAKAVGKVIPSLNGKLTGMSFRVPTADVSVV
DLTARLVNPASYDEIKAAIKSASENEMKGILGYTEKAVVSSDFIGDSHSSIFDAEAGIAL
TDDFVKLVSWYDNEWGYSSRVLDLIEHMVKNE
```

B

```
From:        annep@acs.auc.eun.eg
Date:        Fri 04 Oct 1996 - 16:14.27 - GMT
To:          motiffinder@genome.ad.jp
Subject:
DATALIB MOTIFDIC
>UNKNOWN PROTEIN PHOSPHATASE CATALYTIC SUBUNIT
MNVAINGFGRIGRLVLRASAKNPLINIVAINDPFVSTTYMEYMLEYDTVHGKFDGSLSHD
ETHIFVNGKPIRVFNEMNPENIKWGEEQVQYVVESTGAFTTLEKASTHLKNGVEKVVISA
PSSDAPMFVMGVNHELYEKNMHVVSNASCTTNCLAPLAKVVNDKFGIKEGLMTTVHAVTA
TQKTVDGPSKKDWRGGRGACFNIIPSSTGAAKAVGKVIPSLNGKLTGMSFRVPTADVSVV
DLTARLVNPASYDEIKAAIKSASENEMKGILGYTEKAVVSSDFIGDSHSSIFDAEAGIAL
TDDFVKLVSWYDNEWGYSSRVLDLIEHMVKNE
```

■ Interpreting the analysis results

In contrast to the simplicity of the query message, the results that are reported back by MotifFinder can be complicated. First, if MotifFinder finds any sequence motifs in the query sequence, it returns the information about the motif and the location within the query sequence. In addition, the reply message from the server can also contain up to two sections of results in the PostScript language that graphically depict the matching motifs in relation to the query sequence and the sequences in the databases from which the motifs were originally derived.

Interpreting the first part of the reply message, the identity of the motif and its location on the query sequence, is relatively straightforward. The server lists the entry and its description from the pattern database, then displays the query sequence annotated to show the location of the motif pattern. For most people, this first part of the message will be sufficient, as interpreting the PostScript graphics is a bit more of a challenge. To extract the graphics from this file, several steps must be followed. First, save the entire e-mail reply from the server into a word processor file. Do this by cutting and pasting from the e-mail client program to the word processing program or by saving the e-mail reply as a text document and then opening it in the word processing program. Next, edit the text file in the word processing program to remove the e-mail header section of the reply from the server. This header will be at the beginning of the text file. Also remove the first part of the reply message. This process will leave a results file that contains both the text file showing the location of the motifs and PostScript files that graphically depict the results.

The next step is to locate the individual file headers in the results file. At each of these headers, cut out the individual file and its header and paste it into a new word processing document. Save this document as text files with the names:

- **MotifFind.0, MotifFind.1,** for files showing the location of motifs
- **MotifFeat0.ps, MotifFeat1.ps,** for files showing feature representation
- **Motif3D0.ps, Motif3D1.ps,** for files showing 3-D representation

where the numerals 0, 1, and so on distinguish different motifs found in the query sequence.

The first set of files are text files, but the second and third sets are graphic files. The MotifFeat file contains a schematic diagram of the primary structures showing both the locations of motifs and the locations of functional sites annotated in the database. In this diagram, boxes correspond to motifs, and arrows and triangles mean active/binding sites and intron positions, respectively, as annotated in the feature table of the database. The Motif3D file, if it exists, is a graphical representation of structural entries in the PDB database. It is a stereo drawing of alpha-carbon tracing, with the locations of motifs in the 3D structure marked in bold lines. These files can now be saved and printed to a PostScript-compatible printer or viewed using a PostScript viewer program.

NetGene

The NetGene server is designed to predict splice sites in vertebrate genes using a neural network approach (Brunak et al. 1991). The server assigns scores to each nucleotide in the query sequence on the basis of its relative potential to be a donor or acceptor site in a junction. Nucleotide scores are regulated in part by the predicted transition regions of sequence around introns and exons. These transition regions are identified by Net-Gene based on a separate neural network system that has been trained to recognize exon elements in genomic sequence.

This approach to the prediction of splice sites means that in regions of sudden change in the predicted codon usage, splice sites can be detected with a high degree of confidence even with a low relative score. Conversely, in regions with a more constant pattern of codon usage, potential splice sites must have a higher relative score. This methodology has resulted in a system that has proven more capable in the prediction of splice sites than other approaches. Please note that because of the design of the analysis system, splice sites closer than 225 nucleotides to the ends of the query sequence cannot be determined.

Internet addresses

E-mail:	netgene@cbs.dtu.dk	(Server)
	rapacki@cbs.dtu.dk	(Personal help)
Organization:	Center for Biological Sequence Analysis, Department of Physical Chemistry, Technical University of Denmark, Building 206, DK-2800 Lyngby, Denmark	
Telephone:	+45-4525-2483	
Facsimile:	+45-4593-4808	

■ Accessing the NetGene server

To receive an informational help file back from the server, send a message to the server address containing the word HELP on a single line in the body of the mail message. For questions about the NetGene server that are not covered in the informational help file, send a brief message to Kristoffer Rapacki at the personal help address.

■ Server directives and parameters

The NetGene server has no directives or parameters. The only mandatory component of a query to the server is that the sequence to be analyzed must be in FastA format (see Chapter 3). The structure and an example of a query message to the server are shown in Fig 6.10.

■ **Formatting a query**

The query message is composed of a sequence file in FastA format with some slight modifications. The sequence can only consist of the letters A, T, U, G, C, and N. The letter U is converted to T. If any other letters are used in the query sequence, they are converted to N. All numbering, blank spaces, and other characters are allowed but ignored by the server.

The minimum length of a sequence that can be analyzed by the NetGene server is 451 nucleotides. The maximum length is 100,000 nucleotides. When formatting the sequence in the query message, the maximum length of a line of sequence should be 80 characters.

■ **Example of a query**

An example query message to the NetGene server is shown in Fig. 6.10. Note that the query message is simply a sequence file in FastA format.

Figure 6.10. Example of a Query to the NetGene Server

```
From:      annep@acs.auc.eun.eg
Date:      Wed 25 Sep 1996 - 09:58.16 - GMT
To:        netgene@cbs.dtu.dk
Subject:
>GENOMIC SEQUENCE FROM CLONE 16-98
GGATCCTGAGTACCTCTCCTCCCTGACCTCAGGCTTCCTCCTAGTGTCATTATGAACACCCCCAATCTCC
CAGATGCTGACTGCCCCGATCAGAAGTCAGATCAGATCAGTACAGTCCAGCTAGCAGATCAGTCACAGTC
ACAGATCCAGATCAGCTACAGGAGATATGAGGACAGATCGACAGCTAGACCTTGGCCCCTCTTAGAAGCC
ATTAGGCCCTCAGTTTCTGCAGCGGGGATACAGTCCAGCTAGCAGATCAGTCACGTCAGATCAGATCAGT
CAGTCCAGCTAGCAGATCAGTCACAGTCACAGATCCAGATCAGCTACAGGAGATATGAGGACAGATCGAC
GCTAGACCTTGGCCCCTCTTAGAAGCCAATTAGGCCCTCAGTTTCTGCAGCGGGGATACAGTCCAGCTAG
AGATCAGTCACAGTCACAGATCCAGATCAGCTACAGGAGATATGAGGACAGATCGACAGCTAGAGTAGGA
GTTGTTACACCCCACAGATTAATATGAGTAGGAGTTGTTACACCCCACAGATAGCAGAGTACCCCAGATG
GGGGGGATACAGATACAGTAGTGTACAGATAGTCACAGATCCAGATCAGCTACAGGAGATATGAGGACAG
ATCGACAGCTAGAGTAGGAGTTGTTACACCCCACATCAGTCGATCCGTTTGATCGCAGTACCGAAATGCT
TTAGATAAAACAGATGCAGTCGATGGGCTGACAGCTGACGCTAGACACGATACAGATCCCAGTCA
```

nnPredict

The nnPredict server is designed to predict the secondary structure of protein sequences. To generate the predicted secondary structure, the server calculates the optimal secondary structure type for each residue in a query protein sequence based on a set of predetermined, weighted values (Kneller et al. 1990).

Internet addresses

E-mail:	nnpredict@celeste.ucsf.edu	(Server)
	nnpredict-request@celeste.ucsf.edu	(Personal help)
Web:	http://www.cmpharm.ucsf.edu/~nomi/nnpredict.html	
		(Home page)
Organization:	University of California-San Francisco (UCSF), San Francisco, CA 94143, U.S.A.	

■ Accessing the nnPredict e-mail server

To access the nnPredict server, send a query message containing the server address. To receive a set of instructions on using the server, send a message to the server address with the directive HELP on a single line in the body of the message. For further assistance, send a message with a description of the problem and include the query and any output e-mail messages received back from the server to the personal help address listed above.

■ Server directives and parameters

The nnPredict server has one optional directive, OPTION, which is used to specify the structure class to be used in the analysis of the query sequence. The nnPredict server can use three structure classes in the prediction of a protein secondary structure: all-alpha, all-beta, or alpha/beta, along with a default setting of no structure class. The format for the use of this directive is:

```
OPTION: X
```

where the parameter X is one of four possible parameters. These are:

N	No structure class for prediction; the default setting
A	All-alpha structure class for prediction
B	All-beta structure class for prediction
A/B	Alpha/beta structure class for prediction

If the OPTION directive is not used in a query, then the default setting of N will be used in the analysis.

■ Formatting a query

A query message to the server only needs to contain the sequence to be analyzed. The sequence must be in FastA format (see Chapter 3): the first line of the formatted sequence must be a header line starting with a > character followed by any comments on the same line. The following lines are the query protein sequence data. The sequence data can consist of one-letter amino acid codes (A C D E F G H I K L M N P Q R S T V W Y) or three-letter amino acid codes separated by spaces (ALA CYS ASP GLU PHE GLY HIS ILE LYS LEU MET ASN PRO GLN ARG SER THR VAL TRP TYR). The letters B and Z are not recognized by the server as valid amino acid codes.

Alternatively, the first line of a query to the server can contain the OPTION directive to set the tertiary structure class for the analysis. The sequence would follow on succeeding lines. Note that only one sequence and one structure class can be used in a query.

■ Example of a query

Figure 6.11 shows an example of a query to the nnPredict server. In this query, the OPTION directive is used to set an all-alpha structure class for the protein to be analyzed. In response to this query, the server will return a report that is a secondary structure prediction for each amino acid in the sequence. The predicted type of secondary structure for each amino acid residue will be shown as one of the following values: H for a helix element, E for a beta strand element, or _ for a turn element. If the query sequence contains any codes that are not one of the standard amino acids, question marks will be used in the server reply message to indicate that no structure prediction could be made around the unrecognized amino acid residue.

Figure 6.11. Example of a Query to the nnPredict Server

```
From:      annep@acs.auc.eun.eg
Date:      Wed 25 Sep 1996 - 09:58.16 - GMT
To:        nnpredict@celeste.ucsf.edu
Subject:
OPTION: A
>UNKNOWN PROTEIN PHOSPHATASE REGULATORY SUBUNIT
MNYQPTPEDRFTFGLWTVGWQGRDPFGDATRRALDPVESVQRLAELGAHGVTFHDDDLIPFGSSD
SEREEHVKRFRQALDDTGMKVPMATTNLFTHPVFKDGGFTANDRDVRRYALRKTIRNIDLAVELG
AETYVAWGGREGAESGGAKDVRDALDRMKEAFDLLGEYVTSQGYDIRFIEPKPNEPRGDILLPTVG
HALAFIERLERPELYGVNPEVGHEQMAGLNFPHGIAQALWAGKLFHIDLNGQNGIKYDQDLRFGAG
DLRAAFWLVDLLESAGYSGPRHFDFKPPRTEDFDGVWASAAGCMARNYLILKERAAAFRADPEVQE
ALRASRLDELARPTAADGLQALLDDRSAFSEFDVDAAAARGMAFERLDQLAMDHLLGARG
```

PredictProtein

The PredictProtein server predicts the structure of a submitted protein sequence. A query protein sequence is sent to the server, and PredictProtein returns a multiple sequence alignment and predictions of secondary structure, along with estimates of residue solvent accessibility and helical transmembrane regions. The following additional analysis features are available upon request: (1) fold recognition by a prediction-based threading model in which the protein database is searched for possible remote homologs to the query protein sequence; (2) evaluation of prediction accuracy for a given predicted and observed secondary structure with per-residue and per-segment scores being determined. The prediction method is rated at greater than 70% accuracy for water-soluble, globular protein.

The analysis of the query protein sequence is accomplished in a multistep process. First the SWISS-PROT sequence database is scanned for similar sequences. Next, these sequences are collected by the server and used to generate a multiple sequence alignment by a weighted dynamic programming method called MaxHom. Finally, the resulting multiple alignment is used as input for profile-based neural network predictions using a methodology called PHD.

Internet addresses

E-mail:	predictprotein@embl-heidelberg.de	(Server)
	predict-help@embl-heidelberg.de	(Personal help)
Web:	http://www.embl-heidelberg.de/predictprotein	(Home page)
Organization:	Protein Design Group, European Molecular Biology Laboratory (EMBL), D-69012 Heidelberg, Germany	
Facsimile:	+49-6221-387517	

■ Accessing the PredictProtein server

To submit a protein sequence for analysis, send a query message to PredictProtein at the server address listed above. Detailed questions on the use of the server or other comments should be sent to the personal help address. To get a detailed help and informational file, send an e-mail message to the server address containing the word HELP on the Subject line in the header of the message.

■ Server directives and parameters

For all of the capability of the PredictProtein server, the operation of it is simple because it has no directives or parameters. It does have a strict and unusual format for the query message, which we describe in the next section.

■ Formatting a query

The format for a query message to PredictProtein is different than for most of the other analysis servers we have described in Chapters 5 and 6. The body of the message is composed of several parts. The first part of the query message contains the name of the sender, the institution, and the address of the sender; the first part can occupy one or more lines. Next, the e-mail address of the sender goes on a single line. After this information, one or more lines of relevant information can be entered. Finally, the query protein sequence is entered.

The format for the query protein sequence is similar to FastA format, but with one crucial difference. Instead of using the character > to denote the comment line of the sequence, PredictProtein uses the number character, #. After the comment line, the protein sequence itself is entered in one or more lines of any length. The standard IUB/IUPAC codes (see Appendix D) are supported. No special symbols, numbers, or blank lines are allowed in the sequence.

■ Example of a query

The example shown in Fig. 6.12 demonstrates how to submit a query message to the PredictProtein server. The first part of the body of the message is composed of identifier information on the message sender, along with any other information that is relevant. Following the identifier information comes the query sequence. As previously mentioned, the number character, #, is a crucial component of the query message, as the server interprets anything after this line as a protein sequence. Following the number character comes a one-line description of the protein.

If the submitted protein sequence has at least one nontrivial homolog in the database, the server will send back a large e-mail reply message containing a multiple sequence alignment along with an annotated prediction of the secondary structure of the submitted

Figure 6.12. Example of Query to PredictProtein

```
From:      annep@acs.auc.eun.eg
Date:      Wed 25 Sep 1996 - 09:58.16 - GMT
To:        predictprotein@embl-heidelberg.de
Subject:
Anne Peruski, Biology Section
American University in Cairo
annep@acs.auc.eun.eg
# UNKNOWN PROTEIN PHOSPHATASE REGULATORY SUBUNIT
MNYQPTPEDRFTFGLWTVGWQGRDPFGDATRRALDPVESVQRLAELGAHGVTFHDDDLIPFGSSD
SEREEHVKRFRQALDDTGMKVPMATTNLFTHPVFKDGGFTANDRDVRRYALRKTIRNIDLAVELG
AETYVAWGGREGAESGGAKDVRDALDRMKEAFDLLGEYVTSQGYDIRFIEPKPNEPRGDILLPTVG
HALAFIERLERPELYGVNPEVGHEQMAGLNFPHGIAQALWAGKLFHIDLNGQNGIKYDQDLRFGAG
DLRAAFWLVDLLLESAGYSGPRHFDFKPPRTEDFDGVWASAAGCMARNYLILKERAAAFRADPEVQE
ALRASRLDELARPTAADGLQALLDDRSAFSEFDVDAAAARGMAFERLDQLAMDHLLGARG
```

protein sequence divided into three parts. The first part of the reply message is a multiple sequence alignment of the submitted protein sequence against similar sequences identified from the SWISS-PROT protein sequence database. The second part contains detailed information and explanations about the prediction method used in the analysis. The last part of the reply message displays the actual predicted secondary structure of the query protein sequence.

The prediction accuracy of the algorithms used in the PredictProtein server has been evaluated in several experimental studies. These studies have found that the overall secondary structure prediction has an accuracy of greater than 72% for water-soluble globular proteins (Rost and Sander 1993, 1994). For predictions of solvent accessibility, the expected correlation between observed and predicted relative accessibility is greater than 0.5 (Rost and Sander 1994). Regions of transmembrane helices are predicted with an accuracy of greater than 95% (Rost et al. 1995).

These levels of prediction accuracy are based on finding a cadre of similar sequences in the SWISS-PROT database that can be used as a scaffold on which to base the predictions. If there are only a few weakly homologous protein sequences in the database, then the prediction accuracy drops for structure predictions, including secondary structure, accessibility, and transmembrane helices. In the multiple sequence alignment returned by the server, only homologs with at least 30% identity over 80 or more residues are included in the analysis. This cutoff is five percentage points above the threshold for structural homology (Sander and Schneider 1991) and is set in an attempt to provide high-quality multiple alignments. Further, the server uses an alignment procedure based on globular, water-soluble proteins; this procedure may be of more limited utility with other classes of proteins.

ProDom

The ProDom server was designed to analyze protein and nucleotide sequences for the presence of conserved domain families (Sonnhammer and Kahn 1994). The database of domain families was developed by the application of an algorithm called DOMAINER to locate alignments generated from a comparison of the latest release of the SWISS-PROT database against itself using the BLASTP algorithm. The novelty of the ProDom database is that the modular arrangement of domains in proteins has been taken into account. When domain boundaries are detected, the protein sequence segments are separated to produce consistent families of protein domains. The domain families produced by the DOMAINER algorithm are stored as multiple alignments and also as consensus sequences within the database.

 ### Internet addresses

E-mail:	prodom@toulouse.inra.fr	(Server)
	proquest@toulouse.inra.fr.	(Personal help)
	Daniel.Kahn@toulouse.inra.fr	(Daniel Kahn)
	Jerome.Gouzy@toulouse.inra.fr	(Jerome Gouzy)
	Florence.Corpet@toulouse.inra.fr	(Florence Corpet)
Web:	http://protein.toulouse.inra.fr/prodom.html	
Organization:	Laboratoire de Biologie Moléculaire des Relations Plantes Microorganismes, BP 27, Chemin de Borde Rouge, 31326 Castanet Tolosan, France	
Telephone:	+33-05-61-28-53-29	(Daniel Kahn)
Facsimile:	+33-05-61-28-50-61	(Jerome Gouzy)
	+33-05-61-28-53-08	(Florence Corpet)

Accessing the ProDom server

To receive an informational help file back from the server, send a message to the server address containing the word HELP on a single line in the body of the mail message. If you have questions about the ProDom protein domain server that are not covered in this help file, or if you experience problems using the server, please send a brief message to the personal help address. The developers of the ProDom database and its software, Erik Sonnhammer and Daniel Kahn, may be contacted at their respective e-mail addresses listed above.

Server directives and parameters

The ProDom server has only four directives: BEGIN, END, HELP, and PROGRAM. Use of these directives is summarized in Table 6.9 and demonstrated in the sections that immediately follow.

Table 6.9. ProDom Directives and Parameters

Directive	Attributes	Explanation
BEGIN	Mandatory, none	Specifies the start of the query sequence or query record entry.
END	Optional, none	Indicates the end of a query message.
HELP	Optional, none	Used to retrieve an informational file from the server.
PROGRAM	Mandatory, text	The parameters are ASKDOM, FETCH-DOM, BLASTP, and BLASTX. There is no default setting.

■ Formatting a query

The ProDom server supports three different but related types of query messages. The first is the analysis of a protein or nucleotide sequence against the ProDom database for any similar domain families. Only one query sequence can be submitted to the server in a message of this type. The message is composed of the PROGRAM directive followed by the optional parameter statement. If the query sequence is protein, then the parameter for the PROGRAM directive is BLASTP. If the query sequence is DNA, then the parameter for the PROGRAM directive is BLASTX, which will translate the query sequence in all six frames. These translations will be used in the analysis against the database for matching domains. The next part of the query message is composed of the BEGIN directive, followed by the sequence in FastA format (see Chapter 3).

The second type of query message is used to retrieve information on the domain organization of a protein in the SWISS-PROT database. For this type of query message, up to ten queries are allowed in a single request to the server. The structure of the message consists of the PROGRAM directive followed by the parameter statement ASKDOM. The next part of the query message is the BEGIN directive, followed by the protein name from the SWISS-PROT database identifier (ID) field of the sequence record of interest.

The third type of query message is used to retrieve information on specific domain families that comprise the ProDom database. For this type of query message, up to ten queries are allowed in a single request to the server. The message is composed of the PROGRAM directive followed by the parameter statement FETCHDOM. Next in the query message is the BEGIN directive, followed by the domain family numbers of the ProDom entry of interest.

As with most other servers, a blank line or the directive END should be placed at the conclusion of a query message to prevent the server from interpreting the signature block of the message as part of the query.

■ **Examples of queries**

Examples of the three different types of query messages to the ProDom server are shown in Fig. 6.13. The first example, Fig. 6.13A, demonstrates how to submit a query sequence for analysis against the ProDom database. The protein sequence will be analyzed against the ProDom domain database for matching sequences using the BLASTP algorithm. In the second example, the PROGRAM directive parameter FETCHDOM is used to retrieve information on domain families in the ProDom database. A query message of this sort will get a reply message back from the server that contains a multiple alignment of the protein segments in each domain family and a consensus sequence derived for each of these domain families. In the final example, Fig. 6.13C, the other PROGRAM directive parameter, ASKDOM, is used to retrieve information on domain organizations present in an entry from the SWISS-PROT database. This query message will have the server retrieve the domain organization families from the ProDom database that have been matched to sequence regions of the query SWISS-PROT entry.

Figure 6.13. Examples of Queries to the ProDom Server

A

```
From:       annep@acs.auc.eun.eg
Date:       Wed 25 Sep 1996 - 09:58.16 - GMT
To:         prodom@toulouse.inra.fr
Subject:
PROGRAM BLASTP
BEGIN
>UNKNOWN PROTEIN PHOSPHATASE REGULATORY SUBUNIT
MNYQPTPEDRFTFGLWTVGWQGRDPFGDATRRALDPVESVQRLAELGAHGVTFHDDDLIPFGSSD
SEREEHVKRFRQALDDTGMKVPMATTNLFTHPVFKDGGFTANDRDVRRYALRKTIRNIDLAVELG
AETYVAWGGREGAESGGAKDVRDALDRMKEAFDLLGEYVTSQGYDIRFIEPKPNEPRGDILLPTV
GHALAFIERLERPELYGVNPEVGHEQMAGLNFPHGIAQALWAGKLFHIDLNGQNGIKYDQDLRFG
AGDLRAAFWLVDLLESAGYSGPRHFDFKPPRTEDFDGVWASAAGCMARNYLILKERAAAFRADPE
VQEALRASRLDELARPTAADGLQALLDDRSAFSEFDVDAAAARGMAFERLDQLAMDHLLGARG
```

B

```
From:       annep@acs.auc.eun.eg
Date:       Wed 25 Sep 1996 - 09:58.16 - GMT
To:         prodom@toulouse.inra.fr
Subject:
PROGRAM FETCHDOM
BEGIN
178
2074
103
296
```

C

```
From:       annep@acs.auc.eun.eg
Date:       Wed 25 Sep 1996 - 09:58.16 - GMT
To:         prodom@toulouse.inra.fr
Subject:
PROGRAM ASKDOM
BEGIN
PP2C_HUMA
```

PSORT

The PSORT server is an expert system for the prediction of protein localization sites in cells. Using a submitted protein sequence and the organism in which it is found, the server analyzes the query sequence for compartmentalization and localization based on sequence features of known protein sorting signals. The server then predicts the potential for the query protein sequence at each candidate cellular location.

The PSORT server is organized as an expert analysis system with a knowledge base of "if-then" type rules. About 100 core rules are stored in the knowledge base. Using this expert system, accuracy of the server is approximately 83% based on the analysis of 106 gram-negative proteins with known localization sites (Nakai and Kanehisa 1991). Of the 295 eukaryotic proteins in the initial development of the PSORT server, the compartmentalization of 66% was correctly discriminated (Nakai and Kanehisa 1992). However, caution should be exercised when applying this system to unknown sequences because the prediction accuracy has not been determined.

Internet addresses

E-mail:	psort@nibb.ac.jp	(Server)
	nakai@nibb.ac.jp	(Personal help)
Web:	http://psort.nibb.ac.jp	(Home page)
Organization:	National Institute for Basic Biology, 38 Nishigo-naka, Myodaiji, Okazaki 444, Japan	
Telephone:	+81-564-55-7629	
Facsimile:	+81-564-55-7625	

■ Accessing the PSORT e-mail server

To access the server, send a query message to the server address listed above. A help document on the PSORT server can be obtained by sending a message to the server address with the directive HELP on a single line in the body of the message. Any problems, questions, or comments on the use of the PSORT server should be reported to Kenta Nakai at the address for personal help listed above.

■ Server directives and parameters

A query to PSORT has only two directives, but both of them are mandatory. The first directive is SOURCE, which specifies what class of organism the query protein is from. The second directive is BEGIN, which denotes the start of the query sequence.

The SOURCE line identifies the origin of the sequence data and must contain one of the following five parameters: GRAM-POSITIVE, GRAM-NEGATIVE, YEAST, ANIMAL, or PLANT. Based on the source of the sequence, the candidate localization sites for prediction are as follows.

GRAM-POSITIVE (bacterium)	Cytoplasm, membrane, and outside (protein will be secreted).
GRAM-NEGATIVE (bacterium)	Cytoplasm, inner membrane, periplasm, and outer membrane
YEAST	Cytoplasm, mitochondria (outer membrane, intermembrane space, inner membrane, and matrix space), microbody (peroxisome), nucleus, endoplasmic reticulum (abbreviated as ER), lysosome (lumen and membrane), Golgi body, vacuole, plasma membrane, and outside (protein will be secreted)
ANIMAL	Cytoplasm, mitochondria (outer membrane, intermembrane space, inner membrane, and matrix space), microbody (peroxisome), nucleus, endoplasmic reticulum (lumen and membrane), Golgi body, lysosome (lumen and membrane), plasma membrane, and outside (protein will be secreted)
PLANT	Cytoplasm, mitochondria (outer membrane, intermembrane space, inner membrane, and matrix space), microbody (peroxisome), nucleus, endoplasmic reticulum (lumen and membrane), Golgi body, vacuole, plasma membrane, outside (protein will be secreted), and chloroplast (stroma, thylakoid membrane, and thylakoid space)

■ Formatting a query

Queries consist of an e-mail message with mandatory directives SOURCE and BEGIN to identify the source organism of the query sequence and the query sequence itself. The order for the directives in the query message is first the SOURCE directive and its parameter, followed by the BEGIN directive and the query sequence in FastA format (see Chapter 3).

Only one query sequence is allowed per mail message, and sequence data must be in FastA format. Any characters in the query sequence (except the standard IUB/IUPAC one-letter codes for the 20 amino acids [see Appendix D]) including spaces, numbers, and X will be deleted from the sequence. The server is case-insensitive. All lines of the sequence (including the description line) should be kept to 80 characters or less in length. A blank line should be put at the end of the query sequence to prevent the server from interpreting an electronic signature as part of the query sequence.

Please note that the server expects the query sequence to be a direct translation from the nucleic acid data that contains all information needed by the cell for sorting. Thus, the server will issue a warning message if the protein starts with any amino acid except methionine.

■ **Example of a query**

Figure 6.14 shows an example of a hypothetical query to the PSORT server. The SOURCE directive is set to ANIMAL. It is followed by the BEGIN directive, a single line of comments, and then the protein sequence for analysis.

Figure 6.14. Example of a Query to the PSORT Server

```
From:      annep@acs.auc.eun.eg
Date:      Wed 25 Sep 1996 - 09:58.16 - GMT
To:        psort@nibb.ac.jp
Subject:
SOURCE ANIMAL
BEGIN
>UNKNOWN PROTEIN PHOSPHATASE REGULATORY SUBUNIT
MNYQPTPEDRFTFGLWTVGWQGRDPFGDATRRALDPVESVQRLAELGAHGVTFHDDDLIPFGSSD
SEREEHVKRFRQALDDTGMKVPMATTNLFTHPVFKDGGFTANDRDVRRYALRKTIRNIDLAVELG
AETYVAWGGREGAESGGAKDVRDALDRMKEAFDLLGEYVTSQGYDIRFIEPKPNEPRGDILLPTV
GHALAFIERLERPELYGVNPEVGHEQMAGLNFPHGIAQALWAGKLFHIDLNGQNGIKYDQDLRFG
AGDLRAAFWLVDLLESAGYSGPRHFDFKPPRTEDFDGVWASAAGCMARNYLILKERAAAFRADPE
VQEALRASRLDELARPTAADGLQALLDDRSAFSEFDVDAAAARGMAFERLDQLAMDHLLGARG
```

Results returned from the server are composed of two-part messages. The first part of the message is a summary of the submitted data (for confirmation of the query). The second part of the reply gives the results of analyzing the query sequence for protein sorting signals and calculations used to predict the localization. Calculations are conveniently divided into two reasoning steps. At the end of the message are the prediction results from the server, which include the top five probable localization sites along with an estimation of certainty for the predictions.

SBASE

The SBASE e-mail server accepts a mail message containing a protein query sequence and responds with the list of the most probable domain homologies (Murvai et al. 1996). A database search is performed against the SBASE library of protein domains using the BLAST algorithm; the search results, which contain both a graphic display and annotations if desired, are returned in a reply e-mail message. The SBASE server will do a search using the BLAST algorithm of a submitted protein sequence against the SBASE database of protein sequence domains in order to give clues about the structure and function of the query sequence.

Internet addresses

E-mail:	sbase@icgeb.trieste.it	(Server)
	comment@icgeb.trieste.it	(Personal help)
Gopher:	icgeb.trieste.it	(Gopher server)
Web:	http://www.icgeb.trieste.it	(Home page)
Organization:	International Centre for Genetic Engineering and Biotechnology, Area Science Park, Padriciano 99, 34012 Trieste, Italy	
Telephone:	+39-40-375-7300	
Facsimile:	+39-40-226-555	

Accessing the SBASE server

To access the SBASE server, send a query message to the server address. To receive instructions on using the server, send a message to the server address containing only the directive HELP in the body of the message. For further assistance, send detailed questions to the address for personal help listed above.

Server directives and parameters

Summarized in Table 6.10 are the directives and associated parameters that can be used in a query to the SBASE server. The only mandatory directive that needs to be in a query to the server is BEGIN. Most of the remaining directives are derived from those used by the NCBI BLAST server, so a detailed explanation will not be given here. However, a few of these optional directives require some discussion in order to make optimal use of them in a query.

Table 6.10. SBASE Server Directives and Parameters

Directive	Attribute	Explanation
ALIGNMENTS	Optional, Boolean	The arguments allowed for this directive are either YES or NO. If the argument is YES, then the alignments are added to the search output; the default is YES.
ANNOTATIONS	Optional, Boolean	The arguments allowed for this directive are either YES or NO. If the argument is YES, then annotation lines are added to the search output; the default is NO.
BIAS	Optional, numeric	Sets the cutoff for bias-ratio rejection for the HSPCRUNCH directive. Default is set at 0.75; values can range from 0 to 1.
BEGIN	Mandatory	Indicates to the server that everything after this line is sequence data to be analyzed.
BLASTX	Optional, none	If present, tells the server to run BLASTX on the query; if not present, BLASTP will run on the query.
COVERAGE	Optional, numeric	Sets the coverage limit for the HSPCRUNCH directive. Default is set at 10; values must be a positive integer.
EXPECT PARAMETER	Optional, numeric	Sets the expect parameter for the BLAST search; the default is 25.
GENETIC_CODE	Optional, numeric	Tells BLASTX which genetic code to be used to translate the query sequence; default is 0. Current genetic codes are:

0	Standard or Universal
1	Vertebrate Mitochondrial
2	Yeast Mitochondrial
3	Mold Mitochondrial and Mycoplasma
4	Invertebrate Mitochondrial
5	Ciliate Macronuclear
6	Protozoan Mitochondrial
7	Plant Mitochondrial
8	Echinodermate Mitochondrial

continued

Table 6.10 continued

Directive	Attribute	Explanation
HSPCRUNCH	Optional, none	Tells the server to run the HSPcrunch filter. The default setting is NOT to use the filter.
MATRIX	Optional, text	Sets the scoring matrix for the BLAST algorithm; the default setting is PAM120. The parameter for this directive can be one of the following: PAM120, PAM40, PAM250, BLOSUM62.
NOPICTURE	Optional, none	This directive functions only with the HSPCRUNCH directive. If present, the output style generated by the HSPCRUNCH directive is a list of accepted alignments. If not present, the output will contain a pictoral representation.
OUTPUT PAGES	Optional, numeric	Sets the number of pages of results to report from an analysis; the default is 10. Parameter value must be a positive integer.
SCORE PARAMETER	Optional, numeric	Sets the score parameter for the BLAST search; the default is 35. Parameter value must be a positive integer.
-SORT_BY_COUNT	Optional, none	Search output is sorted from highest to lowest by the number of HSPs found for each database sequence.
-SORT_BY_HIGHSCORE	Optional, none	Search output is sorted from highest to lowest by the score of the highest scoring HSP for each database sequence.
-SORT_BY_PVALUE	Optional, none	Search output is sorted from most to least statistically significant. This is the default setting.
-SORT_BY_TOTALSCORE	Optional, none	Search output is sorted from the highest to the lowest by the sum total score of all HSPs for each database sequence.

First, if the directive ANNOTATIONS is set to YES, it is suggested that the OUTPUT PAGES directive be set to a larger value, such as 20 or 30, since the expanded annotation information will greatly increase the size of the results generated by the server. Failure to make this adjustment may result in the loss of some of the analysis from the server.

The directive BIAS is unique to this server. It determines the relative level of similarity of a sequence segment from the query sequence to a sequence segment from an entry in a database. By altering the value for this directive, the sensitivity of the analysis can be reduced or increased. As a default value, high-scoring pairs (HSPs) with bias-ratio less than 0.75 are considered as biased and rejected.

COVERAGE is another directive that is unique to this server. It can only be used when the HSPCRUNCH directive is used to filter the analysis. The COVERAGE directive looks to see if a region of the sequence has already been matched with another sequence. If so, it compares the scores of the two sequences to the matching sequence. If the score of the new sequence is lower than the original match, it is rejected to avoid redundant data. The default value for COVERAGE is 10, which means that up to ten matching sequences will be allowed. Finally, the family of -SORT_BY_ directives govern how output will be sorted from the server. Only one of these four directives may be used in a query message to the server.

■ Formatting an e-mail message

As mentioned, the only directive that is required in a query message to the SBASE server is BEGIN, followed by the query sequence itself. If any of the optional directives are used in a query, they must be placed prior to the BEGIN directive. As the sequence submitted for analysis must be in FastA format (see Chapter 3), the server reads the sequence up to the first empty line after this directive and no further.

■ Examples of queries

Examples of three query messages are shown in Fig. 6.15. In the first example, Fig. 6.15A, the SBASE server will execute the analysis of the query sequence using the default settings. The BLASTP program is used to perform the analysis using the PAM120 score matrix. The EXPECT PARAMETER directive will be set to its default value of 25. The value for the SCORE PARAMETER will be 35. When reporting the results of the analysis, the server will send back the first ten pages of the output, sorted by *P* value, appended with the sequence alignments.

The second example, Fig. 6.15B, is similar to the first except that BLASTP will use the BLOSUM62 score matrix and the search result will be filtered by the HSPcrunch program. The SBASE server will report results with a graphics output along with the original results from the BLASTP analysis of the query sequence.

The final example of a query to the SBASE server is shown in Fig. 6.15C. In this example, the server will perform the analysis of the query sequence using the BLASTX program and the PAM40 matrix. The genetic code that the server will use will be the Protozoan Mitochondrial code.

Figure 6.15. Examples of Queries to the SBASE Server

A

```
From:       annep@acs.auc.eun.eg
Date:       Wed 25 Sep 1996 - 09:58.16 - GMT
To:         sbase@icgeb.trieste.it
Subject:
BEGIN
>UNKNOWN PROTEIN PHOSPHATASE REGULATORY SUBUNIT
MNYQPTPEDRFTFGLWTVGWQGRDPFGDATRRALDPVESVQRLAELGAHGVTFHDDDLIPFGSSD
SEREEHVKRFRQALDDTGMKVPMATTNLFTHPVFKDGGFTANDRDVRRYALRKTIRNIDLAVELG
AETYVAWGGREGAESGGAKDVRDALDRMKEAFDLLGEYVTSQGYDIRFIEPKPNEPRGDILLPTV
GHALAFIERLERPELYGVNPEVGHEQMAGLNFPHGIAQALWAGKLFHIDLNGQNGIKYDQDLRFG
AGDLRAAFWLVDLLESAGYSGPRHFDFKPPRTEDFDGVWASAAGCMARNYLILKERAAAFRADPE
VQEALRASRLDELARPTAADGLQALLDDRSAFSEFDVDAAAARGMAFERLDQLAMDHLLGARG
```

B

```
From:       annep@acs.auc.eun.eg
Date:       Wed 25 Sep 1996 - 09:58.16 - GMT
To:         sbase@icgeb.trieste.it
Subject:
MATRIX BLOSUM62
HSPCRUNCH
BEGIN
>UNKNOWN PROTEIN PHOSPHATASE REGULATORY SUBUNIT
MNYQPTPEDRFTFGLWTVGWQGRDPFGDATRRALDPVESVQRLAELGAHGVTFHDDDLIPFGSSD
SEREEHVKRFRQALDDTGMKVPMATTNLFTHPVFKDGGFTANDRDVRRYALRKTIRNIDLAVELG
AETYVAWGGREGAESGGAKDVRDALDRMKEAFDLLGEYVTSQGYDIRFIEPKPNEPRGDILLPTV
GHALAFIERLERPELYGVNPEVGHEQMAGLNFPHGIAQALWAGKLFHIDLNGQNGIKYDQDLRFG
AGDLRAAFWLVDLLESAGYSGPRHFDFKPPRTEDFDGVWASAAGCMARNYLILKERAAAFRADPE
VQEALRASRLDELARPTAADGLQALLDDRSAFSEFDVDAAAARGMAFERLDQLAMDHLLGARG
```

C

```
From:       annep@acs.auc.eun.eg
Date:       Wed 25 Sep 1996 - 09:58.16 - GMT
To:         sbase@icgeb.trieste.it
Subject:
MATRIX PAM40
BLASTX
GENETIC_CODE 6
BEGIN
>UNKNOWN PROTEIN PHOSPHATASE REGULATORY SUBUNIT
GAAGTCGTACGTAAGACGAGGCAGGCTCAGGCTACGTCAAATGGTTCAGCCCC
AGTCCCCGGTGGCTGTCAGTCAAAGCAAGCCCGGTTGTTATGACAATGGAAAA
CACTATCAGATAAATCAACAGGACTGTACCTACGATCGCAGTGCATGTGGGAG
CGGACCTACCAGCAATGTACACTGGGAACACTTACCGAGTGGGTACTTGTTAT
GGAGGAAGCCGAGGTTTTAACTGCGAAAGGGGGACTTAAACCTGAAACGCAGC
AGGCTGAAGAGACTTGCTTTGACAAGTGACACTTATGAGCGTCCTAAAGACTC
CATGATCTGGGACTGTTAGGTAATGTGTTGGTTCTACGATCGCAGTGCATCGG
```

Selected readings

For additional information on individual servers we highly recommend that you consult the help documentation from the respective server of interest. Help documentation can usually be obtained by sending an e-mail request to the server with the word HELP on a single line in the body of the message. The server will return the help document in a reply message. If the server is also offered on a Web site, another way to obtain a help file is to browse the Web site. Usually the help documentation is maintained on line and can be accessed via the Web; in many cases it is maintained as a single text-based document that can be downloaded or viewed through a browser.

Journal Articles

Altschul, S. F., W. Gish, W. Miller, E. W. Myers, and D. J. Lipman. 1990. Basic local alignment search tool. *J. Mol. Biol.* **219**:403-410.

Altschul, S. F. 1993. A protein alignment scoring system sensitive at all evolutionary distances. *J. Mol. Evol.* **36**:290-300.

Bairoch, A., and R. Apweiler. 1996. The SWISS-PROT protein sequence data bank and its new supplement TREMBL. *Nucleic Acids Res.* **24**:21-25.

Bairoch, A., P. Bucher, and K. Hofman. 1996. The PROSITE database, its status in 1995. *Nucleic Acids Res.* **24**:189-196.

Benner, S. A., I. Badcoe, M. A. Cohen, and D. L. Gerloff. 1994. Bona fide prediction of aspects of protein conformation: assigning interior and surface residues from patterns of variation and conservation in homologous protein sequences. *J. Mol. Biol.* **235**:926-958.

Blattner, F. R., V. Burland, G. Plunkett, H. J. Sofia, and D. L. Daniels. 1993. Analysis of the *Escherichia coli* genome. IV. DNA sequence of the region from 89.2 to 92.8 minutes. *Nucleic Acids Res.* **21**:5408-5417.

Borodovsky, M., E. V. Koonin, and K. E. Rudd. 1994. New genes in old sequences: a strategy for finding genes in a bacterial genome. *Trends Biochem. Sci.* **19**:309-313.

Borodovsky, M., and J. D. McIninch. 1993. GeneMark: parallel gene recognition for both DNA strands. *Computers Chem.* **17**:123-133.

Borodovsky, M., K. E. Rudd, and E. V. Koonin. 1994. Intrinsic and extrinsic approaches for detecting genes in a bacterial genome. *Nucleic Acids Res.* **22**:4756-4767.

Brunak, S., J. Engelbrecht, and S. Knudsen. 1991. Prediction of human mRNA donor and acceptor sites from the DNA sequence. *J. Mol. Biol.* **220**:49-65.

Feng, D. F., and R. F. Doolittle. 1987. Progressive sequence alignment as a prerequisite to correct phylogenetic trees. *J. Mol. Evol.* **25**:51-360.

Feng, D. F., M. S. Johnson, and R. F. Doolittle. 1984. Aligning amino acid sequences: comparison of commonly used methods. *J. Mol. Evol.* **21**:112-125.

Gerloff, D. L., T. F. Jenny, L. J. Knecht, and S. A. Benner. 1993. A secondary structure prediction of the hemorrhagic metalloprotease family. *Biochem. Biophys. Res. Commun.* **194**:560-565.

Gonnet, G. H., M. A. Cohen, and S. A. Benner. 1993. Exhaustive matching of the entire protein sequence database. *Science* **256**:1443-1445.

Gribskov, M., A. D. McLachlan, and D. Eisenberg. 1987. Profile analysis: detection of distantly related proteins. *Proc. Natl. Acad. Sci. USA* **84**:4355-4358.

Guigo, R., S. Knudsen, N. Drake, and T. Smith. 1992. Prediction of gene structure. *J. Mol. Biol.* **226**:141-157.

Selected readings

183

Hegyi, H., and S. Pongor. 1993. Predicting potential domain-homologies from FASTA search results. *CABIOS* **9**:371-372.

Henikoff, S. 1991. Playing with blocks: some pitfalls of forcing multiple alignments. *New Biol.* **3**:1148-1154.

Henikoff, S. 1992. Detection of *Caenorhabditis* transposon homologs in diverse organisms. *New Biol.* **4**:382-388.

Henikoff, S., and J. G. Henikoff. 1991. Automated assembly of protein blocks for database searching. *Nucleic Acids Res.* **19**:6565-6572.

Henikoff, S., and J. G. Henikoff. 1994. Protein family classification based on searching a database of blocks. *Genomics* **19**:97-107.

Henikoff, S., J. G. Henikoff, W. J. Alford, and S. Pietrokovski. 1995. Automated construction and graphical presentation of protein blocks from unaligned sequences. *Gene* **163**:17-26.

Kneller, D. G., F. E. Cohen, and R. Langridge. 1990. Improvements in protein secondary structure prediction by an enhanced neural network. *J. Mol. Biol.* **214**:171-182.

Murvai, J., A. Gabrielian, P. Fabian, S. Hatsagi, K. Degtyanrenko, H. Hegyi, and S. Pongor. 1996. The SBASE protein domain library, release 4.0: a collection of annotated protein sequence segments. *Nucleic Acids Res.* **24**:210-213.

Nakai, K., and M. Kanehisa. 1992. A knowledge base for predicting protein localization sites in eukaryotic cells. *Genomics* **14**:897-911.

Nakai, K., and M. Kanehisa. 1991. Expert system for predicting protein localization sites in Gram-negative bacteria. *Proteins* **11**:95-110.

Ogiwara, A., I. Uchiyama, Y. Seto, and M. Kanehisa. 1992. Construction of a dictionary of sequence motifs that characterize groups of related proteins. *Protein Eng.* **5**:479-488.

Oliver, S. G. 1996. From DNA sequence to biological function. *Nature* (London) **379**:597-600.

Pappin, D. J. C., P. Hojrup, and A. J. Bleasby. 1993. Rapid identification of proteins by peptide-mass fingerprinting. *Curr. Biol.* **3**:327-332.

Pietrokovski, S., J. G. Henikoff, and S. Henikoff. 1996. The Blocks database—a system for protein classification. *Nucleic Acids Res.* **24**:197-200.

Rost, B., R. Casadio, P. Fariselli, and C. Sander. 1995. Prediction of helical transmembrane segments at 95% accuracy. *Prot. Sci.* **4**:521-533.

Rost, B., C. Sander, and R. Schneider. 1994. PHD—a mail server for protein secondary structure prediction. *CABIOS* **10**:53-60.

Rost, B., and C. Sander. 1993. Prediction of protein secondary structure at better than 70% accuracy. *J. Mol. Biol.* **232**:584-599.

Sander, C., and R. Schneider. 1991. Database of homology-derived structures and the structural meaning of sequence alignment. *Proteins* **9**:56-68.

Sonnhammer, E. L. L., and R. Durbin. 1994. A workbench for large scale sequence homology analysis. *CABIOS* **10**:301-307.

Sonnhammer, E. L. L., and D. Kahn. 1994. Modular arrangement of proteins as inferred from analysis of homology. *Prot. Sci.* **3**:482-492.

Wallace, J. C., and S. Henikoff. 1992. PATMAT: a searching and extraction program for sequence, pattern, and block queries and databases. *CABIOS* **8**:249-254.

Books and Monographs

Gonnet, G. H. 1992. *A Tutorial Introduction to Computational Biochemistry using Darwin.* Informatik E.T.H. Zurich, Switzerland.

Gribskov, M., and J. Devereux (ed.). 1991. *Sequence Analysis Primer.* Stockton Press, New York.

Knecht, L., and G. H. Gonnet. 1992. *Alignment of Nucleotide with Peptide Sequences.* Report 184. Institute for Scientific Computing, E.T.H., Zurich, Switzerland.

Swindell, S. R., R. R. Miller, and G. S. A. Myers (ed.). 1995. *Internet for the Molecular Biologist.* Horizon Scientific Press, Norfolk, England.

Internet Publications

BLAST Manual. 1994. National Center for Biotechnology Information, National Library of Medicine, National Institutes of Health, Bethesda, Md.

Hayden, D. 1994. *Guide to Molecular Biology Databases.* School of Library and Information Studies, University of Alberta, Calgary, Alberta, Canada.

CHAPTER **7** # Software, News, and Information

■ Servers that are archives of software, news, information, and commentary

Servers that are archives of software, news, information, and commentary

The servers that form the basis for Chapter 7 are different from those we described in previous chapters. The servers discussed in Chapters 5 and 6 were designed to retrieve related sequence information and analyze sequence data in order to determine similarities, structure, and function. In contrast, the three servers we describe in this chapter, the BIOSCI Newsgroup Network, EBI's NetServ, and GDB, do not analyze or retrieve sequence information; instead, they give access to informational archives. BIOSCI Newsgroup Network is a worldwide network of newsgroups and archives that allows scientists to join discussion groups and mailing lists related to their fields of interest. The EBI NetServ system is a somewhat chaotic repository of software, data files, and help and informational files, along with references and resources about sequence analysis. The last server we discuss in this chapter, GDB, is perhaps the ultimate repository for all information on the Human Genome Project: sequence, mutations, mapping data, and literature. Like the Query retrieval server described in Chapter 5, it is based on a relational database model; in this case, however, the information contained within this elaborate and extensive database is centered on humans, rather than all organisms.

One of the most useful aspects of these three servers is the flexibility of access that they offer. Like most of the servers discussed in earlier chapters, their resources can be accessed via e-mail or the Web. This flexible access makes it simple to download software or the latest restriction enzyme lists, or to discuss the new and emerging technologies.

In this chapter, we will look first at the BIOSCI Newsgroup Network server and show how to access discussion groups and mailing lists. A discussion of EBI's NetServ will follow, with a look at the archives of software, news, and information that this resource offers the community. Finally, we will look at GDB and how to use this informational system.

BIOSCI Newsgroup Network

BIOSCI Newsgroup Network is a set of electronic communication forums composed of Usenet newsgroups and parallel e-mail lists. This network was developed to allow world-wide communication between biological scientists over a variety of computer networks. By having distribution sites or nodes on each major network, the BIOSCI Newsgroup Network allows its users to contact people anywhere without having to learn a variety of computer addressing tricks. Any user can simply post a message to a regional BIOSCI node and that message will be distributed automatically to all other subscribers on the Internet.

One of the strongest aspects of the BIOSCI Newsgroup Network is that it is driven by the biological sciences community. Biologists decide what should be posted in the moderated newsgroups. Biologists keep track of Internet-based resources to ensure that the archives and libraries are kept both topical and current. The journal tables of contents that can be browsed are from the peer-reviewed literature. These factors strengthen the utility of this resource and foster a sense of community among its users.

 Internet addresses

E-mail:	biosci-server@daresbury.ac.uk	(Europe, Africa, Asia)
	biosci-server@net.bio.net	(Americas, Pacific Rim)
	biosci-help@net.bio.net	(Personal help)
Web:	http://www.bio.net	(Home page)

Additional information on the BIOSCI Newsgroup Network that can be downloaded from the network archives is listed in Table 7.1.

Table 7.1. BIOSCI Information Documents

Document	Directive To Retrieve Document
BIOSCI frequently asked questions (FAQ)	INFO FAQ
BIOSCI info sheet for the Americas & Pacific Rim	INFO USINFO
BIOSCI info sheet for Europe, Africa, & Central Asia	INFO UKINFO
Information on network connections	INFO INTERNET
How to set up Usenet News at your site	INFO USENET1
Usenet Netiquette and other edifying reading	INFO USENET2

Why are there two BIOSCI sites? BIOSCI distribution sites were established on each side of the Atlantic to minimize network e-mail traffic. For example, if a message is posted to the United States site, only one copy is sent on to the U.K. network site before it is distributed to all e-mail subscribers and newsgroups. This system is a more efficient process than sending multiple copies of the same message across the Atlantic. A trade-off for this efficiency is a slight complexity in the distribution network: the mailing lists for each newsgroup are split between two sites. Because of this split, it is imperative that users of this service respect the geographical boundaries of these two sites when subscribing to newsgroups or making an e-mail request for information.

■ BIOSCI Newsgroup Network and the Web

The BIOSCI Newsgroup Network Web pages are visited by thousands of users each week. The site has several separate pages that are linked to a main home page. A listing of the principal pages is given in Table 7.2.

Table 7.2. BIOSCI Newsgroup Network Web Pages

Web Page	URL
Main home page	http://www.bio.net
Newsgroups archives page	http://www.bio.net/archives.html
Journals archive page	http://www.bio.net/bio-journals.html
Employment archive pages	http://www.bio.net/hypermail/employment/
Address database search page	http://www.bio.net/addrsearch.html
Methods newsgroup archive pages	http://www.bio.net/hypermail/methds-reagnts/

The BIOSCI network enables scientists to:

- **Access newsgroups and newsgroup archives.** Using the network Web site as the interface, scientists can browse the files of the BIOSCI newsgroups. This resource includes a complete archive and indexes of all BIOSCI postings. To search for a keyword or phrase occurring in any of the newsgroups, enter the query terms in the search form on the main archive page. To refine a search or narrow it to a single newsgroup, first go to the archive for that newsgroup and then use the search form for that archive. Newsgroups of interest to the molecular and genomic biology community are shown in Table 7.3.

- **Search the contents of journals.** On the network, users can browse the tables of contents of journals distributed on the BIO-JOURNALS newsgroup, or they can search for references on selected topics in those journals. The search system sup-

Table 7.3. Selection of BIOSCI Newsgroups Related to Molecular and Genomic Biology

Newsgroup	Description
ACEDB/bionet.software.acedb	Discussions of genome databases that use ACEDB system.
ADDRESSES/bionet.users.addresses	Listing of who's who in biology.
ARABIDOPSIS/bionet.genome.arabidopsis	Discussions of the *Arabidopsis* genome project.
AUTOMATED-SEQUENCING /bionet.genome.autosequencing	Discussions and support on automated DNA sequencing technology. *Moderated*
BIO-MATRIX/bionet.molbio.bio-matrix	Computer applications for biological databases.
BIO-SOFTWARE/bionet.software	Discussion and information about software for the biological sciences.
BIO-SRS/bionet.software.srs	Discussions about Sequence Retrieval System (SRS) software.
BIO-WWW/bionet.software.www	Information on Web resources for the biological sciences. *Moderated*
BIONEWS/bionet.announce	Announcements of widespread interest to biologists. *Moderated*
CELEGANS/bionet.celegans	Research and information on the model organism *Caenorhabditis elegans*.
CHLAMYDOMONAS /bionet.chlamydomonas	Research and information on the alga *Chlamydomonas* and other green algae. *Moderated*
CHROMOSOMES /bionet.genome.chromosomes	Mapping and sequencing of eukaryote chromosomes.
COMPUTATIONAL-BIOLOGY /bionet.biology.computational	Computer and mathematical applications in the biological sciences. *Moderated*
CSM/bionet.prof-society.csm	Announcements from the Canadian Society of Microbiologists. *Moderated*
DROS/bionet.drosophila	Research and discussion on the model organism *Drosophila*.
EMBL-DATABANK /bionet.molbio.embldatabank	Information on about the EMBL sequence database and related topics.
EMPLOYMENT-WANTED /bionet.jobs.wanted	Requests for employment in the biological sciences.

continued

Table 7.3 continued

Newsgroup	Description
EMPLOYMENT/bionet.jobs.offered	Job position openings in the biological sciences. *Moderated*
GDB/bionet.molbio.gdb	Messages to and from the GDB database staff.
GENBANK-BB /bionet.molbio.genbank	Information about the GenBank sequence database and related topics.
GENETIC-LINKAGE /bionet.molbio.gene-linkage	Research and discussion on genetic linkage analysis.
HUMAN-GENOME-PROGRAM /bionet.molbio.genome-program	Discussions and information on all aspects of the international Human Genome Project. *Moderated*
INFO-GCG/bionet.software.gcg	Information and discussion about the GCG software suite for sequence analysis.
METHODS-AND-REAGENTS /bionet.molbio.methds-reagnts	Requests for information and lab reagents.
MICROBIOLOGY /bionet.microbiology	General information and discussion on the science and profession of microbiology.
MOLECULAR-EVOLUTION /bionet.molbio.evolution	Discussion and information on how genes and proteins have evolved.
PROTEIN-ANALYSIS /bionet.molbio.proteins	Research and information on protein sequence analysis and protein databases.
PSEUDOMONADS /bionet.organisms.pseudomonas	Research on the genus *Pseudomonas*.
RAPD/bionet.molbio.rapd	Research and discussion on randomly amplified polymorphic DNA.
RECOMBINATION /bionet.molbio.recombination	Research on the recombination of DNA or RNA.
RNA/Prototype	Discussions about RNA editing, RNA splicing, and ribozyme activities of RNA.
SCIENCE-RESOURCES /bionet.sci-resources	Information about funding agencies, grants, and fellowships. *Moderated*
STADEN/bionet.software.staden	Information and discussion about the Staden molecular sequence analysis software.
YEAST/bionet.molbio.yeast	Information and discussion about the molecular biology and genetics of yeasts.

ports the Boolean operators AND, OR, and NOT, along with the truncation operator character, *. If multiple search terms are entered without Boolean operators, the OR operator is used as the default, meaning that references containing any one or more of the query terms are returned in the results of the search. For example, the query statement PCR AND SEQUENC* NOT RNA requests references that contain the term PCR and any term starting with SEQUENC only, but no references containing the term RNA. (The order of the terms in the references is not important.) Table 7.4 lists some of the journals that are currently indexed in the network system.

■ **Find addresses of subscribers.** Increasing numbers of biologists are users of the BIOSCI Newsgroup Network. The database of these users can be searched by name, address, research interest, or other related topics through an indexed browsing system. New users of the system can register in the database as well. As with the journal searching system, the indexed browsing system of this database uses Boolean operators. For example, the query JOHN SMITH would search the database and report back address records for JOHN SMITH and for JOHN and for SMITH.

Table 7.4. Selection of Journals Related to Molecular and Genomic Biology Indexed in the BIOSCI Newsgroup Network

Antimicrobial Agents and Chemotherapy	IEEE Engineering in Medicine and Biology
Applications in the Biosciences	Immunogenetics
Applied and Environmental Microbiology	Journal of Bacteriology
Applied Microbiology and Biotechnology	Journal of Biological Chemistry
Archives of Microbiology	Journal of Computational Biology
Biotechniques	Journal of Molecular Biology
Cancer Chemotherapy and Pharmacology	Journal of Molecular Evolution
Cancer Immunology, Immunotherapy	Journal of Theoretical Biology
Chromosoma	Journal of Virology
Current Genetics	Mammalian Genome
Current Microbiology	Molecular and Cellular Biology
Differentiation	Molecular and General Genetics
EMBO Journal	Molecular Microbiology
European Journal of Biochemistry	Nucleic Acids Research
Evolution	Protein Science
Genome	Theoretical and Applied Genetics
Human Genetics	Trends in Biochemical Sciences

■ **Browse for bioinformatics servers.** BIOSCI Newsgroup Network maintains a catalog of useful Web sites that are resources in the general biosciences. The divisions of this bioscience virtual library are shown in Table 7.5. Scientists can browse this virtual library or search for sites of interest with a minimum of effort or time.

Table 7.5. Section Headings in BIOSCI Newsgroup Network Virtual Library

Biology and Medical Sections	Immunology
Agriculture	Instructional Resources in Biology
Biochemistry, Molecular Biology, and Biophysics	Medicine
Biodiversity and Ecology	Microbiology and Virology
Biological Molecules	Mycology (Fungi)
Biotechnology	Model Organisms
Botany	Neurobiology
Developmental Biology	Physiology and Biophysics
Entomology	Whale Watching
Evolution	
Fish and Other Aquatic Animals	*Related Sections of General Interest*
Forestry	Animal Health, Wellbeing, and Rights
Genetics	Electronic Journals
German Biology Sites	History of Science, Technology and Medicine
Herpetology	Publishers

■ Distribution of BIOSCI information

BIOSCI messages are distributed by two means: Usenet news software and e-mail to subscribers. The contents of the Usenet newsgroups and the e-mail distributions are identical: messages sent in by e-mail are forwarded to Usenet, and messages posted to Usenet newsgroups are distributed to e-mail subscribers. However, the Usenet software is preferable to an e-mail subscription for this purpose, for a number of reasons.

Usenet (short for Users Network) is an electronic bulletin board network that uses various public domain versions of software for message transmission. The software can operate over physical networks with modem connections to the Internet and over sophisticated networks with direct connections. Usenet news software has been optimized to transmit messages without loss and to avoid other errors that plague simple e-mail broadcasting of messages.

Usenet news software also keeps messages segregated into their respective newsgroups, making it easier to follow the thread of a discussion. In contrast, if you use e-mail to receive postings from a BIOSCI newsgroup, you receive newsgroup messages and all other messages mixed together at the same e-mail address. Given the volume of messages that a typical newsgroup creates, e-mail subscriptions are not the best way to receive newsgroup messages. Usenet news software allows you to browse the discussion topics quickly and in an organized fashion and to retrieve messages of interest. In addition, Usenet news software recovers messages automatically after network interruptions, whereas e-mail addresses may be removed from the subscription list if the computer network malfunctions and undelivered e-mail starts bouncing back to the BIOSCI administrative sites. As a result, e-mail subscribers to newsgroups then have to retrieve lost messages manually from the archives after subscriptions are suspended. Thus, e-mail participation should be seen as a last resort to be used only if news software cannot be installed locally.

Fortunately, BIOSCI offers advantages for e-mail users too. By having distribution sites or "nodes" on each major network, BIOSCI allows its e-mail users to contact fellow scientists around the world without having to learn a variety of computer addresses. Any user can simply post a message to a newsgroup, and copies of that message will be distributed automatically to all other subscribers.

■ Participating in BIOSCI using news software

If you have access to news client software or a Web browser, you do not need to subscribe to the newsgroups to access the messages. With the client news software or the Web browser, you can read newsgroup messages and post new messages. In most cases, messages submitted to a newsgroup are posted directly without editorial intervention; however, some newsgroups are moderated. Messages to these newsgroups are first reviewed by a moderator who determines whether the message is suitable for the newsgroup.

■ Participating in BIOSCI by e-mail

E-mail subscription and cancellation requests for newsgroups are handled automatically by an e-mail server, although personal assistance is also available through the personal help address. Once an e-mail address is added to a subscription list, mail will be sent to that address automatically every time someone posts a message. This will continue until the address is removed from the subscription list.

To get a listing of all current newsgroups, send a message to the server. In the body of the message, put the directive LISTS. Use the directives SUBSCRIBE or UNSUBSCRIBE to add or delete an e-mail address from the newsgroup mailing lists. The parameter for the above directives is the name of the newsgroup obtained using the LIST directive. Please be sure to send all SUBSCRIBE or UNSUBSCRIBE messages to either the biosci@daresbury.ac.uk or biosci-server@net.bio.net address only (depending on the geographical region), and not to the newsgroup itself.

The list shown in Table 7.3 contains the full names for selected newsgroups. From these full newsgroup names, the parameters to use in the SUBSCRIBE or UNSUB-

SCRIBE directives can be derived. The newsgroup parameter is the first word of the full newsgroup name. To determine the parameter for use with the SUBSCRIBE or UNSUB-SCRIBE directives, begin with the portion of the address to the left of the / character. Use no more than the first eight characters of this name. For example, the e-mail address for the METHODS-AND-REAGENTS mailing list in the Americas is methods@net.bio.net.; the word METHODS is derived from the full name of the newsgroup.

Multiple commands may be placed on separate lines in the same message sent to the server. For example, all requests to either SUBSCRIBE or UNSUBSCRIBE can be in one message, as long as each request is on a separate line. If using multiple requests in a server message, place the directive END as the last directive in a query. This helps to avoid sending any signatures to the server that could be returned in the e-mail message.

Figure 7.1 shows a sample subscription message. The e-mail address, "annep@ acs.auc.eun.eg," would be automatically added to the BIONEWS and HUMAN-GENOME-PROGRAM mailing lists as a result of this message. Note that the list name for HUMAN-GENOME-PROGRAM is HUMAN as derived from the address portion to the left of the / sign in the mailing address for the newsgroup (refer to Table 7.3).

Figure 7.1. Example of a Subscription Message to BIOSCI

```
From:     annep@acs.acu.eun.eg
Date:     02 Dec 1996 22:14:07-GMT
To:       biosci-server@net.bio.net
Subject:
SUBSCRIBE BIONEWS
SUBSCRIBE HUMAN
END
```

As previously mentioned, additional information on the BIOSCI Newsgroup Network can be obtained from a document archive. This archive can be browsed via the Web, and documents can be retrieved via an e-mail request as described below. Table 7.1 summarizes the documents that are kept in this archive and the directives that are necessary to retrieve them by an e-mail query to the server. To construct an e-mail request, put the INFO directive in the body of the message, followed by the document name to be retrieved. Text placed on the Subject line of your message is ignored, so be sure that the directives and parameters are in the body of the e-mail message. Although multiple directives may be placed in one message to the server, it is suggested that each message contain only one directive to avoid having to extract multiple documents from a single message. The server sends back information from a multiple command query in a single response message.

■ **General comments on writing style when posting messages**

Because of the open nature of the BIOSCI Newsgroup Network and the breadth of the audience, careful thought and consideration should be given to the art of preparing messages and documents before posting them to the newsgroups. Keep in mind that the Internet is both an open and an international network. Current or future employers may be reading documents posted to the Internet. So might family, colleagues, and friends whose collective memories will last much longer than it took to compose the message. The following are some general guidelines on writing style and construction of messages to be posted to the BIOSCI network.

First, the writing should be simple in style and structure. Precise writing is just as critical when preparing a document for electronic publication as it is for conventional publication. Paragraphs should be short and sentences shorter still. Select your words with care and spell them correctly. With electronic spell checking being offered in almost every commercial software program, there is little excuse for misspellings. Avoid abbreviations and acronyms and clearly define those that are used in a document. At the same time, watch the context of words and sentences to help keep the message clear. The key principle in crafting a document for the Internet is to make it concise, not cryptic.

To avoid misinterpretation, exercise both caution and judgment when writing. For example, humor and subtlety do not communicate well in a document. This is especially true in an electronic medium such as the Internet. Visual cues such as "smileys"[:-)], "frowns" [:-(], or "winks" [;-)] are useful tools in preventing misunderstandings. With that in mind, do not post an angry or hurtful message on the spur of the moment; take a break, then reread the message before sending it. Not only will this allow for reflection, it will ensure that what was written was intended.

When writing about a topic, remember that references to previous postings on that topic should be made. Include sections of previous messages to provide the necessary context for the reader. Subject lines should be composed and used with precision. Make sure that the text of the Subject line matches what is actually being discussed in the message. Most readers of newsgroups use the Subject line to filter and screen messages for topical items.

When organizing the final message or document for electronic publication, remember that the computer screen offers a different presentation than the printed page. Look carefully at the structure of the message on the screen. Use subheadings to organize content; remember that the size of a typical computer screen display is roughly 30 or 40 lines of text; and finally, white space in a message is not wasted space, but can greatly improve the clarity and organization of the document.

EBI NetServ

As we discussed in Chapter 4, the European Bioinformatics Institute (EBI) replaced the European Molecular Biology Laboratory (EMBL) Data Library in the fall of 1994 (Rodriguez-Tome et al. 1996). EBI's primary function is the development, curation, and distribution of both a comprehensive nucleotide sequence database, the EMBL Nucleotide Sequence database, and other molecular sequence databases such as SWISS-PROT. EBI also maintains an extensive, free series of network services. These include access to the primary and specialized databases maintained by EBI, collections of computer software and documentation, and sequence analysis tools that can be utilized by an e-mail server, Gopher, or the Web.

The EBI network fileserver, NetServ, permits access via e-mail messages, a Gopher server, and the Web to all of the databases, software, and documentation maintained by EBI. The EBI Gopher server uses a simple graphical interface to streamline access to the available network services; the Gopher server also offers links to other genomic biology resources throughout the world. The EBI Web server (described in Chapter 9) is the most wide-reaching of the network services offered by EBI. It permits access to all of EBI's resources including submission of sequence data, as well as database access, similarity analysis, database query and retrieval, and links to other Web resources worldwide.

The information maintained in the NetServ fileserver is extensive. It includes a wide variety of public domain, shareware, and commercial demos of software for DOS/Windows and Macintosh personal computers and for VAX and UNIX systems. Besides computer software, NetServ offers copies of specialty databases for the user community that range from pattern libraries to collections of functionally and structurally related sequences. Extensive documentation such as reference lists, reports, and newsletters is also available. The distribution policy for the files contained in the fileserver archives is liberal: anyone can download files from the archive directories.

Internet addresses

E-mail:	netserv@ebi.ac.uk	(Fileserver)
	nethelp@ebi.ac.uk	(Help system)
Gopher:	gopher.ebi.ac.uk	(EBI Gopher server)
Web:	http://www.ebi.ac.uk	(EBI home page)
Organization:	EMBL Outstation, the EBI, Hinxton Hall, Hinxton, Cambridge CB10 1RQ, United Kingdom	
Telephone:	+44-1223-494400	
Facsimile:	+44-1223-494468	

Accessing the NetServ system

The files that are accessible via the fileserver are organized into directories (shown in Table 7.6). Access the fileserver by sending one or more of the directives described below in a message to the e-mail server address. The requested information will then be returned by the fileserver. For any problems or suggestions, send a message to the help system address.

Table 7.6. NetServ Directories

Directory Name	Description
Databases	
3D_ALI	3D protein alignment database
ALIGN	DNA sequence alignments and consensus sequences
ALU	Alu sequence database
BERLIN	Berlin database of 5S rRNA sequences
BLOCKS	Blocks database of conserved protein patterns
CPGISLE	Database of CpG islands in the human genome
ECD	*E. coli* database
ENZYME	ENZYME database
EPD	Eukaryotic promoter database
HAEMB	Hemophilia B database
HLA	HLA sequence alignments database
IMGT	Immunogenetics database
LISTA	Database of sequences encoding proteins from *Saccharomyces*
NUC	Nucleotide sequences from the EMBL and GenBank databases
P53	Database of p53 somatic mutations
PLMITRNA	Database of higher plant mitochondrial genes
PKCDD	Protein kinase catalytic domain database
PRIMERS	Database of PCR primers
PROSITE	PROSITE pattern database
PROT	SWISS-PROT protein sequence database
PROTEINDATA	Protein structure database
REBASE	Restriction enzyme database
REPBASE	Database of prototypic sequences for human repetitive DNA
RRNA	Database of small ribosomal subunit RNA sequences
SEQANALREF	Sequence analysis bibliographic reference database
SMALLRNA	Database of small RNA sequences

continued

Table 7.6 continued

Directory Name	Description
Databases *(continued)*	
TRNA	tRNA sequence database
FANS_REF	Functional analysis of nucleotide sequences reference database
Software	
BIO_CATAL	Catalog of molecular biological software
DOS_SOFTWARE	Software for IBM PCs and clones
MAC_SOFTWARE	Software for the Apple Macintosh
RELIBRARY	Restriction enzyme lists for various software packages
SOFTWARE	General software and software information
UNIX_SOFTWARE	Software for UNIX systems
VAX_SOFTWARE	Software for VAX/VMS systems
Miscellaneous	
CODONUSAGE	Codon usage tables
DOC	Documents with relevance to molecular biology: submission forms, technical documents, and others
METHYL	Effects of site-specific methylation on methylases and restriction enzymes
XRAY	Information for crystallographers

■ **Server directives and formatting a query**

The directives that can be used to access the NetServ fileserver directories are shown in Table 7.7. The DIR and HELP commands are used to retrieve directory listings and information on retrieving files in a directory. The GET directive is used to retrieve specific files from the server by specifying the directory where the file is located, followed by a colon, and then the name of the file. For example, to retrieve the text file on the submission of sequence data to EBI, DATASUB.TXT, which is stored in the DOC directory of the fileserver, the directive would be: GET DOC:DATASUB.TXT. No spaces are allowed between the directory name, the colon, and the filename. Further help is available from the NetServ file by typing in the directive HELP followed by any of the directory names listed in Table 7.6.

Software, News, and Information

Table 7.7. NetServ Directives

Directive	Explanation
DIR	Retrieves a directory listing of files from the server. Identical in function to the directive, HELP. Example: DIR MAC_SOFTWARE.
GET	Retrieves a specific file from a directory. Example: GET DOC:DATASUB.TXT.
HELP	Retrieves a directory listing of files from the server. Identical in function to the directive DIR. Example: HELP MAC_SOFTWARE.
SIZE	Changes the packet size of split files. Example: SIZE 50, where the value 50 is the size of the packet in kilobytes.

Some of the files available from the NetServ fileserver are larger than 100 kilobytes in size. While the transmission and receipt of such files is generally not a problem with current technology, there are some older systems that cannot handle large files. To accommodate these users, the NetServ fileserver automatically splits files into packets of about 95 kilobytes for transmission over the Internet. The Subject lines of the reply messages tell how many file packets were sent and give the packet numbers. With this information, the user can reassemble the original files by joining the individual file packets after removing the mail headers.

If the server divides a file into smaller packets for transmission, the SIZE directive can be used to change the packet size. The packet size can be increased or decreased by this directive. For example, to set a file packet size of 250 kilobytes, use the directive SIZE 250 in a query message to the server. To reduce the file packet size to 50 kilobytes, use the directive SIZE 50. Remember from Chapter 1 that the TCP/IP software protocol that governs file transfer over the Internet splits data files into smaller packets for transmission. The speed of file transfer is a function of file packet size: the smaller the packet, the faster the transmission. As a result, smaller files will probably be received faster than larger files. Check to see if there are any limitations on the size of files you are allowed to receive.

■ Example of query

Query messages to the NetServ fileserver can be set up to retrieve several different files in a single message to the server. The example query message shown in Fig. 7.2 will retrieve specific help on the nucleotide and protein sequence databases, general help on software, the new nucleotide sequence citation index, and the sequence submission form. In response to this request, the fileserver will first send back a message that reports what was requested and what is being sent back. Subsequent messages from the fileserver will transmit the requested files, each file in a separate message.

Figure 7.2. Example of a Query Request to the NetServ Fileserver

```
From:      annep@acs.auc.eun.eg
Date:      02 Dec 1996 22:14:07-GMT
To:        netserv@ebi.ac.uk
Subject:
HELP NUC
HELP PROT
HELP SOFTWARE
GET NUC:NEWCITATION.NDX
GET DOC:DATASUB.TXT
```

GDB Server

Established at Johns Hopkins University in 1990, the GDB (Genome Data Base) is the official repository for all human genome information. The GDB server allows users to retrieve documents via e-mail from its databases through simple keyword searches.

Internet addresses

E-mail:	mailserv@gdb.org	(Server)
	help@gdb.org	(Personal help)
Web:	http://gdbwww.gdb.org/	(Home page)
Organization:	Division of Biomedical Information Sciences, Johns Hopkins University School of Medicine, Baltimore, MD 21205, U.S.A.	
Telephone:	+1-410-955-9705	
Facsimile:	+1-410-614-0434	

■ Accessing the GDB server

To access the GDB server via e-mail, send a properly formatted query to the server address listed above. To receive a help document that describes the latest information on how to use the server, send a query to the server address with the directive HELP in the body of the message. For assistance with specific problems or for more information, send a query to the personal help address.

■ GDB is not a country-specific site

GDB's organization and operation are overseen by the international scientific community. The scope of this operation can be glimpsed in Table 7.8. Notice that nearly every major geographical region in the world has a regional GDB site. This worldwide community effort is one of the major reasons that GDB has matured into an indispensable tool in the human genome analysis community.

■ Formatting a query

A query to the GDB server can contain an unlimited number of directives and can search a range of available databases. Valid directives and their associated parameters are summarized in Table 7.9. Directives and their associated parameters can be of any length and can even span more than one line of a message. To continue a directive to the server over more than one line, simply place a backslash (\ ) at the end of the line. On the next line of the query, continue with the remainder of the directive as shown below:

```
SEARCH GDB-CONTACT LANDER PAGE WARD\
COHEN CHUMAKOV
```

Table 7.8. GDB Web Sites and User Support Offices Worldwide

The GDB Web server is available directly at the following URLs:

United States	http://gdbwww.gdb.org/
Australia	http://morgan.angis.su.oz.au/gdb/gdbtop.html
France	http://gdb.infobiogen.fr/
Germany	http://gdbwww.dkfz-heidelberg.de/
Israel	http://inherit1.weizmann.ac.il/gdb/docs/gdbhome.html
Japan	http://gdb.gdbnet.ad.jp/gdb/docs/gdbhome.html
Netherlands	http://www-gdb.caos.kun.nl/gdb/gdbtop.html
United Kingdom	http://www.hgmp.mrc.ac.uk/gdb/docs/gdbhome.html
Sweden	http://www.embnet.se/gdb/index.html

GDB User Support Offices

Australia
Ms. Carolyn Bucholtz
ANGIS
Sydney
Phone: + 61-2-692-2948
Facsimile: + 61-2-692-3847
E-mail: bucholtz@angis.su.oz.au

France
Philippe Dessen
INFOBIOGEN
Villejuif
Phone: +33-14559-5241
Facsimile: +33-14559-5250
E-mail: gdb@infobiogen.fr

Germany
Molecular Biophysics Dept.
DKFZ
Heidelberg
Phone: + 49-6221-42-2349
Facsimile: + 49-6221-2333
E-mail: gdb@dkfz-heidelberg.de

Israel
Jaime Prilusky
Weizmann Institute of Science
Rehovot
Phone: +972-8-343456
Facsimile: +972-8-344113
E-mail: lsprilus@weizmann.weizmann.ac.il

Japan
Mika Hirakawa
JICST
Tokyo
Phone: +81-3-5214-8491
Facsimile: +81-3-5214-8470
E-mail: mika@gdb.gdbnet.ad.jp

Netherlands
GDB User Support
CAOS/CAMM Center
Nijmegen
Phone: + 31-80-653391
Facsimile: + 31-80-652977
E-mail: post@caos.caos.kun.nl

Sweden
GDB User Support
Biomedical Center
Uppsala
Phone: + 46-18-174057
Facsimile: + 46-18-524869
E-mail: help@gdb.embnet.se

United Kingdom
Administration
HGMP Resource Centre
Cambridge
Phone: + 44-1223-494511
Facsimile: + 44-1223-494512
E-mail: admin@hgmp.mrc.ac.uk

Table 7.9. GDB Server Directives and Parameters

DONE	Instructs the server to ignore all lines following this one. Recommended if a query message contains a signature at the bottom.
GET [GDB ACCESSION NUMBER]	Retrieves a document from GDB based on accession number. Example: GET G00-000-123. Multiple GET directives can be used in a request. Each must be on a separate line.
HELP	Retrieves a general help and informational document.
MAX-LINES [NUMBER]	Specifies the maximum number of lines in a single mail message returned from the server. If the output from a set of commands exceeds that value, the output will be split into a series of files and mailed separately. The default (and maximum) number of lines before splitting occurs is 3,000.
MAX-RESULTS [NUMBER]	Specifies the maximum number of results returned from a search. The default number of results is 10; the maximum number is 50. Example: MAX-RESULTS 5
SAMPLE [DATABASE]	Retrieves a sample detail entry for the specified database. Example: SAMPLE GDB-MAP
SEARCH [DATABASE] [KEYWORDS]	Searches a database for the occurrence of one or more keywords. Only one database can be specified per line in a query. Example: SEARCH GDB-LOCUS BRAC1
TOTAL [DATABASE]	Retrieves the total number of entries in the specified database. Example: TOTAL GDB-PROBE

As mentioned, the GDB server has several databases available for searching; we list these in Table 7.10.

The GDB server performs a two-step search of the databases for information requested in the query. First, the server searches for documents that contain the specified query text. Next, the server ranks these documents on the frequency of appearance of the query text in the document and whether the query text appears in the title of the document. Because of this searching and ranking strategy, a query to the server should follow these general guidelines.

- **Be specific.** Avoid frequently used, nonspecific terms. As with nearly all molecular and genomic biology databases, the number of records stored in the GDB is increasing rapidly. Because of the number and size of records, common words

Table 7.10. GDB Databases

Directive	Explanation
GDB-CELL-LINE	Clonal mammalian cell lines which are useful for genomic mapping research and information on how to obtain the cell line.
GDB-CITATION	Bibliographic citations for all GDB data.
GDB-CONTACT	Contact information for researchers involved in genome mapping, as well as GDB editors and staff.
GDB-LIBRARY	Collections of DNA clones from specific regions of the genome that are useful for mapping research, and how to obtain them.
GDB-LOCUS	Defined regions of the genome, ranging in size from a single point to an extended region, such as genes, DNA segments, fragile sites, and breakpoints.
GDB-MAP	Genetic or physical maps representing an ordered set of elements in a chromosome. Map descriptions include the method used, a summary locations, information about the loci or probes, and the distances between elements on the map.
GDB-MUTATION	Listing and description of less common sequence variations in loci often responsible for clinical phenotypes.
GDB-POLYM	Common variations in the DNA sequence of loci, which are often useful in mapping experiments.
GDB-PROBE	Reagents and experimental methodologies for detecting genes, DNA segments, and variations.

used as keywords in a nonspecific fashion are often ignored. The statement SEARCH GDB-LOCUS CELL, if used in a query to the server, would not give any results because the word CELL is too general.

- **Use Boolean operators when possible.** In order to make the searches more specific, use Boolean statements. The GDB server, by default, treats query words as if they are separated by an OR statement. The query statements SEARCH GDB-CITATION PAGE LANDER and SEARCH GDB-CITATION PAGE OR LANDER mean the same thing. Both queries would retrieve documents that contain the words PAGE or LANDER. To retrieve documents that contain both words, link them with an AND statement such as SEARCH GDB-CITATION PAGE AND LANDER.

    The Boolean statement NOT can be used to make a query statement even more specific. For example, the query statement SEARCH GDB-CITATION PHOSPHATASE NOT ACID NOT ALKALINE looks for and retrieves documents that include the word PHOSPHATASE but not the words ACID or ALKALINE.

- **Use quotes to narrow a search.** Quotes can be used around the text terms of a query statement to retrieve documents containing that phrase. For example, the search statement SEARCH GDB-CITATION "PROTEIN PHOSPHATASE" would retrieve those documents containing the phrase PROTEIN PHOSPHATASE. Documents that contain just the term PROTEIN or just PHOSPHATASE would not be retrieved. Also, those documents that contain the terms PROTEIN and PHOS-PHATASE, but do not have these words together, would not be retrieved.

- **Use an asterisk to broaden a search.** The asterisk can be used as a wildcard character to broaden a search. This strategy is useful in retrieving relevant records if the correct spelling of a search term is not known, or to retrieve records with a common theme. For example, to retrieve records for genes on the short arm of chromosome 1, use the query statement SEARCH GDB-CITATION 1P*.

    There are two major limitations to the use of an asterisk as a wild card in searches. First, they cannot be used within a phrase that is flanked by quotation marks. This is because all text within the quotation marks is interpreted by the server literally. Second, the asterisk can only be used at the end of a word.

■ Example of a query

A sample query to the GDB server could take the form shown in Fig. 7.3. In this example we are asking the server to report back on all citations pertaining to the short arm of chromosome 1, all phosphatases that are not acid or alkaline, and contact information on Lander, Page, Ward, Cohen, and Chumakov.

Figure 7.3. Example of a Query to the GDB Server

```
From:     annep@acs.auc.eun.eg
Date:     02 Dec 1996 22:14:07-GMT
To:       mailserv@gdb.org
Subject:
SEARCH GDB-CITATION 1P*
SEARCH GDB-CITATION PHOSPHATASE NOT ACID NOT ALKALINE
SEARCH GDB-CONTACT LANDER PAGE WARD\
COHEN CHUMAKOV
```

Selected readings

Journal Articles

Emmert, D. B., P. J. Stoehr, G. Stoeser, and G. N. Cameron. 1994. The European Bioinformatics Institute (EBI) databases. *Nucleic Acids Res.* **22**:3445-3449.

Fasman, K. H., S. I. Letovsky, R. W. Cottingham, and D. T. Kingsbury. 1996. Improvements to the GDB Human Genome Data Base. *Nucleic Acids Res.* **24**:57-63.

Rodriguez-Tome, P., P. J. Stoehr, G. N. Cameron, and T. P. Flores. 1996. The European Bioinformatics Institute (EBI) databases. *Nucleic Acids Res.* **24**:6-12.

Zehetner, G., and H. Lehrach. 1994. The reference library system: sharing biological material and experimental data. *Nature* (London) **367**:489-491.

Books and Monographs

Engst, A. C. 1993. *Internet Starter Kit for Macintosh.* Hayden Books, a division of Prentice Hall Computer Publishing, Indianapolis.

Falk, B. 1994. *The Internet Roadmap.* SYBEX, Alameda, Calif.

Hahn, H. 1996. *The Internet Complete Reference,* 2nd ed. Osbourne McGraw-Hill, Berkeley, Calif.

McKusick, V. A. 1994. *Mendelian Inheritance in Man. Catalogs of Human Genes and Genetic Disorders,* 11th ed. Johns Hopkins University Press, Baltimore.

Swindell, S. R., R. R. Miller, and G. S. A. Myers (ed.). 1995. *Internet for the Molecular Biologist.* Horizon Scientific Press, Norfolk, England.

Internet Publications

Hayden, D. 1994. *Guide to Molecular Biology Databases.* School of Library and Information Studies, University of Alberta, Calgary, Alberta, Canada.

Smith, U. R. 1993. *A Biologist's Guide to Internet Resources.* Yale University, New Haven, Conn.

CHAPTER 8 Analysis of Sequence Data

■ Introduction to sequence analysis

■ Evolution of computer algorithms for sequence analysis

■ DNA sequences: analysis of a nucleotide sequence using BLAST

■ DNA sequences: downloading related sequences using Retrieve

■ Protein sequences: analysis with Blocks, Domain, and MotifFinder

■ Protein sequences: secondary structure prediction using nnPredict and PredictProtein

■ Protein sequences: determination of phylogenetic relationships using MAlign and CBRG

Introduction to sequence analysis

With the increasing pace of genomic sequencing projects, the volume and richness of molecular sequence data is increasing at an exponential level. The analysis and application of the information in these databases requires an understanding of the theory and practice of molecular sequence analysis techniques along with the use of the appropriate computer-based resources. With that in mind, we offer four main reasons for computer-aided analysis of nucleic acid and protein sequences:

- To scan the primary databases for similar sequences in order to determine novelty, function, and homology;

- To scan specialty databases for structural and regulatory motifs, domains, and patterns to determine function, localization, and family relationships;

- To model secondary and super-secondary structures to postulate two- and three-dimensional (2D and 3D) structure;

- To examine hypothetical phylogenetic relationships within a family of query sequences in order to determine evolutionary patterns.

In this chapter, we will be applying the Internet tools and resources we described in Chapters 4, 5, and 6 to the analysis of both a DNA sequence and a protein sequence. Specifically, we will apply these tools to five practical tasks. First, a query nucleotide sequence will be used to search molecular sequence databases for similar sequences. We will examine three different algorithms: BLAST, FastA, and Smith-Waterman. Second, using the information obtained from the database searches, we will download sequence records of interest from the international sequence databases for further analysis. Third, we will look at the protein translation of a query sequence for conserved patterns and motifs using Blocks, Domain, and MotifFinder as aids to determine structural and functional domains. Fourth, we will analyze the query protein sequence for secondary structure patterns using nnPredict and Predict-Protein. We will compare the findings from these two servers to generate a hypothetical consensus prediction. Finally, we will do a multiple alignment using the query protein sequence and those sequences that share the highest relative level of similarity. The results from this process will be used to generate a consensus ancestral sequence and to determine the hypothetical phylogenetic relationship within this family of query sequences.

■ Similarity searching of molecular sequence databases

Making biological sense of tracts of molecular sequence information is of paramount importance in genomic and molecular biology. Sequences by themselves do not yield much in the way of information, so they must be analyzed by comparative methods against the existing databases to deduce function and similarity. For example, an mRNA sequence transcribed at a high level in a cancer cell line may be similar to a known protein phosphatase sequence. Research efforts may then be focused on the role of phosphorylation/dephosphorylation cycles in the regulation of cellular transformation. By using computer-aided sequence analysis, insight may be gained to direct subsequent laboratory investigations into the underlying biological phenomena involved in cellular processes. The analysis examples that follow in this chapter provide a methodological paradigm by which gene and protein sequence similarity searches may be performed.

When checking a query sequence for similar sequences, the computer uses an algorithm to compare the query sequence to every sequence in the database. This is done in a pairwise fashion, which means that each sequence in the database is compared with the query sequence independently of all other database sequences. Each of these comparisons generates a score that reflects the degree of relative similarity between the query sequence and the sequence from the database. Higher scores usually indicate a higher degree of similarity between two sequences. If a global alignment algorithm is used, a scan of a database composed of 100,000 sequences will provide 100,000 scores for ranking and further analysis. If a local alignment method is used, the total number of scores may be much larger because more than one matching sequence may occur within each sequence.

Normally there is some overlap between the genuinely related and unrelated sequences. A number of methods exist for ranking and rescaling similarity scores to discriminate between real matches and those that are artifactual in nature. These methods provide an estimate of the probability of matching a query sequence with an entry from the database by chance alone. However, regardless of the method of ranking, there are nearly always some sequences from the database that are structurally related to the query sequence. In practice, since no method succeeds for all analyses, the aim is to minimize the overlap and ensure that potentially interesting similarities are scored high enough that they will be noticed by the user. Ultimately, what constitutes an interesting match depends on the perspective of the scientist analyzing the data.

Evolution of computer algorithms for sequence analysis

Amino Acid Scoring Schemes

All algorithms used to compare protein sequences rely on some scheme to rank the matches of 210 possible pairs of amino acids: 190 pairs of different amino acids matches and 20 pairs of identical amino acids matches. Most scoring schemes represent these pairs of matches as a matrix of similarities in which amino acids that are identical or similar rank higher than those of different character. Since the first protein sequences were obtained, many different types of scoring schemes have been devised. The most commonly used scoring schemes are those based on the observed substitution of amino acids in related proteins; of these, the matrix called PAM 250 has been the dominant until the last few years. (We will discuss this matrix and other scoring methods in the following sections.)

■ Identity scoring

Identity scoring is the simplest scoring method. Amino acid pairs are classified into two types, identical and nonidentical. Nonidentical pairs are scored 0; identical pairs are given a positive score (usually 1). This scoring scheme is generally considered less effective than schemes that weight nonidentical pairs, particularly for the detection of weak similarities. The normalized sum of identity scores for an alignment is popularly quoted as "percentage identity," but this value can be useful to indicate the overall similarity between two sequences.

■ Substitution scoring

In aligning two protein sequences, some method must be used to score the alignment of one residue against another. Substitution matrices contain such values. There are two widely used substitution matrices in the computer analysis of molecular sequence data, PAM and BLOSUM.

PAM

The late Margaret Dayhoff was a pioneer in the development of protein databases and sequence comparison (Barker et al. 1978; Dayhoff 1965; Dayhoff 1969; Dayhoff 1974; Dayhoff 1976; Dayhoff et al. 1974; Dayhoff et al. 1978). She and her coworkers developed a model of protein evolution that resulted in the development of a set of widely used substitution matrices; these are frequently called Dayhoff, MDM (Mutation Data Matrix), or PAM (Percent Accepted Mutation) matrices (Dayhoff et al. 1978). The key features of this family of matrices are listed below.

- PAM matrices are derived from global alignments of closely related sequences.

- Matrices for greater evolutionary distances are extrapolated from those for lesser ones.

- The matrix value (PAM 40, PAM 100) refers to the evolutionary distance; greater numbers are greater distances.

BLOSUM

Over time it became apparent that the PAM matrices had limitations in sequence analysis and comparison, particularly in the identification of motifs and patterns and the detection of more distant similarities. To address these limitations, the BLOSUM series of matrices was created by Henikoff and colleagues (Henikoff and Henikoff 1992). The key features of these matrices are:

- BLOSUM matrices are derived from local, ungapped alignments of distantly related sequences.

- All matrices are directly calculated; no extrapolations are used.

- The matrix value (BLOSUM 62) refers to the minimum percent identity of the blocks used to construct the matrix; greater numbers are lesser distances.

- The BLOSUM series of matrices generally perform better than PAM matrices for local similarity searches (Henikoff and Henikoff 1993).

From this discussion it should be clear that there is no such thing as a perfect substitution matrix; each matrix has its own limitations. If each matrix has its own limitations, then it should be possible to use multiple matrices so that each one complements the limits of the others.

Sequence Database Searching

The most obvious first stage in the analysis of any new sequence is to perform comparisons with sequence databases to find homologs. These searches can now be performed just about anywhere and on just about any computer. In addition, there are numerous Web servers where one can post or paste a sequence into the server and receive the results via e-mail, as an HTML document, or as a text file downloaded from a Web site.

It is best to search a few different databases in order to find as many homologs as possible. A very important thing to do, and one that is sometimes overlooked, is to compare any new sequence to a database of sequences for which 3D structure information is available. Whether or not a sequence is homologous to a protein of known 3D structure is not obvious in the output from many searches of large sequence databases. Moreover, if the homology is weak, the similarity may not be apparent at all during the search through a larger database.

Because of these reasons, database searches should be conducted using more than one algorithm to perform the analysis. The following are the three most commonly used families of sequence comparison algorithms.

- **BLAST:** very fast, but doesn't allow gaps and can miss distant similarities

- **FastA:** slower than BLAST, but more sensitive in some searches, allows gaps

- **Smith-Waterman:** much slower than either BLAST or FastA, though more sensitive.

The key point to remember when doing sequence comparisons is that different searches can give different results, so it is important to use different analysis algorithms and search parameters to create a thorough list of similar sequences.

■ BLAST (Basic Local Alignment Search Tool) algorithm

BLAST was designed to find the highest scoring locally optimal alignments between a query sequence and a database (Altschul et al. 1990). The important simplification that BLAST makes is not to allow gaps, but the algorithm does allow multiple hits to the same sequence. The BLAST algorithm and the family of programs based on it rely on the principles of ungapped sequence alignments. The BLAST algorithm estimates the probability of obtaining an ungapped alignment or MSP (Maximal Segment Pair) above a particular score. The BLAST algorithm permits nearly all MSPs above a cutoff to be located efficiently in a database. The algorithm operates in three steps:

- For a given word length n (usually three letters for proteins) and score matrix, a list of all words (n-mers) that can score when compared to n-mers from the query is created.

- The database is searched using the list of n-mers to find the corresponding n-mers or hits in the database.

- Each hit is extended to determine if an MSP that includes the n-mer scores the preset threshold score for an MSP. Since pair score matrices typically include negative values, extension of the initial n-mer hit may increase or decrease the score. Accordingly, a parameter defines how great an extension will be tried in an attempt to raise the score of the sequence being compared to the query sequence.

A low value for word length reduces the possibility of missing MSPs with the required score; however, lower values also increase the size of the hit list generated in step 2 and hence the execution time and memory required. For example, the BLASTP program used for protein searches sets compromise values to balance the processor requirements and sensitivity.

BLAST is unlikely to be as sensitive for all protein searches as a full dynamic programming algorithm. However, the underlying statistics provide a direct estimate of the significance of any match found. The program was developed at the NCBI and benefits from strong technical support and continuing refinement. For example, filters have recently been developed to exclude, automatically, regions of the query sequence that have low compositional complexity or short-periodicity internal repeats. The presence of such sequences can yield extremely large numbers of statistically significant but biologically uninteresting MSPs. For example, searching with a sequence that contains a long section of hydrophobic residues will find many proteins with transmembrane helices. In short, BLAST has the following characteristics.

- **Local alignments:** BLAST tries to find patches of regional similarity, rather than trying to find the best alignment between the entire query and an entire database sequence.

- **Ungapped alignments:** Alignments generated with BLAST do not contain gaps. BLAST's speed and statistical model depend on this, but theoretically, it reduces sensitivity. However, BLAST will report multiple local alignments between a query and a database sequence.

- **Explicit statistical theory:** BLAST is based on an explicit statistical theory developed by Karlin and Altschul (Karlin and Altschul 1990). The original theory was later extended to cover multiple weak matches between query and database entry (Karlin and Altschul 1993). However, the repetitive nature of many biological sequences violates the assumptions made in the original theory. Further, the databases are contaminated with numerous artifacts. The use of filters can reduce problems from these sources. Remember that the statistical theory only covers the likelihood of finding a match by chance under particular assumptions; it does not guarantee biological importance.

- **Rapid, but not perfect:** BLAST is extremely fast. BLAST is not guaranteed to find the best alignment between a query and the database; it may miss matches. This is because it uses a strategy that is expected to find most matches, but sacrifices complete sensitivity in order to gain speed. However, in practice, few biologically significant matches are missed by BLAST and those can be found with other sequence search programs. BLAST searches the database in two phases. First, it looks for short subsequences that are likely to produce significant matches; then it tries to extend these subsequences.

- **A substitution matrix is used during all phases of protein searches:** Both phases of the alignment process (scanning and extension) use a substitution matrix to score matches; this is in contrast to FastA, which uses a substitution matrix only for the extension phase. Substitution matrices improve sensitivity.

FastA algorithm

Wilbur and Lipman (Wilbur and Lipman 1983) developed a fast procedure for DNA sequence scans that, in concept, searches for the most significant alignments between two sequences. The initial step in this algorithm is to identify all exact matches of a given length or greater, between the two sequences. This length is known as k-tuples, where k is the length. Speed is achieved by employing a lookup procedure. For example, in the analysis of a protein, if there are 8,000 possible k-tuples in a given sequence, each element of an array of 8,000 is set to represent one of these k-tuples. The query sequence is scanned once, and the location of each k-tuple in it is recorded in the corresponding element of the array. This process generates a lookup table of the query sequence. The sequence database is then scanned with the query sequence, and all k-tuple matches that are common are identified. If two k-tuples are present on the same diagonal, then the difference between their starting position (offset) is also the same and thus the diagonals with the most significant number of matches may be identified. Since runs of identity are relatively rare even between related proteins, Lipman and Pearson (1985) first identified the five diagonals of highest similarity. They then applied the Dayhoff scoring scheme (Dayhoff et al. 1978) to the amino acid pairs over these regions. The region giving the highest score for the protein comparison was used to create a rank order of the sequences from the database for further study by more rigorous procedures.

Pearson and Lipman refined these ideas in the program FastA (Pearson and Lipman, 1988). FastA saves the 10 highest regions of identity, which are then rescored with the PAM250 matrix. If there are several initial regions above a preset cutoff score, those that could form a longer alignment are joined, allowing for gaps, and a score, INITN, is then calculated by subtracting a penalty for each gap. INITN is used to rank the database sequences by similarity. Finally, dynamic programming is used over a narrow region of the high-scoring diagonal to produce an alignment with score OPT. FastA only shows the top-scoring region; it does not locate all high-scoring alignments between two sequences. As a consequence, FastA may not identify direct repeats or multiple domains shared between two proteins. The characteristics of the FastA algorithm include the following.

- **Local alignments:** It tries to find patches of regional similarity, rather than trying to find the best alignment between the entire query and an entire database sequence.

- **Gapped alignments:** Alignments can contain gaps.

- **Rapid but not perfect:** FastA is quite fast, but is not guaranteed to find the best alignment between a query and the database; it may miss matches. This is because it uses a strategy that is expected to find most matches, but sacrifices complete sensitivity in order to gain speed.

- **A substitution matrix is used during the extension phase of protein searches:** This algorithm uses a substitution matrix only for the extension phase. This is in contrast to BLAST, which uses a matrix for both phases. To reduce the penalty of using a substitution matrix for only the second phase, the k-tuple parameter is set to a low value. However, this will incur a significant speed penalty.

■ Smith-Waterman (SW) algorithm

The Smith-Waterman algorithm (Smith and Waterman 1981a, 1981b) is used to search for similarities between one sequence (the query) and a group of sequences (the database). This algorithm calculates the score of the optimal alignment between the query and each sequence in the database and creates a list of the sequences in the database with the best scores. Unlike the two other major search algorithms, FastA and BLAST, the Smith-Waterman algorithm performs no initial filtering of the database. Consequently, distantly related sequences can be discovered, even when there are extensive substitutions and gaps.

Summary

With these lists of characteristics in mind, what is the best method for database scanning? There is no straightforward answer to this question. Attempts have been made to make comparisons, but the process is complicated by the difficulty of designing suitable test cases and the number of adjustable parameters. The most effective method for assessing the success of a scanning technique is to test its ability to find all the members of a known protein family from the database of all known sequences. This can be done by following these steps:

1. Record the identifier codes of all proteins known to be in the family.

2. Select a member to scan with (the query).

3. Perform the scan using the method of choice.

4. Count how many of the known members are found with higher scores than known nonmembers.

A less strict criterion is to count the number of members that score as high as the top 0.5% of the nonmembers in the databank. The best scanning method will give the most members before nonmembers; it detects and reports the fewest false-positives. Of course, evaluation is not as simple as this appears. First, well-characterized protein families must be chosen for the evaluation. The key question at this point is whether all of the members of this protein family have been identified: a high-scoring nonmember may in fact be a previously undiscovered family member. Further difficulties arise for scans where there are many false-negatives. If two methods miss 30 known members, are they missing the same 30? Ideally, evaluation should also explore alternative parameter combinations, but this greatly increases the number of tests that need to be done and complicates the data analysis. For example, when scanning with dynamic programming, there is a choice between the pair-score matrix and gap-penalty, local, or global alignment. The best gap penalty depends on the matrix in use. If both length-dependent and length-independent penalties are used, then the number of alternative combinations increases dramatically. The best combination of matrix and penalty may not be appropriate for other algorithms. BLAST does not consider gaps, so the situation is a little easier; this concept was exploited by Henikoff and Henikoff (Henikoff and Henikoff 1993) to evaluate different substitution matrices.

Given a newly determined sequence, a search using the BLAST or FastA algorithms will quickly reveal the existence of any similar sequence in a database. If this type of analysis does not find any similar sequences, then alternative PAM or BLOSUM matrices should be tried in order to refine the search. When working with PAM matrices, remember that lower matrix values are best for identifying short regions of sequence with very high similarity, while higher PAM matrices are better suited for identifying longer and weaker matches. One possible strategy is to begin an analysis using PAM120, then try PAM250. In each case, vary the gap penalty or INDEL cost around the default value of the matrix. For example, when using PAM250 with the BLITZ server, this default value is 7; values of 6 and 8 are worth investigating (Table 5.10). This process of continual refinement of the analysis still does not preclude the intuitive judgment of apparent matching sequences: consider carefully the biological significance of any match reported back, before treating it as real.

If the sequence contains more than about 500 amino acids, it will almost certainly be divided into discrete functional domains. If possible, it is preferable to split such large proteins up and consider each domain separately. The location of domains can be predicted in several different ways (listed from most to least confident):

- If homology to other sequences occurs only over a portion of the probe sequences, and the other sequences are full sequences, then this provides the strongest evidence for domain structure.

- Regions of low complexity often separate domains into multidomain proteins. Long stretches of repeated residues, particularly proline, glutamine, serine, or threonine, often indicate linker sequences and are usually a good place to split proteins into domains.

- Secondary structure prediction methods will often predict regions of proteins to have different protein structural classes. For example, one region of sequence may be predicted to contain only alpha helices and another to contain only beta sheets. These can often, though not always, suggest likely domain structure.

If a sequence is composed of discrete domains, then it is important to repeat the analysis of the query sequence using each of the separate domains alone. This is because a search of a database with a query sequence composed of several domains may not find all relevant similar sequences, particularly if some of the matching domains are abundantly represented in the database. Instead, highly relevant and biologically significant matches with poorer similarity scores may be masked by matches with many higher-scoring sequences containing the same domain. In this case, analysis of a query sequence using ProDom or SABASE may reveal unusual but very interesting similarities with known protein domains.

With that brief overview of the theory and underlying principles of the analysis of a newly determined sequence, let's now apply this knowledge to the analysis of a query sequence using some of the Internet tools we've described in the preceding chapters. This example analysis will be done using a well-characterized cDNA that encodes the protein PPX, the catalytic subunit of the protein serine/threonine phosphatase complex X, cloned from a human cDNA library. Through the analysis of this putative unknown sequence, a template for the analysis of nearly any sequence will be developed.

DNA sequences: analysis of a nucleotide sequence using BLAST

The first step in the analysis of any new sequence is to determine if any similar sequences exist in the primary international sequence databases. This type of analysis can be accomplished rapidly by a number of servers, such as BLAST or Mail-FastA. In this example, the BLAST server at NCBI will be used for the analysis of the unknown cDNA sequence. In Fig. 8.1A, the query message to the BLAST server contains the unknown nucleotide sequence. Note that the query primarily uses the default parameters of the BLAST server. It is set up so that it will go into the fast queue at the server and will return basic but comprehensive results. The BLASTN program is specified to search the query sequence against the nonredundant nucleotide sequence library. The HISTOGRAM directive is set to NO, so that a histogram will not be reported back by the server.

Reply messages from the BLAST server can be quite large and composed of several pages of alignments and analysis statistics. At first this might seem intimidating, but in reality the structure of, and the information found in, a reply from the BLAST server are straightforward and well organized. Because of the length of the server reply, Fig. 8.1B shows an abridged version of the reply received back from the BLAST. The first part of the message reports back information on the BLAST server itself as well as additional news and information from the staff of NCBI to the general user community. This can include news on modifications to the program itself, such as new directives or new documentation on using the server. This section of the reply also contains specifics on the databases that the server uses for analyses and the version of the BLAST program that was used to perform the analysis.

The second part of the message reports back information on the query sequence and some general observations on the analysis of the sequence (such as its length). Also listed is the database that was used in the analysis as well as any warnings from the server regarding the analysis. The last part of this section of the reply message gives a line listing of the matching sequences and their approximate ranking of significance. From the description of the matching sequences in this line listing, it appears that the query nucleotide sequence is, in fact, a Ser/Thr protein phosphatase. The best match comes from a sequence determined from a human mRNA from the EMBL sequence database with the accession number of X70218 and the locus name of HSPPX. The poorest match reported was from X57115, also from the EMBL database, which encodes calcineurin A determined from rat mRNA.

The third part of the reply message gives the sequence matches. Since the default settings were used, a total of 50 matches are reported back by the server. In this abridged reply, the best match and one of the weaker but still significant matches are shown. These two matches have been highlighted in the line listing of the second part of the message to show their relative ranking in the analysis. As mentioned, the best match is with sequence X70218 (see p. 218), which had complete identity with the query sequence. The other match shown, M36951 (see p. 219), had about 60% similarity in the regions of it that matched the query sequence.

In the comparison of the query sequence with the weaker match sequence M36951, note that BLAST does not allow gaps in the alignment of the two sequences. The alignment that is shown between the two sequences only covers nucleotides 148 to 1009 of the query sequence and nucleotides 100 to 961 of M36951. BLAST aligns the query and database sequences until a gap is reached, and then the alignment is terminated. This does not mean that other alignments between these two sequences are ignored or not reported back from the analysis process. Since BLAST is a tool that generates local alignments, if more than one alignment is identified, then any alignment that exceeds the detection limit of the analysis is reported by the server. In this example, only the single alignment shown met this cutoff.

The fourth part of the reply message gives statistics on the analysis process and determines whether the server actually performed the analysis as requested in the query message. Of those listed, the most important parameter values were for V, B, H, and E. The parameter value of 100 for V means that the server will report a line listing of the 100 best matching sequences. The value of 50 for parameter B means that the server will show the 50 best alignments. Parameter H, which was set to a value of 0, controls the display of a histogram. Since the value is 0, no histogram will be reported. The parameter value for the EXPECT directive, E, was set at 10, meaning that 10 matching sequences would be expected to occur by chance alone. All of these settings match those specified in the original query message.

Figure 8.1. Genomic Sequence Analysis Using BLAST

A: Query message to the Blast server

```
Date:   Tue 16 Jul 1996 - 09:16.02 - GMT
From:   annep@acs.auc.eun.eg
To:     blast@ncbi.nlm.nih.gov
Subj:


PROGRAM BLASTN
DATALIB NR
HISTOGRAM NO
BEGIN
>UNKNOWN NUCLEOTIDE SEQUENCE
CGGCGGCGGCGGTCGAAAGCGGAGTGAAAGAGGGAGGCAGGGAGCCGGAGAGCCGGAACC
GGAGTCGCAGCGGCGGAGACCCCTGTGCGGTGCGGAGGGGGCGGCGGCCCCGACTCTGAC
CCGCGCCGGGGGTGGGCCATGGCGGAGATCAGCGACCTGGACCGGCAGATCGAGCAGCTG
CGTCGCTGCGAGCTCATCAAGGAGAGCGAAGTCAAGGCCCTGTGCGCTAAGGCCAGAGAG
ATCTTGGTAGAGGAGAGCAACGTGCAGAGGGTGGACTCGCCAGTCACAGTGTGCGGCGAC
ATCCATGGACAATTCTATGACCTCAAAGAGCTGTTCAGAGTAGGTGGCGACGTCCCTGAG
AGGAACTACCTCTTCATGGGGGACTTTGTGGACCGTGGCTTCTATAGCGTCGAAACGTTC
CTCCTGCTGCTGGCACTTAAGGTTCGCTATCCTGATCGCATCACACTGATCCGGGGCAAC
CATGAGAGTCGCCAGATCACGCAGGTCTATGGCTTCTACGATGAGTGCCTGCGCAAGTAC
GGCTCGGTGACTGTGTGGCGCTACTGCACTGAGATCTTTGACTACCTCAGCCTGTCAGCC
ATCATCGATGGCAAGATCTTCTGCGTGCACGGGGGCCTCTCCCCCTCCATCCAGACCCTG
GATCAGATTCGGACAATCGACCGAAAGCAAGAGGTGCCTCATGATGGGCCCATGTGTGAC
CTCCTCTGGTCTGACCCAGAAGACACCACAGGCTGGGGCGTGAGCCCGCGCGGGAGCCGGC
TACCTATTTGGCAGTGACGTGGTGGCCCAGTTCAACGCAGCCAATGACATTGACATGATC
TGCCGTGCCCACCAACTGGTGATGGAAGGTTACAAGTGGCACTTCAATGAGACGGTGCTC
ACTGTGTGGTCGGCACCCAACTACTGCTACCGCTGTGGGAATGTGGCAGCCATCTTGGAG
CTGGACGAGCATCTCCAGAAAGATTTCATCATCTTTGAGGCTGCTCCCCAAGAGACACGG
GGCATCCCCTCCAAGAAGCCCGTGGCCGACTACTTCCTGTGACCCCGCCCGGCCCCTGCC
CCCTCCAACCCTTCTGGCCCTCGCACCACTGTGACTCTGCCATCTTCCTCAGACGGAGGC
TGGGGGGGCTGTCCTGGCTCTGCTGTCCCCCAAGAGGGTGCCTTCGAGGGTGAGGACTTC
TCTGGAGAGGCCTGGAGACCTAGCTCCATGTTCCTCCTCCTCTCTCCCCACTTGAACCAT
GAAGTTTCCAATAATTTTTTTTTCTTTTTTTCCTTCTTTTTCTGTTTGTTTTTAGATAAA
AATTTTTGAGAAAAAAAATGAAAAATTCTAATAAAAGAAGAAAAATGGTAAAAAAAAAA
AA
```

continued

Figure 8.1 continued

B: Reply message from the Blast server
First part of the reply message

```
Date:       Tue 16 Jul 1996 - 12:41:11 - GMT
From:       NCBI BLAST E-Mail Server <blast@ncbi.nlm.nih.gov>
To:         annep@acs.auc.eun.eg
Subj:       Results-BLAST Server
Errors-To: <owner-blast@ncbi.nlm.nih.gov>
Reply-To: "NCBI BLAST E-Mail Server" <blast@ncbi.nlm.nih.gov>


=-+===================================================================
 To Obtain Documentation: send an e-mail message to 'blast@ncbi.nlm.nih.gov'
 with the word HELP in the body of the message. The documentation was last
 modified March 18th.



=+====================================================================
 March 11th 1996
 Trying blaster... connected
National Center for Biotechnology Information (NCBI)

Experimental GENINFO(R) BLAST Network Service (Blaster)

Tue Jul 16 12:37:48 EDT 1996, Up 25 days, 14:14, load: 17.39, 17.41, 17.49


PEPTIDE SEQUENCE DATABASES

 nr          Non-redundant GenBank CDS translations+PDB+SwissProt+PIR
 pdb         PDB protein sequences
 kabat       Kabat Sequences of Proteins of Immunological Interest
 alu *       Translations of Select Alu Repeats from REPBASE
 month       All new or revised GenBank CDS translation+PDB+SwissProt+PIR
             sequences released in the last 30 days
 swissprot SwissProt sequences


NUCLEOTIDE SEQUENCE DATABASES

 nr          Non-redundant GenBank+EMBL+DDBJ+PDB sequences (but no EST's or
             STS's)
 est +       Non-redundant Database of GenBank+EMBL+DDBJ EST Division
 sts +       Non-redundant Database of GenBank+EMBL+DDBJ STS Division
 pdb         PDB nucleotide sequences
 vector      Vector subset of GenBank
 mito *      Database of mitochondrial sequences, Rel. 1.0, July 1995
 kabat       Kabat Sequences of Nucleic Acid of Immunological Interest
 epd         Eukaryotic Promotor Database
 alu *+      Select Alu Repeats from REPBASE
 month       All new or revised GenBank+EMBL+DDBJ+PDB sequences released in
             the last 30 days

  * Databases that are not accessible through the NCBI Retrieve E-mail
server.
  + The TBLASTX program is restricted to searching these databases.
```

continued

Figure 8.1 continued

```
==============================================================================
You can obtain the BLAST documentation files, send a message consisting of
 just the word ``help'' (without the quotes) to: blast@ncbi.nlm.nih.gov
 Last modification dates: August 10th 95 for the E-mail server help, January
 19th 94 for the BLAST manual and February 7th 95 for the BLAST FAQ.
==============================================================================
For a free subscription to "NCBI News", the NCBI newsletter, send a request
 along with your name and postal mailing address to: info@ncbi.nlm.nih.gov
==============================================================================
A new GenBank sequence submission tool, called BankIt, is now available
 through the NCBI's home page on the World Wide Web. The URL is
 http://www.ncbi.nlm.nih.gov/
==============================================================================
BLASTN 1.4.9MP [26-March-1996] [Build 14:27:07 Apr  1 1996]
```

Reference: Altschul, Stephen F., Warren Gish, Webb Miller, Eugene W. Myers,
    and David J. Lipman (1990). Basic local alignment search tool. J. Mol.
    Biol.215:403-10.
    Notice: this program and its default parameter settings are optimized to find
    nearly identical sequences rapidly. To identify weak similarities encoded in
    nucleic acid, use BLASTX, TBLASTN or TBLASTX.

Second part of the reply message

```
Query=  Unknown nucleotide sequence data
        (1382 letters)

Database: Non-redundant GenBank+EMBL+DDBJ+PDB sequences
          248,514 sequences; 350,713,451 total letters.
Searching...............................................done

WARNING: -hspmax 100 was exceeded with 5 of the database sequences, with as
         many as 129 HSPs being found at one time.


                                                        Smallest
                                                          Sum
                                                 High  Probability
Sequences producing High-scoring Segment Pairs:  Score P(N)       N

emb|X70218|HSPPX     H. sapiens mRNA for protein phosphat... 6910  0.0         1
gb|S57412|S57412     protein phosphatase X [rabbits, live... 4055  0.0         4
emb|X14031|OCPPX     Rabbit mRNA for protein phosphatase ... 2647  2.3e-249    3
emb|Z22596|ATPROPHOB A.thaliana protein phosphatase mRNA.   1466  2.5e-157    3
emb|X55199|DMPHOS2A  D.melanogaster mRNA for phosphatase  ... 1665  9.6e-131    2
emb|X14087|RNPHO2AB  Rat mRNA for phosphoprotein phosphat... 1617  9.6e-125    1
gb|M23591|RATPP2AB   Rat protein phosphatase 2A-beta cata... 1617  2.1e-124    1
gb|U49113|OSU49113   Oryza sativa protein phosphatase 2A ... 1606  1.7e-123    1
emb|Z67746|MMPHO2AIB M.musculus mRNA for phosphatase 2A c... 1591  4.8e-123    1
emb|X16044|RNP2B2    Rat mRNA for phosphatase 2A catalyti... 1589  4.0e-122    1
emb|Z67745|MMPHO2AIA M.musculus mRNA for phosphatase 2A c... 1556  5.0e-120    1
emb|Z50852|XLPP2ACSB X.laevis mRNA for protein phosphatas... 1531  3.4e-117    1
emb|X14159|RN2APHOS  Rat mRNA for protein phosphatase-2A  ... 1529  4.9e-117    1
emb|X16043|RNP2A2    Rat mRNA for phosphatase 2A catalyti... 1520  2.7e-116    1
gb|M96733|ATHPRPHB   Arabidopsis thaliana protein phospha... 1513  7.0e-116    1
emb|X06087|OCPP2A    Rabbit skeletal muscle mRNA for prot... 1511  2.0e-115    1
dbj|D17531|CHKP2ACS  Chicken mRNA for phosphatase 2A cata... 1501  6.2e-115    1
```

continued

Figure 8.1 continued

```
gb|S66918|S66918        PPV=35 kda protein serine/threonine ...  1474  6.3e-115  2
gb|M20192|PIG2APHA       Porcine protein phosphatase 2A alpha...  1499  1.6e-114  1
gb|M96841|ATHPRPHC       Arabidopsis thaliana protein phospha...  1484  2.5e-113  1
emb|X62114|XLPP2AA       X.laevis mRNA for protein phosphatas...  1478  9.0e-113  1
gb|M96732|ATHPRPHA       Arabidopsis thaliana protein phospha...  1411  2.5e-111  2
emb|X72858|BTPP2A        B.taurus mRNA for protein phosphatas...  1457  5.2e-111  1
emb|X52554|BTPHO2AA      Bovine mRNA for protein phosphatase ...  1457  5.3e-111  1
gb|M74168|TRBPHOAT2A     Trypansoma brucei protein phosphatas...  1451  2.0e-110  1
emb|Y00763|OCPP2AB       Rabbit mRNA for protein phosphatase ...  1447  3.2e-110  1
gb|M36951|HUMLPP2AA      Human protein phosphatase 2A-alpha c...  1439  1.8e-109  1
emb|X12646|HSRPHO2A      Human mRNA for protein phosphatase 2...  1439  1.9e-109  1
gb|M60483|HUMPP2AA       Human protein phosphatase 2A catalyt...  1439  2.2e-109  1
emb|Z22587|ATPROPHSA     A.thaliana protein phosphatase mRNA.     1430  7.7e-109  1
gb|J03805|HUMALPHLB      Human liver phosphatase 2A mRNA, 3' ...  1418  8.5e-108  1
emb|X57439|BNPP2A        B.napus mRNA for phosphatase 2A          1411  2.5e-107  1
emb|X12656|HSPP2A        Human mRNA for protein phosphatase 2...  1409  4.9e-107  1
gb|M60484|HUMPP2AB       Human protein phosphatase 2A catalyt...  1409  6.8e-107  1

gb|M16968|BOVPHO2A       Bovine protein phosphatase type 2A c...  1393  1.2e-105  1
gb|M20193|PIG2APHB       Porcine protein phosphatase 2A beta ...  1365  1.8e-103  1
emb|Z26041|HAPRPH2A      H.annuus mRNA for protein phosphatas...  1337  2.3e-102  2
emb|X77236|RNPPV         R.norvegicus mRNA for protein phosph...  1306  1.7e-98   1
gb|J03804|HUMALPHLA      Human liver phosphatase 2A mRNA, com...  1229  8.0e-95   2
gb|U39568|ATU39568       Arabidopsis thaliana type 2A serine/...  1258  1.8e-94   1
gb|M58518|YSPPPA1        Schizosaccharomyces pombe type 2A pr...  1240  5.4e-93   1
emb|X70399|MSPP2AMS      M.sativa mRNA for phosphoprotein pho...  1233  2.4e-92   1
emb|Z26654|ACMRPP2A      A.cliftonii mRNA for protein phospha...  1215  9.4e-91   1
gb|M58519|YSPPPA2        Schizosaccharomyces pombe type 2A pr...  1112  4.1e-82   1
emb|X56261|SCPPH1G       Yeast PPH1 gene for protein phosphat...  1079  1.9e-79   1
emb|X14832|OCPP1A        Rabbit mRNA for type-1 protein phosp...   564  2.0e-79   3
emb|X58856|SCPPH21       S.cerevisiae PPH21 gene for protein ...  1079  2.6e-79   1
emb|X96876|SCCHRIVFY     S.cerevisiae DNA of cosmid from chro...  1079  3.4e-79   1
emb|X07798|OCCPP1A       Rabbit mRNA for type-1 protein phosp...   555  1.7e-78   3
gb|J04759|HUMPPP1A       Human protein phosphatase I alpha su...   510  2.5e-78   4
emb|X56262|SCPPH22G      Yeast PPH22 gene for protein phospha...  1047  1.1e-76   1
gb|M60317|YSCSIT4A       S.cerevisiae PPH2-alpha protein gene...  1047  1.1e-76   1
emb|X58857|SCPPH22       S.cerevisiae PPH22 gene for protein ...  1047  1.4e-76   1
emb|X83276|SCDNAIV       S.cerevisiae DNA for ORFs from chrom...  1047  1.6e-76   1
emb|X70848|HSPH1CAT      H.sapiens mRNA for phosphatase 1 cat...   510  2.5e-76   4
emb|Y00701|OCPP1         Rabbit mRNA for protein phosphatase-...   530  3.2e-76   3
gb|M63960|HUMPRPHOS1     Human protein phosphatase-1 catalyti...   510  1.4e-75   4
gb|S57501|S57501         protein phosphatase type 1 catalytic...   510  1.6e-75   4
emb|X83593|NCPPH1        N.crassa pph-1 gene                       926  4.0e-73   2
gb|U31445|PTU31445       Paramecium tetraurelia macronuclear ...   542  9.0e-71   3
dbj|D90163|RATPP1AA      Rat PP-1a gene for catalytic subunit...   510  3.7e-69   4
emb|Z18925|SP2APRPH      S.pombe type2A-like protein phosphat...   935  1.6e-67   1
dbj|D13712|YSPPPE1       Yeast ppe1+ gene for protein phospha...   935  1.8e-67   1
gb|S78215|S78215         protein phosphatase 1 alpha [rats, s...   519  1.8e-66   3
emb|X58858|SCPPH3        S.cerevisiae PPH3 gene for type 2A-r...   920  5.3e-66   1
dbj|D00859|RATPP1ACS     Rat PP-1 alpha gene for catalytic su...   510  1.0e-65   3
emb|X82086|SCCHROIV      S.cerevisiae DNA for right arm of ch...   911  8.8e-65   1
emb|Z46796|SC8554        S.cerevisiae chromosome IV cosmid 8554.   911  1.0e-64   2
emb|X56438|DMPP1A1       D.melanogaster PP1-alpha 96A gene fo...   487  1.7e-63   4
emb|X15583|DMPP1A        Fruitfly mRNA for protein phosphatas...   412  9.2e-58   3
emb|X55198|DMPP1         D.melanogaster gene for protein phos...   412  4.0e-57   3
gb|M24395|YSCSIT4C       Saccharomyces cerevisiae protein pho...   800  5.2e-56   1
emb|Z71781|SCCIVL37K     S.cerevisiae chromosome IV left arm ...   800  1.1e-54   3
gb|M58443|RATPPX         Rat protein serine-threonine phospha...   729  3.2e-52   1
emb|X56439|DMPP1B        D.melanogaster PP1-beta 9C gene for ...   429  1.8e-50   3
emb|Z47078|MDPPMD3       M.domestica mRNA for serine/threonin...   613  8.4e-43   1
gb|J05479|MUSCALCAT      Mouse calcineurin catalytic subunit ...   380  3.5e-38   2
gb|U30493|DMU30493       Drosophila melanogaster calcineurin ...   402  3.8e-38   2
emb|X89416|HSRNAPPP5     H.sapiens mRNA for protein phosphata...   319  1.3e-36   3
```

continued

Figure 8.1 continued

```
gb|L07395|HUMPP1G1X    Human protein phosphatase-1 gamma 1 ...   398   5.0e-36   2
emb|X74008|HSPPPICC    H.sapiens mRNA for protein phosphata...   398   5.2e-36   2
gb|U52691|GPU52691     Gonyaulax polyedra putative type-1 s...   551   3.0e-35   1
gb|U25174|PPP5C        Human serine-threonine phosphatase (...   301   3.8e-35   3
emb|X63558|BOBOPP1     B.oleracea mRNA for type 1 protein s...   363   2.9e-32   2
gb|U00063|CELF56C9     Caenorhabditis elegans cosmid F56C9       346   5.3e-32   2
gb|M60215|MZEZMPP1     Z.mays protein phosphatase-1 (ZmPP1)...   473   1.0e-28   1
gb|M27067|EMEBIMG      Aspergillus nidulans phosphoprotein ...   466   4.2e-28   1
dbj|D90165|RATPP1G1    Rat PP-1g1 gene for catalytic subuni...   255   3.2e-27   3
emb|Z73974|CEF29F11    Caenorhabditis elegans cosmid F29F11      283   3.3e-27   2
dbj|D90166|RATPP1G2    Rat PP-1g2 gene for catalytic subuni...   246   1.0e-24   3
emb|X77237|RNPPT       R.norvegicus mRNA for protein phosph...   290   1.7e-23   2
gb|U12203|RNU12203     Rattus norvegicus phosphoprotein pho...   290   1.7e-23   2
gb|M27071|MUSDIS2M1A   Mus musculus protein phosphatase typ...   246   5.0e-23   2
dbj|D85137|MUSPIM1K    House mouse; Musculus domesticus bra...   246   8.6e-23   2
gb|U53456|MMU53456     Mus musculus protein phosphatase 1cg...   246   9.1e-23   2
emb|Z49886|CEC06A1     Caenorhabditis elegans cosmid C06A1       394   4.8e-22   1
gb|M81475|MUSPPPPA     Murine phosphoprotein phosphatase mR...   379   1.3e-21   2
emb|Z46996|CEC34C12    Caenorhabditis elegans cosmid C34C12      217   1.9e-20   4
gb|M29275|RATCNRA      Rat calcineurin A mRNA, complete cds...   371   3.6e-20   1
emb|X57115|RNCALCA     Rat mRNA for calcineurin A                371   3.6e-20   1
```

```
WARNING: Descriptions of 1703 database sequences were not reported due to
the limiting value of parameter V = 100.
```

Third part of the reply message

```
>emb|X70218|HSPPX H. sapiens mRNA for protein phosphatase X
          Length = 1382

  Plus Strand HSPs:

 Score = 6910 (1909.4 bits), Expect = 0.0, P = 0.0
 Identities = 1382/1382 (100%), Positives = 1382/1382 (100%), Strand = Plus
/ Plus

Query:    1 CGGCGGCGGCGGTCGAAAGCGGAGTGAAAGAGGGAGGCAGGGAGCCGGAGAGCCGGAACC 60
            ||||||||||||||||||||||||||||||||||||||||||||||||||||||||||||
Sbjct:    1 CGGCGGCGGCGGTCGAAAGCGGAGTGAAAGAGGGAGGCAGGGAGCCGGAGAGCCGGAACC 60

Query:   61 GGAGTCGCAGCGGCGGAGACCCCTGTGCGGTGCGGAGGGGGCGGCGGCCCCGACTCTGAC 120
            ||||||||||||||||||||||||||||||||||||||||||||||||||||||||||||
Sbjct:   61 GGAGTCGCAGCGGCGGAGACCCCTGTGCGGTGCGGAGGGGGCGGCGGCCCCGACTCTGAC 120

Query:  121 CCGCGCCGGGGGTGGGCCATGGCGGAGATCAGCGACCTGGACCGGCAGATCGAGCAGCTG 180
            ||||||||||||||||||||||||||||||||||||||||||||||||||||||||||||
Sbjct:  121 CCGCGCCGGGGGTGGGCCATGGCGGAGATCAGCGACCTGGACCGGCAGATCGAGCAGCTG 180

Query:  181 CGTCGCTGCGAGCTCATCAAGGAGAGCGAAGTCAAGGCCCTGTGCGCTAAGGCCAGAGAG 240
            ||||||||||||||||||||||||||||||||||||||||||||||||||||||||||||
Sbjct:  181 CGTCGCTGCGAGCTCATCAAGGAGAGCGAAGTCAAGGCCCTGTGCGCTAAGGCCAGAGAG 240

Query:  241 ATCTTGGTAGAGGAGAGCAACGTGCAGAGGGTGGACTCGCCAGTCACAGTGTGCGGCGAC 300
            ||||||||||||||||||||||||||||||||||||||||||||||||||||||||||||
Sbjct:  241 ATCTTGGTAGAGGAGAGCAACGTGCAGAGGGTGGACTCGCCAGTCACAGTGTGCGGCGAC 300

Query:  301 ATCCATGGACAATTCTATGACCTCAAAGAGCTGTTCAGAGTAGGTGGCGACGTCCCTGAG 360
            ||||||||||||||||||||||||||||||||||||||||||||||||||||||||||||
Sbjct:  301 ATCCATGGACAATTCTATGACCTCAAAGAGCTGTTCAGAGTAGGTGGCGACGTCCCTGAG 360
```

continued

Figure 8.1 continued

```
Query:  361 AGGAACTACCTCTTCATGGGGGACTTTGTGGACCGTGGCTTCTATAGCGTCGAAACGTTC 420
            ||||||||||||||||||||||||||||||||||||||||||||||||||||||||||||
Sbjct:  361 AGGAACTACCTCTTCATGGGGGACTTTGTGGACCGTGGCTTCTATAGCGTCGAAACGTTC 420

Query:  421 CTCCTGCTGCTGGCACTTAAGGTTCGCTATCCTGATCGCATCACACTGATCCGGGGCAAC 480
            ||||||||||||||||||||||||||||||||||||||||||||||||||||||||||||
Sbjct:  421 CTCCTGCTGCTGGCACTTAAGGTTCGCTATCCTGATCGCATCACACTGATCCGGGGCAAC 480

Query:  481 CATGAGAGTCGCCAGATCACGCAGGTCTATGGCTTCTACGATGAGTGCCTGCGCAAGTAC 540
            ||||||||||||||||||||||||||||||||||||||||||||||||||||||||||||
Sbjct:  481 CATGAGAGTCGCCAGATCACGCAGGTCTATGGCTTCTACGATGAGTGCCTGCGCAAGTAC 540

Query:  541 GGCTCGGTGACTGTGTGGCGCTACTGCACTGAGATCTTTGACTACCTCAGCCTGTCAGCC 600
            ||||||||||||||||||||||||||||||||||||||||||||||||||||||||||||
Sbjct:  541 GGCTCGGTGACTGTGTGGCGCTACTGCACTGAGATCTTTGACTACCTCAGCCTGTCAGCC 600

Query:  601 ATCATCGATGGCAAGATCTTCTGCGTGCACGGGGGCCTCTCCCCCTCCATCCAGACCCTG 660
            ||||||||||||||||||||||||||||||||||||||||||||||||||||||||||||
Sbjct:  601 ATCATCGATGGCAAGATCTTCTGCGTGCACGGGGGCCTCTCCCCCTCCATCCAGACCCTG 660

Query:  661 GATCAGATTCGGACAATCGACCGAAAGCAAGAGGTGCCTCATGATGGGCCCATGTGTGAC 720
            ||||||||||||||||||||||||||||||||||||||||||||||||||||||||||||
Sbjct:  661 GATCAGATTCGGACAATCGACCGAAAGCAAGAGGTGCCTCATGATGGGCCCATGTGTGAC 720

Query:  721 CTCCTCTGGTCTGACCCAGAAGACACCACAGGCTGGGGCGTGAGCCCGCGCGGAGCCGGC 780
            ||||||||||||||||||||||||||||||||||||||||||||||||||||||||||||
Sbjct:  721 CTCCTCTGGTCTGACCCAGAAGACACCACAGGCTGGGGCGTGAGCCCGCGCGGAGCCGGC 780

Query:  781 TACCTATTTGGCAGTGACGTGGTGGCCCAGTTCAACGCAGCCAATGACATTGACATGATC 840
            ||||||||||||||||||||||||||||||||||||||||||||||||||||||||||||
Sbjct:  781 TACCTATTTGGCAGTGACGTGGTGGCCCAGTTCAACGCAGCCAATGACATTGACATGATC 840

Query:  841 TGCCGTGCCCACCAACTGGTGATGGAAGGTTACAAGTGGCACTTCAATGAGACGGTGCTC 900
            ||||||||||||||||||||||||||||||||||||||||||||||||||||||||||||
Sbjct:  841 TGCCGTGCCCACCAACTGGTGATGGAAGGTTACAAGTGGCACTTCAATGAGACGGTGCTC 900

Query:  901 ACTGTGTGGTCGGCACCCAACTACTGCTACCGCTGTGGGAATGTGGCAGCCATCTTGGAG 960
            ||||||||||||||||||||||||||||||||||||||||||||||||||||||||||||
Sbjct:  901 ACTGTGTGGTCGGCACCCAACTACTGCTACCGCTGTGGGAATGTGGCAGCCATCTTGGAG 960

Query:  961 CTGGACGAGCATCTCCAGAAAGATTTCATCATCTTTGAGGCTGCTCCCCAAGAGACACGG 1020
            ||||||||||||||||||||||||||||||||||||||||||||||||||||||||||||
Sbjct:  961 CTGGACGAGCATCTCCAGAAAGATTTCATCATCTTTGAGGCTGCTCCCCAAGAGACACGG 1020

Query: 1021 GGCATCCCCTCCAAGAAGCCCGTGGCCGACTACTTCCTGTGACCCCGCCCGGCCCCTGCC 1080
            ||||||||||||||||||||||||||||||||||||||||||||||||||||||||||||
Sbjct: 1021 GGCATCCCCTCCAAGAAGCCCGTGGCCGACTACTTCCTGTGACCCCGCCCGGCCCCTGCC 1080

Query: 1081 CCCTCCAACCCTTCTGGCCCTCGCACCACTGTGACTCTGCCATCTTCCTCAGACGGAGGC 1140
            ||||||||||||||||||||||||||||||||||||||||||||||||||||||||||||
Sbjct: 1081 CCCTCCAACCCTTCTGGCCCTCGCACCACTGTGACTCTGCCATCTTCCTCAGACGGAGGC 1140

Query: 1141 TGGGGGGGCTGTCCTGGCTCTGCTGTCCCCCAAGAGGGTGCCTTCGAGGGTGAGGACTTC 1200
            ||||||||||||||||||||||||||||||||||||||||||||||||||||||||||||
Sbjct: 1141 TGGGGGGGCTGTCCTGGCTCTGCTGTCCCCCAAGAGGGTGCCTTCGAGGGTGAGGACTTC 1200

Query: 1201 TCTGGAGAGGCCTGGAGACCTAGCTCCATGTTCCTCCTCCTCTCTCCCCACTTGAACCAT 1260
            ||||||||||||||||||||||||||||||||||||||||||||||||||||||||||||
Sbjct: 1201 TCTGGAGAGGCCTGGAGACCTAGCTCCATGTTCCTCCTCCTCTCTCCCCACTTGAACCAT 1260

Query: 1261 GAAGTTTCCAATAATTTTTTTTTTCTTTTTTTCCTTCTTTTTCTGTTTGTTTTTAGATAAA 1320
            ||||||||||||||||||||||||||||||||||||||||||||||||||||||||||||
Sbjct: 1261 GAAGTTTCCAATAATTTTTTTTTTCTTTTTTTCCTTCTTTTTCTGTTTGTTTTTAGATAAA 1320
```

continued

Figure 8.1 continued

```
Query:  1321 AATTTTTGAGAAAAAAAATGAAAAATTCTAATAAAAGAAGAAAAATGGTAAAAAAAAAA 1380
             ||||||||||||||||||||||||||||||||||||||||||||||||||||||||||
Sbjct:  1321 AATTTTTGAGAAAAAAAATGAAAAATTCTAATAAAAGAAGAAAAATGGTAAAAAAAAAA 1380

Query:  1381 AA 1382
             ||
Sbjct:  1381 AA 1382


>gb|M36951|HUMLPP2AA Human protein phosphatase 2A-alpha catalytic subunit
mRNA,
                complete cds.
                Length = 2034

  Plus Strand HSPs:

  Score = 1439 (397.6 bits), Expect = 1.8e-109, P = 1.8e-109
  Identities = 543/862 (62%), Positives = 543/862 (62%), Strand = Plus / Plus

Query:   148 ATCAGCGACCTGGACCGGCAGATCGAGCAGCTGCGTCGCTGCGAGCTCATCAAGGAGAGC 207
             | ||   || |||||||| |  ||||||||||||||     ||| |||   |   ||| |
Sbjct:   100 ACCAAGGAGCTGGACCAGTGGATCGAGCAGCTGAACGAGTGCAAGCAGCTGTCCGAGTCC 159

Query:   208 GAAGTCAAGGCCCTGTGCGCTAAGGCCAGAGAGATCTTGGTAGAGGAGAGCAACGTGCAG 267
             | |||||| |||  |||| |||||  ||| ||| ||| ||  | | || ||||||||||
Sbjct:   160 CAGGTCAAGAGCCTCTGCGAGAAGGCTAAAGAAATCCTGACAAAAGAATCCAACGTGCAA 219

Query:   268 AGGGTGGACTCGCCAGTCACAGTGTGCGGCGACATCCATGGACAATTCTATGACCTCAAA 327
             |||   |   |||||| || |||| || | ||| | ||||| ||||| |||| ||| |||
Sbjct:   220 GAGGTTCGATGTCCAGTTACTGTCTGTGGAGATGTGCATGGGCAATTTCATGATCTCATG 279

Query:   328 GAGCTGTTCAGAGTAGGTGGCGACGTCCCTGAGAGGAACTACCTCTTCATGGGGGACTTT 387
             || ||||| |||   | |||||  || ||||||| || |||| || |||  |||| ||
Sbjct:   280 GAACTGTTTAGAATTGGTGGCAAATCACCAGATACAAATTACTTGTTTATGGGAGATTAT 339

Query:   388 GTGGACCGTGGCTTCTATAGCGTCGAAACGTTCCTCCTGCTGCTGGCACTTAAGGTTCGC 447
             || || | ||| |||  ||| || ||||| |||  | |||||  || ||||||||||||
Sbjct:   340 GTTGACAGAGGATATTATTCAGTTGAAACAGTTACACTGCTTGTAGCTCTTAAGGTTCGT 399

Query:   448 TATCCTGATCGCATCACACTGATCCGGGGCAACCATGAGAGTCGCCAGATCACGCAGGTC 507
             || | ||| |||||||| ||| ||| ||| | |  |||||||| | | ||||||| ||
Sbjct:   400 TACCGTGAACGCATCACCATTCTTCGAGGGAATCATGAGAGCAGACAGATCACACAAGTT 459

Query:   508 TATGGCTTCTACGATGAGTGCCTGCGCAAGTACGGCTCGGTGACTGTGTGGCGCTACTGC 567
             ||||| |||| |||||   | ||| |  | ||  | || || |   |  ||| || |  |
Sbjct:   460 TATGGTTTCTATGATGAATGTTTAAGAAAAATATGGAAATGCAAATGTTTGGAAATATTTT 519

Query:   568 ACTGAGATCTTTGACTACCTCAGCCTGTCAGCCATCATCGATGGCAAGATCTTCTGCGTG 627
             || || ||| ||||||| || | |  |||  || || |||| |||  ||||| |||  |
Sbjct:   520 ACAGATCTTTTTGACTATCTTCCTCTCACTGCCTTGGTGGATGGGCAGATCTTCTGTCTA 579

Query:   628 CACGGGGGCCTCTCCCCCTCCATCCAGACCCTGGATCAGATTCGGACAATCGACCGAAAG 687
             || ||  | ||| ||||| |||  |  ||  ||| || ||  ||  || || |  | |
Sbjct:   580 CATGGTGGTCTCTCGCCATCTATAGATACACTGGATCATATCAGAGCACTTGATCGCCTA 639

Query:   688 CAAGAGGTGCCTCATGATGGGCCCATGTGTGACCTCCTCTGGTCTGACCCAGAAGACACC 747
             |||||| | |||  |||| |||  | |||||||||| || ||  |  |||  ||| ||
Sbjct:   640 CAAGAAGTTCCCCATGAGGGTCCAATGTGTGACTTGCTGTGGTCAGATCCAGATGACCGT 699

Query:   748 ACAGGCTGGGGCGTGAGCCCGCGCGGAGCCGGCTACCTATTTGGCAGTGACGTGGTGGCC 807
             || || ||||| |   || || |||||| | ||||| |||||  ||     ||| |  |
Sbjct:   700 GGTGGTTGGGGGTATATCTCCTCGAGGAGCTGGTTACACCTTTGGGCAAGATATTTCTGAG 759
```

continued

Figure 8.1 continued

```
Query:   808 CAGTTCAACGCAGCCAATGACATTGACATGATCTGCCGTGCCCACCAACTGGTGATGGAA 867
             ||  ||     ||||||| | |     ||| |    | || ||||| || ||||||||
Sbjct:   760 ACATTTAATCATGCCAATGGCCTCACGTTGGTGTCTAGAGCTCACCAGCTAGTGATGGAG 819

Query:   868 GGTTACAAGTGGCACTTCAATGAGACGGTGCTCACTGTGTGGTCGGCACCCAACTACTGC 927
             || || || |||  |    |    || ||  | ||  || |      || || ||||| ||
Sbjct:   820 GGATATAACTGGTGCCATGACCGGAATGTAGTAACGATTTTCAGTGCTCCAAACTATTGT 879

Query:   928 TACCGCTGTGGGAATGTGGCAGCCATCTTGGAGCTGGACGAGCATCTCCAGAAAGATTTC 987
             ||  || ||||| ||    || || ||| |||| || |||||     ||| |   |  ||||
Sbjct:   880 TATCGTTGTGGTAACCAAGCTGCAATCATGGAACTTGACGATACTCTAAAATACTCTTTC 939

Query:   988 ATCATCTTTGAGGCTGCTCCCC 1009
             |      |||||  |  || || |
Sbjct:   940 TTGCAGTTTGACCCAGCACCTC 961
```

Fourth part of the reply message

```
WARNING: HSPs involving 1753 database sequences were not reported due
to the limiting value of parameter B = 50.
Parameters:
  V=100
  B=50
  H=0
  -qtype
  P=4
  -ctxfactor=2.00
  E=10
```

DNA sequences: downloading related sequences using Retrieve

Based on the results of the comparison of the query sequence to the nonredundant nucleotide sequence database using the BLASTN program, several interesting matching and putative similar sequences were identified. Some of these sequences were clones from humans and some belonged to the same family of sequences as the query sequence (namely the protein ser/thr phosphatase catalytic subunits). For further comparative analysis of this family of sequences, the Retrieve server was used to download several entries for clones of these enzymes identified in humans. The query message to the server and a portion of the reply message are shown in Fig. 8.2A.

In this message, six sequence records, X70218, M36951, M63960, U25174, S57501, and M60483, are retrieved from the GenBank database by using the accession number field of the records as the search criterion. Note that some of these sequence records were originally submitted to sequence databases other than GenBank, but because of the exchange of data between the international sequence databases, the same sequence records can be found in all databases. An example of one of the returned records from the server is shown in Fig. 8.2B.

Figure 8.2. Retrieval of Genomic Sequence Information Using Retrieve

A: Query message to the Retrieve server

```
Date:   Tue 16 Jul 1996 - 09:16.02 - GMT
From:   annep@acs.auc.eun.eg
To:     retrieve@ncbi.nlm.nih.gov
Subj:

DATALIB GENBANK
MAXDOCS 100
MAXLINES 10000
BEGIN
X70218 [ACC]
M36951 [ACC]
M63960 [ACC]
U25174 [ACC]
S57501 [ACC]
M60483 [ACC]
```

B: Reply message from the Retrieve server

```
Date:      Tue 16 Jul 1996 - 12:41:11 - EDT
To:        annep@auc-acs.eun.eg
From:      RETRIEVE E-Mail Server <retrieve@ncbi.nlm.nih.gov>
Subject:   Results-RETRIEVE Server:
Reply-To:  Retrieve Server <retrieve@ncbi.nlm.nih.gov>


===========================================================================
To Obtain Help Documentation: send e-mail to 'retrieve@ncbi.nlm.nih.gov'
   with the word 'help' in the body of the mail message.

Note: GenBank retrieval and submission tools are available through
   the World Wide Web at the URL: http://www.ncbi.nlm.nih.gov/  For more
   information contact User Services at: info@ncbi.nlm.nih.gov
===========================================================================

Database: GenBank Updates (95.0+, 07/15/96)
Query: x70218 [acc] m36951 [acc] m63960 [acc] u25174 [acc] s57501 [acc]
m60483 [acc]
Parse status: OK: 0 documents retrieved.
//
Database: GenBank (95.0, 6/15/96)
Query: x70218 [acc] m36951 [acc] m63960 [acc] u25174 [acc] s57501 [acc]
m60483 [acc]
Parse status: OK: 9 documents retrieved.
Documents selected: 1-9  (up to 10000 lines)


LOCUS       HSPPX        1382 bp    RNA              PRI      30-JUN-1993
DEFINITION  H. sapiens mRNA for protein phosphatase X.
ACCESSION   X70218 S55208
NID         g312813
KEYWORDS    protein phosphatase X.
```

continued

Figure 8.2 continued

```
SOURCE       human.
  ORGANISM   Homo sapiens
             Eukaryotae; mitochondrial eukaryotes; Metazoa; Chordata;
             Vertebrata; Eutheria; Primates; Catarrhini; Hominidae; Homo.
REFERENCE    1  (bases 1 to 1382)
  AUTHORS    Brewis,N.D. and Cohen,P.T.
  TITLE      Protein phosphatase X has been highly conserved during mammalian
             evolution
  JOURNAL    Biochim. Biophys. Acta 1171 (2), 231-233 (1992)
  MEDLINE    93129628
FEATURES             Location/Qualifiers
     source          1..1382
                     /organism="Homo sapiens"
                     /cell_type="teratocarcnoma"
                     /clone_lib="teratocarcnoma cDNA lambda 10"
     CDS             139..1062
                     /gene="PPX"
                     /codon_start=1
                     /product="protein phosphatase X"
                     /db_xref="PID:g312814"
                     /db_xref="SWISS-PROT:P33172"
                     /translation="MAEISDLDRQIEQLRRCELIKESEVKALCAKAREVEESNVQR
                     VDSPVTVCGDIHGQFYDLKELFRVGGDVPERNYLFMGDFVDRGFYSVETFLLLLALKV
                     RYPDRITLIRGNHESRQITQVYGFYDECLRKYGSVTVWRYCTEIFDYLSLSAIIDGKI
                     FCVHGGLSPSIQTLDQIRTIDRKQEVPHDGPMCDLLWSDPEDTTGWGVSPRGAGYLFG
                     SDVVAQFNAANDIDMICRAHQLVMEGYKWHFNETVLTVWSAPNYCYRCGNVAAILELD
                     EHLQKDFIIFEAAPQETRGIPSKKPVADYFL"
     primer_bind     666..688
                     /gene="PPX"
     primer_bind     807..836
                     /gene="PPX"
     polyA_signal    1350..1355
BASE COUNT      302 a     386 c     400 g     294 t
ORIGIN
        1 cggcggcggc ggtcgaaagc ggagtgaaag agggaggcag ggagccggag agccgaacc
       61 ggagtcgcag cggcggagac ccctgtgcgg tgcggagggg gcggcggccc cgactctgac
      121 ccgcgccggg ggtgggccat ggcggagatc agcgacctgg accggcagat cgagcagctg
      181 cgtcgctgcg agctcatcaa ggagagcgaa gtcaaggccc tgtgcgctaa ggccagagag
      241 atcttggtag aggagagcaa cgtgcagagg gtggactcgc cagtcacagt gtgcggcgac
      301 atccatggac aattctatga cctcaaagag ctgttcagag taggtggcga cgtccctgag
      361 aggaactacc tcttcatggg ggactttgtg gaccgtggct tctatagcgt cgaaacgttc
      421 ctcctgctgc tggcacttaa ggttcgctat cctgatcgca tcacactgat ccggggcaac
      481 catgagagtc gccagatcac gcaggtctat ggcttctacg atgagtgcct gcgcaagtac
      541 ggctcggtga ctgtgtggcg ctactgcact gagatctttg actacctcag cctgtcagcc
      601 atcatcgatg gcaagatctt ctgcgtgcac gggggcctct cccctccat ccagaccctg
      661 gatcagattc ggacaatcga ccgaaagcaa gaggtgcctc atgatgggcc catgtgtgac
      721 ctcctctggt ctgacccaga agacaccaca ggctggggcg tgagcccgcg cggagccggc
      781 tacctatttg gcagtgacgt ggtggcccag ttcaacgcag ccaatgacat tgacatgatc
      841 tgccgtgccc accaactggt gatggaaggt tacaagtggc acttcaatga cggtgctc
      901 actgtgtggt cggcacccaa ctactgctac cgctgtggga atgtggcagc catcttggag
      961 ctggacgagc atctccagaa agatttcatc atctttgagg ctgctcccca agagacacgg
     1021 ggcatcccct ccaagaagcc cgtggccgac tacttcctgt gaccccgccc ggcccctgcc
     1081 ccctccaacc cttctggccc tcgcaccact gtgactctgc catcttcctc agacggaggc
     1141 tgggggggct gtcctggctc tgctgtcccc caagagggtg ccttcgaggg tgaggacttc
     1201 tctggagagg cctggagacc tagctccatg ttcctcctcc tctctcccca cttgaaccat
     1261 gaagtttcca ataatttttt tttctttttt tccttctttt tctgtttgtt tttagataaa
     1321 aatttttgag aaaaaaaatg aaaaattcta ataaaagaag aaaaatggta aaaaaaaaaa
     1381 aa
//
```

The sequence record is composed of more than just the raw nucleotide sequence. The first line of the record gives the locus name, HSPPX; the length, 1,382 base pairs; the source of the clone, RNA; the database division, primate (PRI); and the date that the sequence was entered into the database, 30-JUN-1993. The next line is composed of a single descriptive line of text about the sequence, followed by the accession number or numbers for the entry on the next line. The next line gives the NID term, a unique identifier for the sequence itself. After that come keywords and details on the source of the clone. Information on the authors and a primary literature citation take up the lines just before the feature table of the sequence record.

The feature table gives information on the source of the sequence, the coding region, other related entries in other databases, and a translation of the sequence. The last part of the feature table gives details of primer binding sites and the polyadenylation signal site.

The last part of the sequence record is the actual sequence itself as reported by the authors. From this record, the protein sequence and nucleotide sequence can be extracted using a word processor for further analysis. Alternatively, FastA format sequences can be obtained directly using the Query server (see Fig. 5.25B) via e-mail or the Web.

Protein sequences: analysis with Blocks, Domain, and MotifFinder

Chapter 6 detailed the range of servers that can be used for pattern and domain analysis of protein sequences. Analysis of the cDNA sequence using BLAST tentatively identified the query sequence as that of the catalytic subunit of human PPX. Through the use of the Retrieve server, the sequence for the human PPX catalytic subunit was downloaded and the predicted protein sequence was extracted using a word processor. In the examples below, this PPX protein sequence is analyzed for potential pattern matches that may give additional clues to domains and motifs that are involved in structure and function.

As we mentioned, there is no one correct strategy for the analysis of a protein sequence. One suggested methodology is to first use the Blocks server to look for patterns and conserved domains. One powerful aspect of Blocks is that the results from it are keyed not only to the Blocks database, but the highly annotated PROSITE database as well, allowing the user to extract information from both sources. In the analysis of the query sequence started with BLAST, the Blocks server will be used to begin the annotation of the predicted protein sequence.

■ Blocks

The query to the Blocks server with the PPX protein sequence is shown in Fig. 8.3A. As we described in Chapter 6, the query format for Blocks is simple. In this query, the sequence is submitted in FastA format; the body of the message is composed of a single line of comments, followed immediately by the sequence itself. The reply from the server, shown in Fig. 8.3B, is a bit more complex and will be explained below.

Before looking in detail at the results of the query sequence from the Blocks analysis, it is helpful to understand how the Blocks server works. When performing a comparison analysis, Blocks compares the query sequence to a database of protein sequence blocks that are shared in a family of proteins. In brief, when Blocks analyzes the query sequence, the first position of the query sequence is aligned with the first position of the first block in the database, and an alignment score is determined. Scores are summed over the width of the alignment, and then the block is aligned with the next position of the

Figure 8.3. Pattern Analysis Using Blocks

A: Query message to the Blocks server

```
Date:   Tue 16 Jul 1996 - 09:16.02 - GMT
From:   annep@acs.auc.eun.eg
To:     blocks@howard.fhcrc.org
Subj:

>PPX CATALYTIC SUBUNIT, HUMAN, ACC NO. X70218, S55208
MAEISDLDRQIEQLRRCELIKESEVKALCAKAREILVEESNVQRVDSPVTVCGDIHGQFYDLKELFRVGGDV
PERNYLFMGDFVDRGFYSVETFLLLLALKVRYPDRITLIRGNHESRQITQVYGFYDECLRKYGSVTVWRYCT
EIFDYLSLSAIIDGKIFCVHGGLSPSIQTLDQIRTIDRKQEVPHDGPMCDLLWSDPEDTTGWGVSPRGAGYL
FGSDVVAQFNAANDIDMICRAHQLVMEGYKWHFNETVLTVWSAPNYCYRCGNVAAILELDEHLQKDFIIFEA
APQETRGIPSKKPVADYFL
```

B: Results from the Blocks server

```
Date:   Sun 07 Jul 96 - 11:19:25 - PDT
From:   blocks@howard.fhcrc.org (Block Server)
To:     annep@acs.auc.eun.eg
Subj:   Result of Blocks Search

Query= Human PPX catalytic subunit ,
 Size=307 Amino Acids
Database=mats.dat, Blocks Searched=3179


1.--------------------------------------------------------------------
Block      Rank Frame Score Strength   Location (aa) Description
BL00125A     1    0   1805  1835          48-     87 Serine/threonine specifi
BL00125A    23    0    991  1835          65-    104 Serine/threonine specifi
BL00125B     2    0   1619  1578         154-    200 Serine/threonine specifi
BL00125B   176    0    947  1578          58-    104 Serine/threonine specifi

1805=100.00th percentile of anchor block scores for shuffled queries
P<0.00045 for BL00125B in support of BL00125A
                         |--- 133 amino acids---|
    BL00125 AAAAAAAA::::::::::::.............................BBBBBBBBB
            AAAAAAAA:::::::::::::BBBBBBBBB
            <  AAAAAAAA
            <  BBBBBBBBB


    BL00125A   <->A   (38,446):47
     PPX_HUMAN 48     PVTVCGDIHGQFYDLKELFRVGGDVPERNYLFMGDFVDRG
                      ||||||||||||||||||||||||||||||||||||||||
              48     PVTVCGDIHGQFYDLKELFRVGGDVPERNYLFMGDFVDRG
```

continued

Figure 8.3 continued

```
BL00125B    A<->B    (65,231):66
  PPX_HUMAN 154       AIIDGKIFCVHGGLSPSIQTLDQIRTIDRKQEVPHDGPMCDLLWSDP
                      |||||||||||||||||||||||||||||||||||||||||||||||
            154       AIIDGKIFCVHGGLSPSIQTLDQIRTIDRKQEVPHDGPMCDLLWSDP


2.-------------------------------------------------------------------
Block       Rank Frame Score Strength    Location (aa) Description
BL00134B       3    0  1031  1079            203-     209 Serine proteases, trypsi

1031=26.53th percentile of anchor block scores for shuffled queries
P not calculated for single block BL00134B
                       |--- 115 amino acids---|
     BL00134 AAAA:.....................BB::::::::::...........CC:...DDDDD
             <:::::::::::::::::::::::::::BB
BL00134B    <->B    (37,4474):202



CTR2_CANFA 156        TTGWGLT
                      |||||
           203        TTGWGVS

5.-----------------------------------------------------------Block
Rank Frame Score Strength    Location (aa) Description
BL01143B       6    0  1025  1814             47-      79 Ribosomal protein L31 pr

1025=19.23th percentile of anchor block scores for shuffled queries
P not calculated for single block BL01143B
                        |---    27 amino acids---|
     BL01143 AAAAAAAAAAAAAAAAAAAAAAAAAAA::::..BBBBBBBBBBBBBBBBBBBBBBBBBBB
             <:::::::::::::::::::::::::::::::::BBBBBBBBBBBBBBBBBBBBBBBBBBB


BL01143B    <->B    (30,32):46
RL31_ECOLI 32        LNLDVCSKCHPFFTGKQRDVATGGRVDRFNKRF
                     ||   |   |           ||   |   | |
           47        spVtVCGdiHgqFydlkelfrvGGdVperNylF


10 possible hits reported
```

query sequence. This procedure is carried out exhaustively for all positions of the sequence, for all blocks in the database, and the best alignments between a sequence and entries in the Blocks database are noted. Results for a DNA query are analyzed similarly to those for a protein query, except that here it is assumed that multiple block alignments with a single sequence might be detected in different frames because of frameshift errors in the sequence.

If a particular block scores highly, it is possible that the sequence is related to the group of sequences the block represents. Typically, a group of proteins has more than one region in common and their relationship is represented as a series of blocks separated by unaligned regions. If a second block for a group also scores highly in the search, the evidence that the sequence is related to the group is strengthened; it is further strengthened if a third block also scores it highly, and so on for all of the blocks that are members of that pattern family.

At the conclusion of this analysis process, Blocks ranks the 400 highest scoring blocks. The basis for this rank ordering is a score that is obtained by dividing the raw alignment score described above by a 99.5 percentile calibration score for the block and multiplying by 1,000. The purpose of the calibration score is to allow blocks to be compared to one another, even though they are quite heterogeneous in width (4 to 60 amino acids), in number of sequences (2 to >300), and in the degree of similarity between sequences in each block. A score above 1,000 is expected for 0.5% of the blocks in the search using a protein query of average size. Since there are about 2,000 blocks, a protein of average size is expected to score about 10 blocks above 1,000 by chance alone. Up to 10 blocks with scores of over 1,000, called anchor blocks, are reported. If additional supporting blocks for one of the PROSITE groups represented by these top 10 blocks are also detected among the highest scoring blocks, these are reported as well.

Each numbered result consists of one or more blocks from a PROSITE group found in the query sequence. The program then selects one of the highest scoring blocks for analysis. The blocks are placed in the correct order and separated by distances based on information that is extrapolated from the Blocks database. If this set includes multiple blocks, the server estimates the probability that the lower scoring blocks support the highest scoring block. Maps of the database blocks and query sequence are shown in which:

AAA	represents the first block of a family, roughly in proportion to its length;
:	represents the minimum distance between the blocks of a family;
.	represents the maximum distance between the blocks of a family;
< >	indicates that the sequence has been truncated to fit the page.

Next, the query map is aligned on the highest scoring block. Multiple block hits that are consistent with the highest scoring block are separated by colons. Block hits that are not consistent with the consensus map are listed below the block hits that are consistent. The next part of the response from the server shows the distance between detected blocks. This is listed as the minimum and maximum distances for the database entry, followed by the distance for that block found in the query sequence. Uppercase letters in the query indicate at least one occurrence of the residue in that column of the block. The distance of the anchor block from the amino-terminal end of the protein is indicated by the symbol ->. Distances between blocks within a family are indicated by the symbol <->. In the example shown in Fig. 8.3B, BL00125A can be between 38 and 447 amino acids from the amino-terminal end of the protein, and 47 amino acids is typical.

Note that the highest ranking blocks belong to the Ser/Thr protein phosphatase family and are correctly spaced. Striking regions of identity with the closest segment in the block are seen, further evidence that this query sequence is a member of this protein family. The fact that for all blocks the scores are close to the respective strengths indicates that this predicted protein is about as distant from other members of the family as is the typical member.

The server detected nine other potential alignments. Two of these potential alignments are shown in Fig. 8.3B. However, unlike the blocks of the Ser/Thr phosphatase family, these blocks are clearly poorer matches at the sequence level, and also the alignments cannot be extended to other blocks within the same family. These alignments are therefore the result of chance and not true homology. An example of one of these potential alignments is the second matching block family, the serine proteases. While the block shown in the results is a good match for a region in the query sequence (unlike the blocks that compose the Ser/Thr phosphatase family), the single matching block in this family cannot be extended further to identify supporting blocks in this family (as is the case with the Ser/Thr phosphatase block family). This indicates that this match is probably artifact. The same is true for the other matching block family shown, a ribosomal protein.

Detection of repeats

Sometimes a single block will be reported to align at multiple positions with high scores. These might represent repeats in the query sequence. If repeats are characteristic of a family, as documented in PROSITE, then the Blocks Searcher will show the putative repeats aligned. Exceptions are (1) cases in which the block aligns with overlapping positions in the query; (2) cases in which the number of putative repeats detected exceeds the maximum number of repeats seen for a member of the PROSITE group (MAX-REPEAT value), in which case only the maximum number of alignments is shown; and (3) cases in which no putative repeats are detected for the highest ranking block in the group.

There are two cautions that should be kept in mind when interpreting putative repeats reported by Blocks. One is that compositional bias can inflate the score of the occasional block with a similar compositional bias. This will often appear as a repeat, with the block aligned at multiple positions along the compositionally biased segment. Another is that a chance hit will be within a repeated region of the query, leading to a report of multiple alignments within the query. The repeat is real; homology is not.

■ Domain

Another analysis tool to look for pattern matches in a query sequence is Domain. Domain uses sequences derived from the SWISS-PROT protein sequence database. This analysis is based on a BLAST search, which returns those database sequences that present the best level of similarity. From this analysis, the most likely domains are determined and ranked based on the relative strength of the match.

As with the results from Blocks, the list of domain matches provided by Domain should be considered as only potential homologies. Whether or not the reported matches are biologically relevant is another question. The relevance of these results depends on several factors, including the quality of the annotations for each sequence in the SWISS-PROT database, the quality of the BLAST search, and the effectiveness of the ranking algorithm, FTHOM, which is used by Domain to analyze matching domains. Therefore, there is no guarantee that all matching domains found are biologically correct.

What this means in practical terms is that if your query sequence includes a domain that is not annotated in the SWISS-PROT database, then the list of best matching sequences reported by the Domain server may be meaningless. However, one advantage of the Domain server over a server such as Blocks is that the parameters for the BLAST analysis performed by Domain can be modified to search for more distant (or stringent) similarities. Thus, failure to find a domain type annotated in SWISS-PROT might suggest that the sensitivity of the BLAST search be adjusted.

Figure 8.4 shows both the query to and the results from an analysis of the PPX protein sequence using the Domain server. In this example, the default settings are used to simplify the query message (Fig. 8.4A). The sequence is in FastA format (as with the query to the Blocks server). The reply (Fig. 8.4B) is also straightforward and has some of the features as the reply from the BLAST server. The reply first lists the domain features that have the highest relative similarity with the query sequence. The information in the line descriptions of the similar sequences can be rather cryptic and may not be of much use. In this example, a specific domain called "CATALYTIC" clearly has the highest overall score of the seven matching domains reported back by the server, but obviously this description tells little about the biological meaning of the match: catalytic for what? The next two parts of the reply message, the line listing of the best matching sequences that were identified in the BLAST search of the SWISS-PROT database and the actual alignments themselves, are of more use in discerning the meaning of the analysis.

Figure 8.4. Pattern Analysis Using Domain

A: Query message to the Domain server

```
Date:   Sat 12 Oct 1996 - 12:49.19 - GMT
From:   annep@acs.auc.eun.eg
To:     domain@hubi.abc.hu
Subj:


BEGIN
>PPX CATALYTIC SUBUNIT, HUMAN, ACC NO. X70218, S55208
MAEISDLDRQIEQLRRCELIKESEVKALCAKAREILVEESNVQRVDSPVTVCGDIHGQFYDLKELFRVGGDV
PERNYLFMGDFVDRGFYSVETFLLLLALKVRYPDRITLIRGNHESRQITQVYGFYDECLRKYGSVTVWRYCT
EIFDYLSLSAIIDGKIFCVHGGLSPSIQTLDQIRTIDRKQEVPHDGPMCDLLWSDPEDTTGWGVSPRGAGYL
FGSDVVAQFNAANDIDMICRAHQLVMEGYKWHFNETVLTVWSAPNYCYRCGNVAAILELDEHLQKDFIIFEA
APQETRGIPSKKPVADYFL
```

B: Reply message from the Domain server

```
Date: Sun, 13 Oct 1996 11:02:40 +0200 (MET DST)
From: FTHOM domain homology server <domain@abc.hu>
To: ANNEP@acs.auc.eun.eg
Subject: Fthom24818

Number of entries:70

Query: PPX
```

continued

Figure 8.4 continued

```
Feature                                             FT    score
name                                                freq  sum

CATALYTIC                                            23   1996
DNA-BINDING; ZN-FINGER                                2     68
DNA-BINDING; ZN-FINGER; C2HC                          2     56
CALPAIN-3-DOMAIN                                      1     30
SIGNAL                                                1     22
PRPP-BINDING                                          1     14
CALCINEURIN-B-BINDING                                 4      4


The FTHOM search is based on the following BLAST result:


Query=  PPX
        (307 letters)


Database: swissprot.FASTA
           49,340 sequences; 17,385,503 total letters.
Searching................................................done


                                                          Smallest
                                                          Poisson
                                                   High   Probability
Sequences producing High-scoring Segment Pairs:    Score  P(N)       N

PPX_HUMAN  SERINE/THREONINE PROTEIN PHOSPHATASE PP-X (EC ... 1705  1.4e-247  1
PPX_RABIT  SERINE/THREONINE PROTEIN PHOSPHATASE PP-X (EC ... 1687  6.2e-245  1
P2A1_ARATH SERINE/THREONINE PROTEIN PHOSPHATASE PP2A-1 CA... 1149  1.1e-165  1
P2A2_ARATH SERINE/THREONINE PROTEIN PHOSPHATASE PP2A-2 CA... 1138  4.7e-164  1
P2AB_HUMAN SERINE/THREONINE PROTEIN PHOSPHATASE PP2A-BETA... 1107  1.7e-159  1
P2AB_RABIT SERINE/THREONINE PROTEIN PHOSPHATASE PP2A-BETA... 1107  1.7e-159  1
P2AA_HUMAN SERINE/THREONINE PROTEIN PHOSPHATASE PP2A-ALPH... 1106  2.4e-159  1
P2AA_RAT   SERINE/THREONINE PROTEIN PHOSPHATASE PP2A-ALPH... 1106  2.4e-159  1
P2A_DROME  SERINE/THREONINE PROTEIN PHOSPHATASE PP2A (EC ... 1098  3.7e-158  1
P2A_BRANA  SERINE/THREONINE PROTEIN PHOSPHATASE PP2A CATA... 1097  5.2e-158  1
P2A2_SCHPO MAJOR SERINE/THREONINE PROTEIN PHOSPHATASE PP2... 1096  6.9e-158  1
P2A3_ARATH SERINE/THREONINE PROTEIN PHOSPHATASE PP2A-3 CA... 1096  7.3e-158  1
P2A_MEDSA  SERINE/THREONINE PROTEIN PHOSPHATASE PP2A CATA... 1089  7.7e-157  1
P2A1_SCHPO MINOR SERINE/THREONINE PROTEIN PHOSPHATASE PP2... 1076  6.4e-155  1
P2AB_PIG   SERINE/THREONINE PROTEIN PHOSPHATASE PP2A-BETA... 1065  2.8e-153  1
P2A3_YEAST SERINE/THREONINE PROTEIN PHOSPHATASE PPH3 (EC ... 1037  3.6e-149  1
P2A2_YEAST SERINE/THREONINE PROTEIN PHOSPHATASE PP2A-2 CA... 1025  1.7e-147  1
P2A1_YEAST SERINE/THREONINE PROTEIN PHOSPHATASE PP2A-1 CA... 1025  1.7e-147  1
PPE1_SCHPO SERINE/THREONINE PROTEIN PHOSPHATASE PPE1 (EC ...  930  2.1e-133  1
PP11_YEAST SERINE/THREONINE PROTEIN PHOSPHATASE PP1-1 (EC...  716  6.9e-102  1
P2A4_YEAST SERINE/THREONINE PROTEIN PHOSPHATASE PP2A-LIKE...  430  1.3e-89   3
PP1_ARATH  SERINE/THREONINE PROTEIN PHOSPHATASE PP1 (EC 3...  374  1.1e-82   2
PP1B_DROME SERINE/THREONINE PROTEIN PHOSPHATASE PP1-87B (...  390  9.1e-50   1
PP1G_HUMAN SERINE/THREONINE PROTEIN PHOSPHATASE PP1-GAMMA...  388  2.0e-49   1
PP1G_RAT   SERINE/THREONINE PROTEIN PHOSPHATASE PP1-GAMMA...  388  2.0e-49   1
PP1A_HUMAN SERINE/THREONINE PROTEIN PHOSPHATASE PP1-ALPHA...  388  2.1e-49   1
PP1G_XENLA SERINE/THREONINE PROTEIN PHOSPHATASE PP1-GAMMA...  385  5.8e-49   1
PP12_RABIT SERINE/THREONINE PROTEIN PHOSPHATASE PP1-ALPHA...  382  1.5e-48   1
PP11_TRYBB SERINE/THREONINE PROTEIN PHOSPHATASE PP1(4.8) ...  269  4.6e-48   3
PP1_EMENI  SERINE/THREONINE PROTEIN PHOSPHATASE PP1 (EC 3...  379  4.6e-48   1
PP12_TRYBB SERINE/THREONINE PROTEIN PHOSPHATASE PP1(5.9) ...  269  1.3e-47   3
PP1B_HUMAN SERINE/THREONINE PROTEIN PHOSPHATASE PP1-BETA ...  366  4.3e-46   1
PP11_SCHPO SERINE/THREONINE PROTEIN PHOSPHATASE PP1-1 (EC...  364  8.6e-46   1
PP12_YEAST SERINE/THREONINE PROTEIN PHOSPHATASE PP1-2 (EC...  362  1.6e-45   1
PP1_MAIZE  SERINE/THREONINE PROTEIN PHOSPHATASE PP1 (EC 3...  360  3.3e-45   1
PPY_DROME  SERINE/THREONINE PROTEIN PHOSPHATASE PP-Y (EC ...  240  2.2e-44   2
P2B1_HUMAN SERINE/THREONINE PROTEIN PHOSPHATASE 2B CATALY...  195  5.0e-44   2
P2B2_HUMAN SERINE/THREONINE PROTEIN PHOSPHATASE 2B CATALY...  287  5.1e-44   2
```

continued

Figure 8.4 continued

```
P2B2_RAT    SERINE/THREONINE PROTEIN PHOSPHATASE 2B CATALY...   287   5.1e-44   2
PPZ2_YEAST  SERINE/THREONINE PROTEIN PHOSPHATASE PP-Z2 (EC...   276   7.8e-44   2
PPZ1_YEAST  SERINE/THREONINE PROTEIN PHOSPHATASE PP-Z1 (EC...   283   1.2e-42   2
PP12_SCHPO  SERINE/THREONINE PROTEIN PHOSPHATASE PP1-2 (EC...   343   1.2e-42   1
P2B1_RAT    SERINE/THREONINE PROTEIN PHOSPHATASE 2B CATALY...   301   3.1e-42   2
PPQ1_YEAST  SERINE/THREONINE PROTEIN PHOSPHATASE PPQ (EC 3...   290   1.3e-41   2
P2B2_YEAST  SERINE/THREONINE PROTEIN PHOSPHATASE 2B CATALY...   255   8.7e-37   2
P2B1_YEAST  SERINE/THREONINE PROTEIN PHOSPHATASE 2B CATALY...   255   2.4e-35   2
PP1_BRANA   SERINE/THREONINE PROTEIN PHOSPHATASE PP1 (EC 3...   262   1.4e-30   1
YK84_CAEEL  HYPOTHETICAL 43.0 KD PROTEIN C30A5.4 IN CHROMO...   117   1.3e-21   3
RDGC_DROME  SERINE/THREONINE PROTEIN PHOSPHATASE RDGC (EC ...   124   1.1e-12   2
YL39_CAEEL  HYPOTHETICAL 25.0 KD PROTEIN F44B9.9 IN CHROMO...    68   6.5e-05   2
MEDB_GIALA  MEDIAN BODY PROTEIN.                                 47   0.44      2
PYP2_SCHPO  PROTEIN-TYROSINE PHOSPHATASE 2 (EC 3.1.3.48) (...    63   0.60      1
KPRS_ECOLI  RIBOSE-PHOSPHATE PYROPHOSPHOKINASE (EC 2.7.6.1...    43   0.998     3
CAN3_PIG    CALPAIN P94, LARGE (CATALYTIC) SUBUNIT (EC 3.4...    57   0.999     1
UHPT_ECOLI  HEXOSE PHOSPHATE TRANSPORT PROTEIN.                  57   0.999     1
UHPT_SALTY  HEXOSE PHOSPHATE TRANSPORT PROTEIN.                  57   0.999     1
CP51_YEAST  CYTOCHROME P450 L1 (14DM) (LANOSTEROL 14-ALPHA...    57   0.9990    1
ILV2_TOBAC  ACETOLACTATE SYNTHASE II PRECURSOR (EC 4.1.3.1...    56   0.99995   1
ILV1_TOBAC  ACETOLACTATE SYNTHASE I PRECURSOR (EC 4.1.3.18...    56   0.99995   1
YMR1_CAEEL  PUTATIVE ATP-DEPENDENT DNA HELICASE K02F3.1 IN...    55   0.999999  1
RRPL_SV41   RNA POLYMERASE BETA SUBUNIT (EC 2.7.7.48) (LAR...    40   1.000000  3
EAR_ASFB7   APOPTOSIS REGULATOR BCL-2 HOMOLOG PRECURSOR.         54   1.000000  1
EAR_ASFE4   APOPTOSIS REGULATOR BCL-2 HOMOLOG PRECURSOR.         54   1.000000  1
EAR_ASFM2   APOPTOSIS REGULATOR BCL-2 HOMOLOG PRECURSOR.         54   1.000000  1
CP51_SCHPO  PUTATIVE CYTOCHROME P450 L1 (LANOSTEROL 14-ALP...    45   1.000000  2
CP51_CANAL  CYTOCHROME P450 L1 (P450-L1A1) (LANOSTEROL 14-...    54   1.000000  1
DHSA_ECOLI  SUCCINATE DEHYDROGENASE FLAVOPROTEIN SUBUNIT (...    54   1.000000  1
TBP2_HAEIN  PROBABLE TRANSFERRIN BINDING PROTEIN 2 PRECURSOR.    54   1.000000  1
HEMA_IADMA  HEMAGGLUTININ PRECURSOR (FRAGMENT).                  53   1.000000  1
CNBP_HUMAN  CELLULAR NUCLEIC ACID BINDING PROTEIN (CNBP).        44   1.000000  2


>PPX_HUMAN SERINE/THREONINE PROTEIN PHOSPHATASE PP-X (EC 3.1.3.16).
          Length = 307
 Score = 1705 (834.4 bits), Expect = 1.4e-247, P = 1.4e-247
 Identities = 307/307 (100%), Positives = 307/307 (100%)


Query:    1 MAEISDLDRQIEQLRRCELIKESEVKALCAKAREILVEESNVQRVDSPVTVCGDIHGQFY 60
            MAEISDLDRQIEQLRRCELIKESEVKALCAKAREILVEESNVQRVDSPVTVCGDIHGQFY
Sbjct:    1 MAEISDLDRQIEQLRRCELIKESEVKALCAKAREILVEESNVQRVDSPVTVCGDIHGQFY 60


Query:   61 DLKELFRVGGDVPERNYLFMGDFVDRGFYSVETFLLLLALKVRYPDRITLIRGNHESRQI 120
            DLKELFRVGGDVPERNYLFMGDFVDRGFYSVETFLLLLALKVRYPDRITLIRGNHESRQI
Sbjct:   61 DLKELFRVGGDVPERNYLFMGDFVDRGFYSVETFLLLLALKVRYPDRITLIRGNHESRQI 120


Query:  121 TQVYGFYDECLRKYGSVTVWRYCTEIFDYLSLSAIIDGKIFCVHGGLSPSIQTLDQIRTI 180
            TQVYGFYDECLRKYGSVTVWRYCTEIFDYLSLSAIIDGKIFCVHGGLSPSIQTLDQIRTI
Sbjct:  121 TQVYGFYDECLRKYGSVTVWRYCTEIFDYLSLSAIIDGKIFCVHGGLSPSIQTLDQIRTI 180


Query:  181 DRKQEVPHDGPMCDLLWSDPEDTTGWGVSPRGAGYLFGSDVVAQFNAANDIDMICRAHQL 240
            DRKQEVPHDGPMCDLLWSDPEDTTGWGVSPRGAGYLFGSDVVAQFNAANDIDMICRAHQL
Sbjct:  181 DRKQEVPHDGPMCDLLWSDPEDTTGWGVSPRGAGYLFGSDVVAQFNAANDIDMICRAHQL 240


Query:  241 VMEGYKWHFNETVLTVWSAPNYCYRCGNVAAILELDEHLQKDFIIFEAAPQETRGIPSKK 300
            VMEGYKWHFNETVLTVWSAPNYCYRCGNVAAILELDEHLQKDFIIFEAAPQETRGIPSKK
Sbjct:  241 VMEGYKWHFNETVLTVWSAPNYCYRCGNVAAILELDEHLQKDFIIFEAAPQETRGIPSKK 300


Query:  301 PVADYFL 307
            PVADYFL
Sbjct:  301 PVADYFL 307
```

continued

Figure 8.4 continued

```
>P2AA_HUMAN SERINE/THREONINE PROTEIN PHOSPHATASE PP2A-ALPHA, CATALYTIC SUBUNIT
           Length = 309

 Score = 1106 (541.3 bits), Expect = 2.4e-159, P = 2.4e-159
 Identities = 194/286 (67%), Positives = 239/286 (83%)

Query:    6 DLDRQIEQLRRCELIKESEVKALCAKAREILVEESNVQRVDSPVTVCGDIHGQFYDLKEL 65
            +LD+ IEQL  C +  ES+VK+LC KA+EIL  ESNVQ V  PVTVCGD+HGQF DL EL
Sbjct:    9 ELDQWIEQLNECKQLSESQVKSLCEKAKEILTKESNVQEVRCPVTVCGDVHGQFHDLMEL 68

Query:   66 FRVGGDVPERNYLFMGDFVDRGFYSVETFLLLLALKVRYPDRITLIRGNHESRQITQVYG 125
            FR+GG  P+ NYLFMGD+VDRG+YSVET  LL+ALKVRY +RIT++RGNHESRQITQVYG
Sbjct:   69 FRIGGKSPDTNYLFMGDYVDRGYYSVETVTLLVALKVRYRERITILRGNHESRQITQVYG 128

Query:  126 FYDECLRKYGSVTVWRYCTEIFDYLSLSAIIDGKIFCVHGGLSPSIQTLDQIRTIDRKQE 185
            FYDECLRKYG+  VW+Y T++FDYL+L+A++DG IFC+HGGLSPSI+TLD+IR++DR QE
Sbjct:  129 FYDECLRKYGNANVWKYFTDLFDYLPLTALVDGQIFCLHGGLSPSIDTLDHIRALDRLQE 188

Query:  186 VPHDGPMCDLLWSDPEDTTGWGVSPRGAGYLFGSDVVAQFNAANDIDMICRAHQLVMEGY 245
            VPH+GPMCDLLWSDP+D  GWG+SPRGAGY FG D+   FN AN + ++ RAHQLVMEGY
Sbjct:  189 VPHEGPMCDLLWSDPDDRGGWGISPRGAGYTFGQDISETFNHANGLTLVSRAHQLVMEGY 248

Query:  246 KWHFNETVLTVWSAPNYCYRCGNVAAILELDEHLQKDFIIFEAAPQ 291
            +W +  V+T+ SAPNYCYRCGN AAI+ELD+ L   F+ F++AP+
Sbjct:  249 NWCHDRNVVTIFSAPNYCYRCGNQAAIMELDDTLKYSFLQFDPAPR 294
```

The results from the Domain server show again that the query sequence is the PPX catalytic subunit. Two of the alignments contained in the reply message are displayed in Fig. 8.4B. The first alignment of the query sequence is with PPX from the Domain server database. The match is perfect, with the query sequence being completely identical to the entry from the database. In contrast, the other match shown (this one with PP2A is 63% identical and 83% conserved) suggests that these two proteins are similar and perhaps homologous.

■ MotifFinder

The last pattern analysis of the predicted protein sequence was done with MotifFinder. As described in Chapter 6, the MotifFinder server is a useful tool to screen sequences for entries in the PROSITE database. The query message shown in Fig. 8.5A sends the sequence to the server for analysis against the default PROSITE library. As expected, the server reports back that the only matching pattern in the PROSITE database is for a signature sequence associated with a Ser/Thr protein phosphatase. A portion of this reply message is shown in Fig. 8.5B.

Figure 8.5. Pattern Analysis Using MotifFinder

A: Query message to the MotifFinder server

```
Date:   Fri 05 Jul 1996 - 09:16.02 - GMT
From:   annep@acs.auc.eun.eg
To:     motif@genome.ad.jp
Subj:

>PPX CATALYTIC SUBUNIT, HUMAN, ACC NO. X70218, S55208
MAEISDLDRQIEQLRRCELIKESEVKALCAKAREILVEESNVQRVDSPVTVCGDIHGQFYDLKELFRVGGDV
PERNYLFMGDFVDRGFYSVETFLLLLALKVRYPDRITLIRGNHESRQITQVYGFYDECLRKYGSVTVWRYCT
EIFDYLSLSAIIDGKIFCVHGGLSPSIQTLDQIRTIDRKQEVPHDGPMCDLLWSDPEDTTGWGVSPRGAGYL
FGSDVVAQFNAANDIDMICRAHQLVMEGYKWHFNETVLTVWSAPNYCYRCGNVAAILELDEHLQKDFIIFEA
APQETRGIPSKKPVADYFL
```

B: Reply message from the MotifFinder server

```
Date:           Sat 06 Jul 96 - 23:01:57 - JST
From:           MotifFinder@genome.ad.jp
To:             annep@acs.auc.eun.eg
Subj:           result of MotifFinder
Errors-To:      MotifFinder-Manager@genome.ad.jp
Reply-To:       MotifFinder-Manager@genome.ad.jp


--------------------<Cut Here>--------------------
#! /bin/csh
# Result of PrositeFind
################################################################
cat <<'==PRO_END==' >MotifFind.0
PROSITE_ID:     SER_THR_PHOSPHATASE
Accession:      PS00125
Definition:     Serine/threonine specific protein phosphatases signature.
Pattern:        [LIVM]-R-G-N-H-E.


Sequence:
        MAEISDLDRQIEQLRRCELIKESEVKALCAKAREILVEESNVQRVDSPVTVCGDIHGQFY

        DLKELFRVGGDVPERNYLFMGDFVDRGFYSVETFLLLLALKVRYPDRITLIRGNHESRQI
                                                              ++++++
        TQVYGFYDECLRKYGSVTVWRYCTEIFDYLSLSAIIDGKIFCVHGGLSPSIQTLDQIRTI

        DRKQEVPHDGPMCDLLWSDPEDTTGWGVSPRGAGYLFGSDVVAQFNAANDIDMICRAHQL

        VMEGYKWHFNETVLTVWSAPNYCYRCGNVAAILELDEHLQKDFIIFEAAPQETRGIPSKK

        PVADYFL


'==PRO_END=='
```

Protein sequences: secondary structure prediction using nnPredict and Predict Protein

Based on the results from the initial analysis of the query sequence using BLAST and supported by the results from Blocks, Domain, and MotifFinder, it is clear that the query sequence used in the example in this chapter is a protein Ser/Thr phosphatase catalytic subunit, PPX. The next question that is to be addressed is the predicted secondary structure of this protein. To address this, the nnPredict and PredictProtein servers can be used in parallel to generate a consensus picture of the protein.

Secondary structure prediction has been around for almost a quarter of a century. The early methods suffered from a lack of data on which to base predictions. As a result, most predictions were based on single sequences rather than families of homologous sequences. In addition, there were relatively few known 3D structures of protein from which to derive a framework to base prediction rules. Because of these limitations, the first prediction methods were no more than 50% to 60% accurate.

The availability of large families of homologous sequences revolutionized secondary structure prediction. Traditional methods, when applied to a family of proteins rather than a single sequence, proved much more accurate at identifying core secondary structure elements. The combination of sequence data with sophisticated computing techniques (such as neural networks) led to prediction accuracies in excess of 70%. Though this seems a small percentage of increase, these predictions are actually much more useful, since they tend to predict the core of the protein structure accurately. Moreover, the limit of 70% to 80% may be a function of secondary structure variation within homologous proteins rather than a limitation of the actual prediction method.

■ nnPredict

As mentioned, two servers are highly useful in the prediction of secondary structure, nnPredict and PredictProtein. For the first step in determining a hypothetical secondary structure for PPX, the protein sequence is sent for analysis to nnPredict. Figure 8.6 shows the query message sent to the server and the reply that was received back after the completion of the analysis.

Figure 8.6. Secondary Structure Prediction Using nnPredict

A: Query message to the nnPredict server

```
Date:   Sat 12 Oct 1996 - 12:49.19 - GMT
From:   annep@acs.auc.eun.eg
To:     nnpredict@celeste.ucsf.edu
Subj:
>PPX CATALYTIC SUBUNIT, HUMAN, ACC NO. X70218, S55208
MAEISDLDRQIEQLRRCELIKESEVKALCAKAREILVEESNVQRVDSPVTVCGDIHGQFYDLKELFRVGGDV
PERNYLFMGDFVDRGFYSVETFLLLLALKVRYPDRITLIRGNHESRQITQVYGFYDECLRKYGSVTVWRYCT
EIFDYLSLSAIIDGKIFCVHGGLSPSIQTLDQIRTIDRKQEVPHDGPMCDLLWSDPEDTTGWGVSPRGAGYL
FGSDVVAQFNAANDIDMICRAHQLVMEGYKWHFNETVLTVWSAPNYCYRCGNVAAILELDEHLQKDFIIFEA
APQETRGIPSKKPVADYFL
```

continued

Figure 8.6 continued

B: Reply message from the nnPredict server

```
Date:   Sat 12 Oct 1996 - 12:49.19 - GMT
From:   nnpredict@celeste.ucsf.edu (the nnpredict server)
To:     annep@acs.auc.eun.eg
Subj:   Reply to your nnpredict query

Sequence:
>PPX CATALYTIC SUBUNIT, HUMAN, ACC NO. X70218, S55208
MAEISDLDRQIEQLRRCELIKESEVKALCAKAREILVEESNVQRVDSPVTVCGDIHGQFYDLKELFRVGGDV
PERNYLFMGDFVDRGFYSVETFLLLLALKVRYPDRITLIRGNHESRQITQVYGFYDECLRKYGSVTVWRYCT
EIFDYLSLSAIIDGKIFCVHGGLSPSIQTLDQIRTIDRKQEVPHDGPMCDLLWSDPEDTTGWGVSPRGAGYL
FGSDVVAQFNAANDIDMICRAHQLVMEGYKWHFNETVLTVWSAPNYCYRCGNVAAILELDEHLQKDFIIFEA
APQETRGIPSKKPVADYFL

Secondary structure prediction
(option n):
-------HHHHHHH-HHH-H-HHHHHHHHHHHHHHHHHH----------EEE-------HHHHHHHH-----
----HHE----------EHHHHHHHHHHH-------EEEE------HHEEE----HHHHH-----EEEEEEH
HHHHHH---EEE---EEEE--------------H----------------------------------EE
E---HHHHH-----HHHHHHHHHHHHHHH-------HEEE-------E-------HHHHHHHHH--HEEE--
------------------

(H = helix, E = strand, - = no prediction)
```

Replies from the nnPredict server are compact and concise. The query sequence is echoed back to show what was analyzed by the server. This is followed by the prediction of the secondary structure. The predicted type will be either H (a helix element); E (a beta-strand element); or - (a turn element). Note that if the query sequence were to contain any characters that are not standard amino acids, ?s would be used in the reply from the server to indicate that no prediction could be made in that region of the query sequence.

■ PredictProtein

After analysis with the nnPredict server, the same sequence was sent to the Predict-Protein server for prediction of secondary structure. Figure 8.7 shows the query message sent to the server and a small portion of the reply message received back after the analysis had been completed.

Figure 8.7. Secondary Structure Prediction Using PredictProtein

A: Query message to the PredictProtein server

```
Date:   Sat 12 Oct 1996 - 12:49.19 - GMT
From:   annep@acs.auc.eun.eg
To:     predictprotein@embl-heidelberg.de
Subj:
```

continued

Figure 8.7 continued

```
Anne H. Peruski
Biology Section, American University in Cairo
annep@acs.auc.eun.eg
# PPX CATALYTIC SUBUNIT, HUMAN, ACC NO. X70218, S55208
MAEISDLDRQIEQLRRCELIKESEVKALCAKAREILVEESNVQRVDSPVTVCGDIHGQFYDLKELFRVGGDV
PERNYLFMGDFVDRGFYSVETFLLLLALKVRYPDRITLIRGNHESRQITQVYGFYDECLRKYGSVTVWRYCT
EIFDYLSLSAIIDGKIFCVHGGLSPSIQTLDQIRTIDRKQEVPHDGPMCDLLWSDPEDTTGWGVSPRGAGYL
FGSDVVAQFNAANDIDMICRAHQLVMEGYKWHFNETVLTVWSAPNYCYRCGNVAAILELDEHLQKDFIIFEA
APQETRGIPSKKPVADYFL
```

B: Reply message from the PredictProtein server

```
Date:    Sun, 13 Oct 1996 11:05:45 GMT
From:    <phd@EMBL-Heidelberg.de>
To:      ANNEP@acs.auc.eun.eg
Subject: Predict-Protein

The following information has been received by the server:
~~~~~~~~~~~~~~~~~~~~~~~~~~~~~~~~~~~~~~~~~~~~~~~~~~~~~~~

_____
reference predict_e25968 (Sun Oct 13 12:58:08 MDT 1996)
from ANNEP@acs.auc.eun.eg
resp MAIL
orig MAIL
Anne H. Peruski
Biology Section, American University in Cairo
annep@acs.auc.eun.eg
# PPX CATALYTIC SUBUNIT, HUMAN, ACC NO. X70218, S55208
 MAEISDLDRQIEQLRRCELIKESEVKALCAKAREILVEESNVQRVDSPVTVCGDIHGQFYDLKELFRVGGDV
PERNYLFMGDFVDRGFYSVETFLLLLALKVRYPDRITLIRGNHESRQITQVYGFYDECLRKYGSVTVWRYCT
EIFDYLSLSAIIDGKIFCVHGGLSPSIQTLDQIRTIDRKQEVPHDGPMCDLLWSDPEDTTGWGVSPRGAGYL
FGSDVVAQFNAANDIDMICRAHQLVMEGYKWHFNETVLTVWSAPNYCYRCGNVAAILELDEHLQKDFIIFEA
APQETRGIPSKKPVADYFL
_____

The sequence had been interpreted as being:
~~~~~~~~~~~~~~~~~~~~~~~~~~~~~~~~~~~~~~~~~~~~~

_____
>P1; /home/phd/server/work/predict_e25968
(#)  ppx catalytic subunit, human, acc no. x70218, s55208
MAEISDLDRQIEQLRRCELIKESEVKALCAKAREILVEESNVQRVDSPVTVCGDIHGQFY
DLKELFRVGGDVPERNYLFMGDFVDRGFYSVETFLLLLALKVRYPDRITLIRGNHESRQI
TQVYGFYDECLRKYGSVTVWRYCTEIFDYLSLSAIIDGKIFCVHGGLSPSIQTLDQIRTI
DRKQEVPHDGPMCDLLWSDPEDTTGWGVSPRGAGYLFGSDVVAQFNAANDIDMICRAHQL
VMEGYKWHFNETVLTVWSAPNYCYRCGNVAAILELDEHLQKDFIIFEAAPQETRGIPSKK
PVADYFL
_____


PHD output for your protein
~~~~~~~~~~~~~~~~~~~~~~~~~~~~

Sun Oct 13 13:05:23 1996
Jury on:      10    different architectures (version   5.94_317 ).
Note: differently trained architectures, i.e., different versions can
result in different predictions.
```

continued

Figure 8.7 continued

```
About the protein
~~~~~~~~~~~~~~~~~


HEADER        /home/phd/server/work/predict_e25968_240
COMPND
SOURCE
AUTHOR
SEQLENGTH   307
NCHAIN          1 chain(s) in predict_e25968_24068 data set
NALIGN         84 (=number of aligned sequences in HSSP file)


Abbreviations: PHDsec
~~~~~~~~~~~~~~~~~~~~~


sequence:
AA : amino acid sequence
secondary structure:
HEL: H=helix, E=extended (sheet), blank=other (loop)
PHD: Profile network prediction HeiDelberg
Rel: Reliability index of prediction (0-9)
detail:
prH: 'probability' for assigning helix
prE: 'probability' for assigning strand
prL: 'probability' for assigning loop
note: the 'probabilites' are scaled to the interval 0-9, e.g.,
prH=5 means, that the first output node is 0.5-0.6
subset:
SUB: a subset of the prediction, for all residues with an expected
average accuracy > 82% (tables in header)
note: for this subset the following symbols are used:
L: is loop (for which above " " is used)
".": means that no prediction is made for this residue, as the
reliability is: Rel < 5


Abbreviations: PHDacc
~~~~~~~~~~~~~~~~~~~~~


solvent accessibility:
3st: relative solvent accessibility (acc) in 3 states:
b = 0-9%, i = 9-36%, e = 36-100%.
PHD: Profile network prediction HeiDelberg
Rel: Reliability index of prediction (0-9)
P_3: predicted relative accessibility in 3 states
note: for convenience a blank is used intermediate (i).
10st:relative accessibility in 10 states:
= n corresponds to a relative acc. of n*n %
subset:
SUB: a subset of the prediction, for all residues with an expected
 average correlation > 0.69 (tables in header)
note: for this subset the following symbols are used:
"I": is intermediate (for which above " " is used)
".": means that no prediction is made for this residue, as the
reliability is: Rel < 4


Abbreviations: PHDhtm
~~~~~~~~~~~~~~~~~~~~~


secondary structure:
HL: T=helical transmembrane region, blank=other (loop)
PHD: Profile network prediction HeiDelberg
PHDF:filtered prediction, i.e., too long transmembrane segments
```

continued

Figure 8.7 continued

```
are split, too short ones are deleted
Rel: Reliability index of prediction (0-9)
detail:
prH: 'probability' for assigning helical transmembrane region
prL: 'probability' for assigning loop
note: the 'probabilites' are scaled to the interval 0-9, e.g.,
prH=5 means, that the first output node is 0.5-0.6
subset:
SUB: a subset of the prediction, for all residues with an expected
average accuracy > 82% (tables in header)
note: for this subset the following symbols are used:
L: is loop (for which above " " is used)
".": means that no prediction is made for this residue, as the
 reliability is: Rel < 5




protein:        predict        length      307


                  ....,....1....,....2....,....3....,....4....,....5....,....6
           AA   |MAEISDLDRQIEQLRRCELIKESEVKALCAKAREILVEESNVQRVDSPVTVCGDIHGQFY|
           PHD sec |    HHHHHHHHHH    HHHHHHHHHHHHHHHHHH    EE    EEEE   HHHH|
           Rel sec |9997632899999953267659999999999999999964884253899267653213699|
detail:
           prH sec |0001235899999966421129999999999999999873101110000000011346799|
           prE sec |0000000000000000000000000000000000000000000002463100378762000000|
           prL sec |9998764000000023577770000000000000000000268853258985111125543100|
subset: SUB sec |LLLLL..HHHHHHHH..LLLLHHHHHHHHHHHHHHHHHH.LL..E.LLL.EEEE....HHH|

ACCESSIBILITY
 3st:   P_3 acc |eee eebeebbeebeeeeebeeeebeebbeebeebbee  eebeebebbbebbbbbbbbb |
 10st:  PHD acc |99757706700760677770777606700770670067576067070000600000000004|
           Rel acc |36403431423523162551454152353544244415141214453152877051516161|
subset: SUB acc |.ee..e..e..e....e.ee.eee.b..b.eeb.ebb.e.e...ebe..b.bbb.b.b.|
                  ....,....7....,....8....,....9....,....10....,....11....,....12
           AA   |DLKELFRVGGDVPERNYLFMGDFVDRGFYSVETFLLLLALKVRYPDRITLIRGNHESRQI|
           PHD sec |HHHHHHHH     EEEEE      HHHHHHHHHHHHHHHH    EEEEEE    HHHHH|
           Rel sec |9999999985899988623663314345433589998889876324776887478335778|
detail:
           prH sec |9999999821001100000001223226568898888987754200000000000366778|
           prE sec |0000000000000014577554210000011000001101111111178888610000010|
           prL sec |0000001789988753223345567633200000000010236872110138853211|
subset: SUB sec |HHHHHHHHHLLLLLLL..EE......L...HHHHHHHHHHHH...LLEEEEE.LL..HHHH|

ACCESSIBILITY
 3st:   P_3 acc |bbbebbeebee eebbbbbbbbbbeebeebbebbbbbbbbebebeebbbb b bebeeb|
 10st:  PHD acc |0006007709756700000000000660660060000000060706760000405070670|
           Rel acc |17538652135103114677302411422341695575541340142535512103413|
subset: SUB acc |.bb.bbe...e.....bbbb...b..b...b.bbbbbbbb..e..e.b.bb....b...|
                  ....,....13....,....14....,....15....,....16....,....17....,....18
           AA   |TQVYGFYDECLRKYGSVTVWRYCTEIFDYLSLSAIIDGKIFCVHGGLSPSIQTLDQIRTI|
           PHD sec |HHHH HHHHHHH   HHHHHHHHHHHH   EEEEE   EEEEEE    HHHHHHHHHHHH|
           Rel sec |7522452899987427713679999986166424431324899747998347569999981|
detail:
           prH sec |7644115899886411456789898874211222111100000000001668679999985|
           prE sec |0122210000000001211000000000024555423688876100000000000000000|
           prL sec |11226631001135874211000000247763112455210012899833122000014|
subset: SUB sec |HH...L.HHHHH..LL..HHHHHHHH.LL.........EEEE.LLLL..HHHHHHHHH.|

ACCESSIBILITY
 3st:   P_3 acc |bebbbb eebeee eebebbeebbebbebbbbbbbebebbbbbbbbbbeebeebeeb eb|
 10st:  PHD acc |0600004660667367060070060060000000070600000000770760760560|
           Rel acc |12532611271250040164415011612535583430189943313134241231822|
```

continued

Figure 8.7 continued

```
subset: SUB acc  |..b..b...b..e..e..bbe.b...b..b.bbb.b..bbbb......e.e....b...|
                 ....,....19...,....20...,....21...,....22...,....23...,....24
          AA     |DRKQEVPHDGPMCDLLWSDPEDTTGWGVSPRGAGYLFGSDVVAQFNAANDIDMICRAHQL|
          PHD sec|         EEEEE            E E HHHHHHHHHH HHHHHHHHHH|
          Rel sec|89989999996214432499979999979986321247489999621472899999898|
detail:
          prH sec|00010000000000000000010000000000000001689998875531589998898|
          prE sec|00000000013466552000000000100123545200000000000000000000000|
          prL sec|89989999987542333699989999989987534367200000014468310000001|
subset: SUB sec  |LLLLLLLLLLL.......LLLLLLLLLLLLLLL.....L.HHHHHHH...L.HHHHHHHHH|

ACCESSIBILITY
  3st:   P_3 acc |e eeeb eebbbbbbbbb eeeeee eeeeebbbbbbbeebbeebbeeeebebbbbbbbeb|
 10st:   PHD acc |75777059800000000057778795767760000000770067006776060000060|
         Rel acc |41354106520451663112566440303312432023252414401531425760821 6|
subset: SUB acc  |e..ee..ee..bb.bb....eeeee.....b.....e.b.eb..e..b.bbb.b..b|
                 ....,....25...,....26...,....27...,....28...,....29...,....30
          AA     |VMEGYKWHFNETVLTVWSAPNYCYRCGNVAAILELDEHLQKDFIIFEAAPQETRGIPSKK|
          PHD sec|HHHHHHHH    EEEEEE            EEEEE HHHHHHH          |
          Rel sec|99968842153189997288511122443235765111223112127799999999998 9|
detail:
          prH sec|99978854300000000000024445332321111223445543422000000000000 0|
          prE sec|00000022223589997400000000100125676632111233333100000000000 0|
          prL sec|00021112465410001588644435665432111444332222347899999999988 8|
subset: SUB sec  |HHHHHH...L..EEEEE.LLL..........EEEE...........LLLLLLLLLLLLLLL|

ACCESSIBILITY
  3st:   P_3 acc |beebbebbeeeebbbbbbbbbbbbeebebebbbbbebeeebeebbebbeeeeeeeeeeeee|
 10st:   PHD acc |07700600777700000000000066060700007067706600600777877977777787|
         Rel acc |43451210344457686671014112003476444044311141045426776764346 6|
subset: SUB acc  |b.eb.....eeebbbbbbb...b......bbbbbeb.ee....b..bee.eeeeee.eee|
                 ....,....31...,....32...,....33...,....34...,....35...,....36
          AA     |PVADYFL|
          PHD sec|       |
          Rel sec|9987669|
detail:
          prH sec|0001100|
          prE sec|0000110|
          prL sec|9887679|
subset: SUB sec  |LLLLLLL|

ACCESSIBILITY
  3st:   P_3 acc |e ee bb|
 10st:   PHD acc |7577400|
         Rel acc |4136111|
subset: SUB acc  |e..e...|
```

Unlike the clean, compact nature of the reply from nnPredict, reply messages from PredictProtein are voluminous. While much of the reply does contain useful statistics on the actual analysis (along with a nice multiple sequence alignment of the query sequence with homologous sequences), the reply message is still overly lengthy. Much of the information could be put into a special help file that could be requested from the server if needed. The truly critical part of the analysis, the actual prediction itself, is near the end of the reply message.

Figure 8.7B shows only a portion of the reply message. When merging the results from both nnPredict and PredictProtein to create a consensus secondary structure for PPX, caution in the interpretation of the results is crucial. Pooling the secondary structure predictions from nnPredict and PredictProtein can be done by pasting the results from the reply messages into a word processor file. This file can be edited to generate a consensus prediction for the tendencies of a given amino acid in the sequence to assume a given structure or an approximation of the consensus secondary structure prediction. Such a consensus is shown in Fig. 8.8.

Figure 8.8. Consensus Secondary Structure Tendencies for PPX

```
No                ........10........20........30........40........50........60
AA                MAEISDLDRQIEQLRRCELIKESEVKALCAKAREILVEESNVQRVDSPVTVCGDIHGQFY
PHD                  HHHHHHHHHH      HHHHHHHHHHHHHHHHHH     EE      EEEE     HHHH
nnP               -------HHHHHHH-HHH-H-HHHHHHHHHHHHHHHHHHH----------EEE-------H
Con               -------HHHHHH-H-----HHHHHHHHHHHHHHHHHH----------EEE-------H

No                ........70........80........90.......100......110......120
AA                DLKELFRVGGDVPERNYLFMGDFVDRGFYSVETFLLLLALKVRYPDRITLIRGNHESRQI
PHD               HHHHHHHH        EEEEE          HHHHHHHHHHHHHHHHH     EEEEE      HHHHH
nnP               HHHHHHH---------HHE----------EHHHHHHHHHHH-------EEEE------HH
Con               HHHHHHH-----------E----------HHHHHHHHHHH-------EEEE------HH

No                .......130.......140.......150.......160.......170.......180
AA                TQVYGFYDECLRKYGSVTVWRYCTEIFDYLSLSAIIDGKIFCVHGGLSPSIQTLDQIRTI
PHD               HHHH  HHHHHHH     HHHHHHHHHHH     EEEEE  EEEEEE       HHHHHHHHHHHHH
nnP               EEE----HHHHH-----EEEEEEHHHHHHH---EEE---EEEE--------------H-
Con               -------HHHHH-----------HHHHH-----EEE---EEEE--------------H-

No                .......190.......200.......210.......220.......230.......240
AA                DRKQEVPHDGPMCDLLWSDPEDTTGWGVSPRGAGYLFGSDVVAQFNAANDIDMICRAHQL
PHD                         EEEEE                   E E  HHHHHHHHHH  HHHHHHHHHH
nnP               -------------------------------------EEE---HHHHH-----HHHHHHHHHH
Con               -------------------------------------E----HHHHH-----HHHHHHHHHH

No                .......250.......260.......270.......280.......290.......300
AA                VMEGYKWHFNETVLTVWSAPNYCYRCGNVAAILELDEHLQKDFIIFEAAPQETRGIPSKK
PHD               HHHHHHHH  EEEEE               EEEEE  HHHHHHH
nnP               HHHHH-------HEEE-------E-------HHHHHHHHH--HEEE--------------
Con               HHHHH-------EEE-------------------HHH--H----------------

No                .......
AA                PVADYFL
PHD
nnP               -------
Con               -------
```

The first line is the numbering system for the amino acids. The next line is the amino acid sequence of PPX itself, with the predictions from PredictProtein (PHD) and nnPredict immediately under that. The last line is the consensus results.

The take-home message from this analysis is to use as many state-of-the-art prediction approaches as possible and combine this with some human insight to obtain a consensus prediction for the family. Align all predictions, including multiple sequence alignments, to get an overall picture of the structure. In the process of looking for similarity patterns, a prediction of the secondary structure can be built for most regions of the protein. Note that most methods will agree for many regions of the alignment. By combining the results of several prediction methods, one can create a consensus picture of the secondary structure.

Protein sequences: determination of phylogenetic relationships using MAlign and CBRG

■ Multiple sequence alignment

Because the previous analyses suggest that PPX is a member of a family of homologous sequences, a multiple sequence alignment containing the query sequence and all the putative homologs is critical for this determination. Alignments can provide several kinds of information that are useful in categorizing sequences, such as information on protein domain structure, residues likely to be involved in protein function, residues likely to be buried in the protein core or exposed to solvent, and information useful in homology modeling and secondary structure prediction.

In Chapter 6, we discussed two servers that could be used for the analysis of a set of protein sequences in order to determine family relationships. One is the MAlign server (maintained by DDBJ) and the other is the CBRG server in Zurich. In Fig. 8.9, sequences of all of the known, homologous classes of protein Ser/Thr phosphatases catalytic subunits isolated from humans are submitted to the MAlign server for the creation of a multiple sequence alignment. These sequences were obtained using the Retrieve server. Based on previous reports from the peer-reviewed literature, this family of functional and sequence homologs includes PP1, PP2A, PP2B, and PPX.

Figure 8.9. Multiple Alignment of Sequences Using MAlign

A: Query message to the MAlign server

```
Date:   Sat 12 Oct 1996 - 12:49.19 - GMT
From:   annep@acs.auc.eun.eg
To:     malign@nig.ac.jp
Subj:

ANCES
MOLTYPE PROTEIN
TREE
BEGIN
>PPX sequence 1
MAEISDLDRQIEQLRRCELIKESEVKALCAKAREILVEESNVQRVDSPVTVCGDIHGQFYDLKELFRVGGDVPERNYL
FMGDFVDRGFYSVETFLLLLALKVRYPDRITLIRGNHESRQITQVYGFYDECLRKYGSVTVWRYCTEIFDYLSLSAII
DGKIFCVHGGLSPSIQTLDQIRTIDRKQEVPHDGPMCDLLWSDPEDTTGWGVSPRGAGYLFGSDVVAQFNAANDIDMI
CRAHQLVMEGYKWHFNETVLTVWSAPNYCYRCGNVAAILELDEHLQKDFIIFEAAPQETRGIPSKKPVADYFL
//
>PP1a sequence 2
MSDSEKLNLDSIIGRLLEGSRVLTPHCAPVQGSRPGKNVQLTENEIRGLCLKSREIFLSQPILLELEAPLKICGDIHG
QYYDLLRLFEYGGFPPESNYLFLGDYVDRGKQSLETICLLLAYKIKYPENFFLLRGNHECASINRIYGFYDECKRRYN
IKLWKTFTDCFNCLPIAAIVDEKIFCCHGGLSPDLQSMEQIRRIMRPTDVPDQGLLCDLLWSDPDKDVQGWGENDRGV
SFTFGAEVVAKFLHKHDLDLICRAHQVVEDGYEFFAKRQLVTLFSAPNYCGEFDNAGAMMSVDETLMCSFQILKPADK
NKGKYGQFSGLNPGGRPITPPRNSAKAKK
//
```

continued

Figure 8.9 continued

```
>
PP2Aa sequence 3
MDEKVFTKELDQWIEQLNECKQLSESQVKSLCEKAKEILTKESNVQEVRCPVTVCGDVHGQFHDLMELFRIGGKSPDT
NYLFMGDYVDRGYYSVETVTLLVALKVRYRERITILRGNHESRQITQVYGFYDECLRKYGNANVWKYFTDLFDYLPLT
ALVDGQIFCLHGGLSPSIDTLDHIRALDRLQEVPHEGPMCDLLWSDPDDRGGWGISPRGAGYTFGQDISETFNHANGL
TLVSRAHQLVMEGYNWCHDRNVVTIFSAPNYCYRCGNQAAIMELDDTLKYSFLQFDPAPRRGEPHVTRRTPDYFL
//
>PP2Ba sequence 4
MSEPKAIDPKLSTTDRVVKAVPFPPSHRLTAKEVFDNDGKPRVDILKAHLMKEGRLEESVALRIITEGASILRQEKNL
LDIDAPVTVCGDIHGQFFDLMKLFEVGGSPANTRYLFLGDYVDRGYFSIECVLYLWALKILYPKTLFLLRGNHECRHL
TEYFTFKQECKIKYSERVYDACMDAFDCLPLAALMNQQFLCVHGGLSPEINTLDDIRKLDRFKEPPAYGPMCDILWSD
PLEDFGNEKTQEHFTHNTVRGCSYFYSYPAVCEFLQHNNLLSILRAHEAQDAGYRMYRKSQTTGFPSLITIFSAPNYL
DVYNNKAAVLKYENNVMNIRQFNCSPHPYWLPNFMDVFTWSLPFVGEKVTEMLVNVLNICSDDELGSEEDGFDGATAA
ARKEVIRNKIRAIGKMARVFSVLREESESVLTLKGLTPTGMLPSGVLSGGKQTLQSATVEAIEADEAIKGFSPQHKIT
SFEEAKGLDRINERMPPRRDAMPSDANLNSINKALTSETNGTDSNGSNSSNIQ
//
```

B: Reply message from the MAlign server

```
ALIGNMENT OF SEQUENCES:


PP1a   seq MSD---SEKLNLD------SIIGRLLEGSRVLTPHCAPVQGSRPG---------KNVQLT   42   1
           MSE---SKALDPD------SIKARPLPPSHRLTPHCAPDQGGKPR---------KNGQLT   42   4
PPX    seq MAE---ISDLD---------------------;-----RQIEQLR---------RCELIK   21   0
           MAE---TKDLD----------------------------QQIEQLR---------RCEQLS   21   5
PP2Aa  seq MDEKVFTKELD--------------------------QWIEQLN---------ECKQLS   24   2
           MSE---SKALD--------------------------DQGGQPR---------RCGQLS   21   6
PP2Ba  seq MSE---PKAIDPKLSTTDRVVKAVPFPPSHRLTAKEVFDNDGKPRVDILKAHLMKEGRLE   57   3
                *


PP1a   seq ENEIRGLCLKSREIFLSQPILLELEAPLKICGDIHGQYYDLLRLFEYGGFPPESNYLFLG  102  1
           ESEARGLCTKSREILRQEPNLLEIDAPVTVCGDIHGQFYDLMKLFEVGGSPPETNYLFLG  102  4
PPX    seq ESEVKALCAKAREILVEESNVQRVDSPVTVCGDIHGQFYDLKELFRVGGDVPERNYLFMG   81   0
           ESEVKSLCAKAREILTEESNVQEVDSPVTVCGDIHGQFYDLMELFRVGGDSPETNYLFMG   81   5
PP2Aa  seq ESQVKSLCEKAKEILTKESNVQEVRCPVTVCGDVHGQFHDLMELFRIGGKSPDTNYLFMG   84   2
           ESEARSLCTKSREILTEESNLQEIDSPVTVCGDIHGQFYDLMELFEVGGSSPETNYLFMG   81   6
PP2Ba  seq ESVALRIITEGASILRQEKNLLDIDAPVTVCGDIHGQFFDLMKLFEVGGSPANTRYLFLG  117  3
                 *          *           *   *** ***  **   **   **      *** *


PP1a   seq DYVDRGKQSLETICLLLAYKIKYPENFFLLRGNHECASINRIYGFYDECKRRY-NIKLWK  161  1
           DYVDRGYFSIETVLLLLALKIKYPETLFLLRGNHECRHITEYYGFYDECKRKY-SEKVWK  161  4
PPX    seq DFVDRGFYSVETFLLLLALKVRYPDRITLIRGNHESRQITQVYGFYDECLRKYGSVTVWR  141  0
           DYVDRGYYSVETVLLLLALKVRYPERITLLRGNHESRQITQVYGFYDECLRKYGSATVWK  141  5
PP2Aa  seq DYVDRGYYSVETVTLLVALKVRYRERITILRGNHESRQITQVYGFYDECLRKYGNANVWK  144  2
           DYVDRGYFSIETVLLLLALKIRYPETITLLRGNHECRQITEVYGFYDECKRKY-SATVWK  140  6
PP2Ba  seq DYVDRGYFSIECVLYLWALKILYPKTLFLLRGNHECRHLTEYFTFKQECKIKY-SERVYD  176  3
               * ****  *  *    * * *  *        *****      *   **    *


PP1a   seq TFTDCFNCLPIAAIVDEKIFCCHGGLSPDLQSMEQIRRIMRPTDVPDQGLLCDLLWSDPD  221  1
           TCTDCFDCLPLAAIVDEQIFCVHGGLSPDIQTLDQIRKIDRFKEVPDQGPMCDLLWSDPD  221  4
PPX    seq YCTEIFDYLSLSAIIDGKIFCVHGGLSPSIQTLDQIRTIDRKQEVPHDGPMCDLLWSDPE  201  0
           YCTDIFDYLPLSAIVDGQIFCVHGGLSPSIQTLDQIRTIDRLQEVPHDGPMCDLLWSDPD  201  5
PP2Aa  seq YFTDLFDYLPLTALVDGQIFCLHGGLSPSIDTLDHIRALDRLQEVPHEGPMCDLLWSDPD  204  2
           TCTDCFDCLPLSAIVDGQIFCVHGGLSPSIQTLDQIRTIDRLQEVPDDGPMCDLLWSDPD  200  6
PP2Ba  seq ACMDAFDCLPLAALMNQQFLCVHGGLSPEINTLDDIRKLDRFKEPPAYGPMCDILWSDPL  236  3
                   * *        * * ****** **   *   *   ** *****


PP1a   seq KDVQGWG-------ENDRGVSFTFGAEVVAKFLHKHDLDLICRAHQVVEDGYEFFAKRQ-  273  1
           EDVGGWG-------ETVRGVSYTFGAEVVAKFLHKNDLDLICRAHQAVDDGYRMFRKRQ-  273  4
PPX    seq -DTTGWG-------VSPRGAGYLFGSDVVAQFNAANDIDMICRAHQLVMEGYKWHFNET-  252  0
           -DTGGWG-------VSPRGAGYTFGSDVVATFNHANDLDLICRAHQLVMEGYKWCHNRT-  252  5
PP2Aa  seq -DRGGWG-------ISPRGAGYTFGQDISETFNHANGLTLVSRAHQLVMEGYNWCHDRN-  255  2
           -DTGGWG-------ESPRGASYTFGSDVVATFNHANDLDLICRAHQAVDDGYRMCHNRT-  251  6
PP2Ba  seq EDFGNEKTQEHFTHNTVRGCSYFYSYPAVCEFLQHNNLLSILRAHEAQDAGYRMYRKSQT  296  3
                *           **         *          ***      **
```

continued

Figure 8.9 continued

```
PP1a    seq  -----LVTLFSAPNYCGEFDNAGAMMSVDETLMC------------------------- 302 1
             -----LVTIFSAPNYCGEFDNAAAMMKYDETLMN------------------------- 302 4
PPX     seq  -----VLTVWSAPNYCYRCGNVAAILELDEHLQK------------------------- 281 0
             -----VVTIFSAPNYCYRCGNVAAIMELDETLQK------------------------- 281 5
PP2Aa   seq  -----VVTIFSAPNYCYRCGNQAAIMELDDTLKY------------------------- 284 2
             -----LVTIFSAPNYCGECGNAAAMMELDETLQN------------------------- 280 6
PP2Ba   seq  TGFPSLITIFSAPNYLDVYNNKAAVLKYENNVMNIRQFNCSPHPYWLPNFMDVFTWSLPF 356 3
                  *   *****       *   *

PP1a    seq  ----------------------------------------------------------- 302 1
             ----------------------------------------------------------- 302 4
PPX     seq  ----------------------------------------------------------- 281 0
             ----------------------------------------------------------- 281 5
PP2Aa   seq  ----------------------------------------------------------- 284 2
             ----------------------------------------------------------- 280 6
PP2Ba   seq  VGEKVTEMLVNVLNICSDDELGSEEDGFDGATAAARKEVIRNKIRAIGKMARVFSVLREE 416 3

PP1a    seq  --------------------------------SFQILKPADKNKGKYGQF---------- 320 1
             --------------------------------SFQIIEPADKNKGKSPQH---------- 320 4
PPX     seq  --------------------------------DFIIFEAAPQE---------------- 292 0
             --------------------------------SFLIFEPAPRE---------------- 292 5
PP2Aa   seq  --------------------------------SFLQFDPAPRR---------------- 295 2
             --------------------------------SFQIIEPAPRN---------------- 291 6
PP2Ba   seq  SESVLTLKGLTPTGMLPSGVLSGGKQTLQSATVEAIEADEAIKGFSPQHKITSFEEAKGL 476 3

PP1a    seq  -----------------SGLNPGGRPITPPRNSAKAK-------K 341 1
             -----------------SGLNSGGRPITSERNSADSN-------Q 341 4
PPX     seq  ---------------------TRGIPSKKPVADYF-------L 307 0
             ---------------------GRPIPSKRTVADYF-------L 307 5
PP2Aa   seq  ---------------------GEPHVTRRT-PDYF-------L 309 2
             ---------------------GRPITSERTSADSN-------Q 306 6
PP2Ba   seq  DRINERMPPRRDAMPSDANLNSINKALTSETNGTDSNGSNSSNIQ 521 3

number of completely conserved sites: 79
  Tree:
0:      PPX    sequence 1
1:      PP1a   sequence 2
2:      PP2Aa  sequence 3
3:      PP2Ba  sequence 4
((1:232,(0:110,2:171)5:272)4:226,3:669)6:0;
```

CAUTION: This tree is rooted assuming a clock, which might not be justified. The exact branch lengths can be found in the nested parenthesis representation just above. Lengths are roughly proportional to vertical lines.

In Fig. 8.9A, the query to the MAlign server requests that the submitted protein sequences be aligned to give a phylogenetic tree of the sequences and a putative ancestral sequence. In response, the server returns a set of results, a portion of which is shown in Fig. 8.9B. MAlign has created the best possible alignment of the four phosphatase sequences. The query sequences are echoed back by the server, each on its own line, with gaps inserted as necessary for the alignment. Between each pair of query sequences, a consensus sequence is given. These consensus sequences are based on the relative degree of similarity shared between pairs of query sequences. The two pairs of numbers at the end of each sequence line are the number of the last amino acid in the sequence listed in that line and the identifying number that the server has assigned each sequence, query, and consensus. Amino acids that are identical between all of the sequences are shown with a * underneath them. Finally, the server uses these results to generate a hypothetical tree showing the phylogenetic relationship of these query and consensus sequences. The last part of Fig. 8.9B shows this tree. The numbers at the branch tips of the tree correspond to the identifying numbers assigned to each sequence by the server during the analysis process. Based on this tree, PPX and PP2A are the closest sequences evolutionarily, sharing consensus sequence 5 as an ancestor. PP1 and consensus sequence 5 share consensus sequence 4 as an ancestor, while PP2B is the most distant member of this protein family. These results are in excellent agreement with the current model for this protein family.

Another analysis approach is taken with the AllAll program of the CBRG server. Figure 8.10A shows a query to the server using the same phosphatase sequences as before. A multiple alignment is requested, along with the construction of an ancestral sequence and data showing evolutionary distances between the query sequences.

Figure 8.10. Phylogenetic Analysis of Sequences Using CBRG

A: Query message to the CBRG server

```
Date:    Sat 12 Oct 1996 - 12:49.19 - GMT
From:    annep@acs.auc.eun.eg
To:      cbrg@inf.ethz.ch
Subj:
ALLALL
MSDSEKLNLDSIIGRLLEVQGSRPGKNVQLTENEIRGLCLKSREIFLSQPILLELEAPLKICGDIHGQYYDL
LRLFEYGGFPPESNYLFLGDYVDRGKQSLETICLLLAYKIKYPENFFLLRGNHECASINRIYGFYDECKRRY
NIKLWKTFTDCFNCLPIAAIVDEKIFCCHGGLSPDLQSMEQIRRIMRPTDVPDQGLLCDLLWSDPDKDVQGW
GENDRGVSFTFGAEVVAKFLHKHDLDLICRAHQVVEDGYEFFAKRQLVTLFSAPNYCGEFDNAGAMMSVDET
LMCSFQILKPADKNKGKYGQFSGLNPGGRPITPPRNSAKAKK
,

MDEKVFTKELDQWIEQLNECKQLSESQVKSLCEKAKEILTKESNVQEVRCPVTVCGDVHGQFHDLMELFRIG
GKSPDTNYLFMGDYVDRGYYSVETVTLLVALKVRYRERITILRGNHESRQITQVYGFYDECLRKYGNANVWK
YFTDLFDYLPLTALVDGQIFCLHGGLSPSIDTLDHIRALDRLQEVPHEGPMCDLLWSDPDDRGGWGISPRGA
GYTFGQDISETFNHANGLTLVSRAHQLVMEGYNWCHDRNVVTIFSAPNYCYRCGNQAAIMELDDTLKYSFLQ
FDPAPRRGEPHVTRRTPDYFL
,

MSEPKAIDPKLSTTDRVVKAVPFPPSHRLTAKEVFDNDGKPRVDILKAHLMKEGRLEESVALRIITEGASIL
RQEKNLLDIDAPVTVCGDIHGQFFDLMKLFEVGGSPANTRYLFLGDYVDRGYFSIECVLYLWALKILYPKTL
FLLRGNHECRHLTEYFTFKQECKIKYSERVYDACMDAFDCLPLAALMNQQFLCVHGGLSPEINTLDDIRKLD
RFKEPPAYGPMCDILWSDPLEDFGNEKTQEHFTHNTVRGCSYFYSYPAVCEFLQHNNLLSILRAHEAQDAGY
RMYRKSQTTGFPSLITIFSAPNYLDVYNNKAAVLKYENNVMNIRQFNCSPHPYWLPNFMDVFTWSLPFVGEK
VTEMLVNVLNICSDDELGSEEDGFDGATAAARKEVIRNKIRAIGKMARVFSVLREESESVLTLKGLTPTGML
PSGVLSGGKQTLQSATVEAIEADEAIKGFSPQHKITSFEEAKGLDRINERMPPRRDAMPSDANLNSINKALT
SETNGTDSNGSNSSNIQ
,

MAEISDLDRQIEQLRRCELIKESEVKALCAKAREILVEESNVQRVDSPVTVCGDIHGQFYDLKELFRVGGDV
PERNYLFMGDFVDRGFYSVETFLLLLALKVRYPDRITLIRGNHESRQITQVYGFYDECLRKYGSVTVWRYCT
EIFDYLSLSAIIDGKIFCVHGGLSPSIQTLDQIRTIDRKQEVPHDGPMCDLLWSDPEDTTGWGVSPRGAGYL
FGSDVVAQFNAAANDIDMICRAHQLVMEGYKWHFNETVLTVWSAPNYCYRCGNVAAILELDEHLQKDFIIFEA
APQETRGIPSKKPVADYFL
```

continued

Figure 8.10 continued

```
                .
             MULALIGNMENT
             PROBANCESTRAL
             SPLITDATA
             SPLITDATA
             PAMDATA
```

B: Reply message from the CBRG server

```
        From: Computational Biochemistry <cbrg@inf.ethz.ch>
        Date: Sun, 13 Oct 1996 11:42:53 +0200
        To:   ANNEP@acs.auc.eun.eg

        Cross reference
        a -- (entry 1)  Input sequence number 1
        b -- (entry 2)  Input sequence number 2
        c -- (entry 3)  Input sequence number 3
        d -- (entry 4)  Input sequence number 4

        Estimated pam distances and variances between sequence pairs:
        a-b = 80.2,   53.8   a-c = 95.8,   80.1   a-d = 81.2,   55.3
        b-c = 96.1,   76.4   b-d = 41.5,   20.0
        c-d = 93.3,   77.8
        tree fitting index 0.24 (poor is >1, good is <1)

        Distances are 100.0 % splittable
        54.0  {c}
        39.9  {a}
        20.2  {b}
        20.0  {b,d}
        19.3  {d}
         1.9  {c,d}


           1 ..74
        d - DLDRQIEQLRRCELIKESEVKALCAKAREILVEESNVQRVDSPVTVCGDIHGQFYDLKELFRVGGDVPERNYLF
        b - ELDQWIEQLNECKQLSESQVKSLCEKAKEILTKESNVQEVRCPVTVCGDVHGQFHDLMELFRIGGKSPDTNYLF
        a -          QLTENEIRGLCLKSREIFLSQPILLELEAPLKICGDIHGQYYDLLRLFEYGGFPPESNYLF
        c -          ----------------ILRQEKNLLDIDAPVTVCGDIHGQFFDLMKLFEVGGSPANTRYLF
            ..  .... ...   ... .. ...   .*. ....***.***..**. **..** .....***

          75 ..148
        d - MGDFVDRGFYSVETFLLLLALKVRYPDRITLIRGNHESRQITQVYGFYDECLRKYGSVTVWRYCTEIFDYLSLS
        b - MGDYVDRGYYSVETVTLLVALKVRYRERITILRGNHESRQITQVYGFYDECLRKYGNANVWKYFTDLFDYLPLT
        a - LGDYVDRGKQSLETICLLLAYKIKYPENFFLLRGNHECASINRIYGFYDECKRRY_NIKLWKTFTDCFNCLPIA
        c - LGDYVDRGYFSIECVLYLWALKILYPKTLFLLRGNHECRHLTEYFTFKQECKIKYSE_RVYDACMDAFDCLPLA
            .**.****..* .*.....*.*.*..*.. ...***** ........*..**...*...*... ... *..*..

         149 ..222
        d - AIIDGKIFCVHGGLSPSIQTLDQIRTIDRKQEVPHDGPMCDLLWSDPE_DTTGWGVSP_____RGAGYLFGS
        b - ALVDGQIFCLHGGLSPSIDTLDHIRALDRLQEVPHEGPMCDLLWSDPD_DRGGWGISP_____RGAGYTFGQ
        a - AIVDEKIFCCHGGLSPDLQSMEQIRRIMRPTDVPDQGLLCDLLWSDPDKDVQGWGEND_____RGVSFTFGA
        c - ALMNQQFLCVHGGLSPEINTLDDIRKLDRFKEPPAYGPMCDILWSDPLEDFGNEKTQEHFTHNTVRGCSYFYSY
            *...  ...*.*****. ......** ..*. ..* *..**.*****. *  ....  ......**......

         223 ..296
        d - DVVAQFNAANDIDMICRAHQLVMEGYKWHFNET_____VLTVWSAPNYCYRCGNVAAILELDEHLQKDFIIFE
        b - DISETFNHANGLTLVSRAHQLVMEGYNWCHDRN_____VVTIFSAPNYCYRCGNQAAIMELDDTLKYSFLQFD
        a - EVVAKFLHKHDLDLICRAHQVVEDGYEFFAKRQ_____LVTLFSAPNYCGEFDNAGAMMSVDETLMCSFQILK
        c - PAVCEFLQHNNLLSILRAHEAQDAGYRMYRKSQTTGFPSLITIFSAPNYLDVYNNKAAVL------
            ... *.. ........***. ...** . .............*..*****.  *  .*.........  ..
```

continued

Figure 8.10 continued

```
         297 ..370
    d  -  AAP
    b  -  PAP
    a  -  PA
    c  -
         ...



Probability vectors of the most ancestral sequence
Sun Oct 13 11:39:01 1996
  (printing only probabilities larger than 5%)

    1   D=42.2 E=38.9
    2   L=80.5
    3   D=77.6 E=6.6
    4   R=40.7 Q=31.0 K=8.5
    5   W=43.1 Q=28.2
    6   I=72.2 V=10.7 L=8.0
    7   E=74.4 D=6.1
    8   Q=68.1 E=5.7
    9   L=80.5
   10   R=35.9 N=31.6 K=8.9
```

CBRG returns the results shown in Fig. 8.10B. Because of the length of the reply, only a part of the results are shown. First, the server reports on the distances and variances found between each pairwise comparison of the query sequences. These findings indicate that the query sequences do, in fact, constitute a family of related sequences, which is in good agreement with both the results from the MAlign server and the published literature.

Second, the server shows the multiple sequence alignment, with – indicating gaps. Identical amino acids are shown with an * and conserved amino acids are shown with a ·. While not identical to the alignment returned by MAlign, the overall pattern and conservation are similar. CBRG takes a different approach to generating an ancestral sequence. A portion of this sequence is shown in the last part of the figure. Here, CBRG estimates the probability of an amino acid being present at each position of the ancestral sequence. A column listing is generated, with the first number in the column being the number of the amino acid in the ancestral sequence, followed by the single letter code for the amino acid or acids that might occupy that position and the probability that they will occupy that position. From this listing, a set of hypothetical sequences can be assembled and used to search the sequence databases for other, more distantly related proteins.

Finally, it should be noted that, like the results reported by the MAlign server, the results from the CBRG server are in excellent agreement with current models for the relationship between these sequences.

■ Documenting the analysis

Upon the completion of the computer analysis of the sequence data, it is important that the method of analysis be recorded. Unfortunately, many published sequence comparisons do not meet these standards. When documenting the analysis, the following parameters should be reported:

- **Algorithm or algorithms** used in the analysis.

- **Substitution matrix** used to create the initial analysis. (All modern search programs use substitution matrices.) The choice of substitution matrix can greatly affect search results; therefore, document which matrix (or matrices) were used.

- **Gap penalty.** For algorithms that use gap penalties, such as Mail-FastA, it is critical to state the gap penalty.

- **Name of database or databases.** Specify the name of the database used in the search and the version of it. Databases are changing very rapidly, much faster than the publication cycle and frequently faster than local system administrators can handle.

- **Computer.** The computer used to execute the analysis should be identified.This is actually the least important parameter to state, because the same algorithm with the same database and parameters should produce the same results on every machine. However, if an off-site computer system was used (such as an e-mail or Internet server), then it is standard scientific courtesy to identify the server and credit its maintainers.

■ Summary

The following are our observations and recommendations on sequence analysis and alignments.

- Don't take every matching sequence found in the initial database and use it to generate a multiple sequence alignment. Initial searches will almost always report sequence matches that are not a significant sequence similarity. Screen the results from the server carefully and discard those matches that do not appear to be a member of the sequence family. Inclusion of nonmembers in an alignment will confuse things and likely lead to analysis errors later.

- Remember that the programs for aligning sequences aren't perfect, and they do not always provide the best alignment. This is particularly so for large families of proteins with low sequence identities. If observation shows a better way to manually align the sequences, then edit the alignment manually.

- Use BLAST or an analogous server to identify similar sequences in the nonredundant databases, followed by Retrieve or a like server to download identical or related sequences files.

- Search either the nucleotide sequence or the predicted protein sequence of the query for pattern matches using a combination of servers such as Blocks, Domain, and MotifFinder.

■ Use the default settings for the initial analysis with all servers. If this does not return satisfactory results, then adjust the sensitivity of the analysis if possible. It should be noted that the servers used in this analysis are not the only choices. Mail-FastA or BioSCAN could be substituted for BLAST. GeneMark or GeneID might be more useful tools for the analysis of a genomic DNA sequence. Domains, motifs, and patterns could have been identified through the use of another combination of servers such as ProDom, SBASE, and Domain.

Selected readings

Journal Articles

Altschul, S. F. 1991. Amino acid substitution matrices from an information theoretic perspective. *J. Mol. Biol.* **219**:555-565.

Altschul, S. F. 1993. A protein alignment scoring system sensitive at all evolutionary distances. *J. Mol. Evol.* **36**:290-300.

Altschul, S. F., W. Gish, W. Miller, E. W. Myers, and D. J. Lipman. 1990. Basic local alignment search tool. *J. Mol. Biol.* **219**:403-410.

Altschul, S. F., and D. J. Lipman. 1990. Protein database searches for multiple alignments. *Proc. Natl. Acad. Sci. USA* **87**:5509-5513.

Altschul, S. F., M. S. Boguski, W. Gish, and J. C. Wootton. 1994. Issues in searching molecular sequence databases. *Nat. Genet.* **6**:119-129.

Barker, W. C., L. K. Ketcham, and M. O. Dayhoff. 1978. A comprehensive examination of protein sequences for evidence of internal gene duplication. *J. Mol. Evol.* **10**:265-281.

Barton, G. J. 1994. SCOP: structural classification of proteins. *Trends Biochem. Sci.* **19**:519-559.

Barton, G. J. 1995. Protein secondary structure prediction. *Curr. Opin. Struc. Biol.* **5**:372-376.

Benner, S. A., I. Badcoe, M. A. Cohen, and D. L. Gerloff. 1994. Bona fide prediction of aspects of protein conformation: assigning interior and surface residues from patterns of variation and conservation in homologous protein sequences. *J. Mol. Biol.* **235**:926-958.

Bork, P., C. Ouzounis, C. Sander, M. Scharf, R. Schneider, and E. Sonnhammer. 1992. What's in a genome? *Nature* (London) **358**:287.

Brunak, S., J. Engelbrecht, and S. Knudsen. 1991. Prediction of human mRNA donor and acceptor sites from the DNA sequence. *J. Mol. Biol.* **220**:49-65.

Chothia, C., and A. M. Lesk. 1986. The relation between the divergence of sequence and structure in proteins. *EMBO J.* **5**:823-826.

Collins, J. F., and A. F. W. Coulson. 1990. Significance of protein sequence similarities. *Methods Enzymol.* **183**:474-486.

Dayhoff, M. O. 1965. Computer aids to protein sequence determination. *J. Theor. Biol.* **8**:97-112.

Dayhoff, M. O. 1969. Computer analysis of protein evolution. *Sci. Am.* **221**: 86-95.

Dayhoff, M. O. 1974. Computer analysis of protein sequences. *Fed. Proc.* **33**:2314-2316.

Dayhoff, M. O. 1976. The origin and evolution of protein superfamilies. *Fed. Proc.* **35**:2132-2138.

Dayhoff, M. O., W. C. Barker, and L. T. Hunt. 1983. Establishing homologies in protein sequences. *Methods Enzymol.* **9**:524-545.

Dayhoff, M. O., W. C. Barker, and P. J. McLaughlin. 1974. Inferences from protein and nucleic acid sequences: early molecular evolution, divergence of kingdoms and rates of change. *Orig. Life* **5**:311-330.

Devereux, J., P. Haeberli, and O. Smithies. 1984. A comprehensive set of sequence analysis programs for the VAX. *Nucleic Acids Res.* **12**:387-395.

Doolittle, R. F., D. F. Feng, M. S. Johnson, and M. A. McClure. 1986. Relationships of human protein sequences to those of other organisms. *Cold Spring Harbor Symp. Quant. Biol.* **51**(Part 1):447-455.

Doolittle, R. F. 1981. Similar amino acid sequences: chance or common ancestry? *Science* **214**:149-159.

Doolittle, R. F. 1989. Similar amino acid sequences revisited. *Trends Biochem. Sci.* **14**:244-245.

Doolittle, R. F. 1990. Searching through sequence databases. *Methods Enzymol.* **183**:99-110.

Doolittle, R. F. 1992. Stein and Moore Award address. Reconstructing history with amino acid sequences. *Protein Sci.* **1**:191-200.

Doolittle, R. F., and D. F. Feng. 1990. Nearest neighbor procedure for relating progressively aligned amino acid sequences. *Methods Enzymol.* **183**:659-669.

Feng, D. F., and R. F. Doolittle. 1987. Progressive sequence alignment as a prerequisite to correct phylogenetic trees. *J. Mol. Evol.* **25**:351-360.

Feng, D. F., and R. F. Doolittle. 1990. Progressive alignment and phylogenetic tree construction of protein sequences. *Methods Enzymol.* **183**:375-387.

Feng, D. F., M. S. Johnson, and R. F. Doolittle. 1984. Aligning amino acid sequences: comparison of commonly used methods. *J. Mol. Evol.* **21**:112-125.

Geourjon, C., and G. Deleage. 1994. SOPM: a self optimised prediction method for protein secondary structure prediction. *Protein Eng.* **7**:157-216.

Gerloff, D. L., T. F. Jenny, L. J. Knecht, and S. A. Benner. 1993. A secondary structure prediction of the hemorrhagic metalloprotease family. *Biochem. Biophys. Res. Commun.* **194**:560-565.

Gonnet, G. H., M. A. Cohen, and S. A. Benner. 1993. Exhaustive matching of the entire protein sequence database. *Science* **256**:1443-1445.

Gribskov, M., A. D. McLachlan, and D. Eisenberg. 1987. Profile analysis: detection of distantly related proteins. *Proc. Natl. Acad. Sci. USA* **84**:4355-4358.

Guigo, R., S. Knudsen, N. Drake, and T. Smith. 1992. Prediction of gene structure. *J. Mol. Biol.* **226**:141-157.

Hegyi, H., and S. Pongor. 1993. Predicting potential domain-homologies from FASTA search results. *CABIOS* **9**:371-372.

Henikoff, S. 1991. Playing with blocks: some pitfalls of forcing multiple alignments. *New Biol.* **3**:1148-1154.

Henikoff, S. 1992. Detection of *Caenorhabditis* transposon homologs in diverse organisms. *New Biol.* **4**:382-388.

Henikoff, S., and J. C. Wallace. 1988. Detection of protein similarities using nucleotide sequence databases. *Nucleic Acids Res.* **16**:6191-6204.

Henikoff, S., and J. G. Henikoff. 1991. Automated assembly of protein blocks for database searching. *Nucleic Acids Res* **19**:6565-6572.

Henikoff, S., and J. G. Henikoff. 1992. Amino acid substitution matrices from protein blocks. *Proc. Natl. Acad. Sci. USA* **89**:10915-10919.

Henikoff, S., and J. G. Henikoff. 1993. Performance evaluation of amino acid substitution matrices. *Proteins* **17**:49-61.

Henikoff, S., and J. G. Henikoff. 1994. Position-based sequence weights. *J. Mol. Biol.* **243**:574-578.

Henikoff, S., and J. G. Henikoff. 1994. Protein family classification based on searching a database of blocks. *Genomics* **19**:97-107.

Henikoff, S., J. C. Wallace, and J. P. Brown. 1990. Finding protein similarities with nucleotide sequence databases. *Methods Enzymol.* **183**:111-132.

Hunt, L. T., D. G. George, L. S. Yeh, and M. O. Dayhoff. 1984. Evolution of prokaryote and eukaryote lines inferred from sequence evidence. *Orig. Life* **14**:657-664.

Johnson, M. F., and R. F. Doolittle. 1986. A method for the simultaneous alignment of three or more amino acid sequences. *J. Mol. Evol.* **23**:267-278.

Jones, D. T., W. R. Taylor, and J. M. Thornton. 1992. The rapid generation of mutation data matrices from protein sequences. *Comput. Appl. Biosci.* **8**:275-282.

Karlin, S., and S. F. Altschul. 1990. Methods for assessing the statistical significance of molecular sequence features by using general scoring schemes. *Proc. Natl. Acad. Sci. USA* **87**:2264-2268.

Karlin, S., and S. F. Altschul. 1993. Applications and statistics for multiple high-scoring segments in molecular sequences. *Proc. Natl. Acad. Sci. USA* **90**:5873-5877.

Koonin, E. V., P. Bork, and C. Sander. 1994. Yeast chromosome III: new gene functions. *EMBO J.* **13**:493-503.

Koonin, E., R. Tatusov, and K. Rudd. 1995. Sequence similarity analysis of Escherichia coli proteins: functional and evolutionary implications. *Proc. Natl. Acad. Sci. USA* **92**:11921-11925.

Kyte, J., and R. F. Doolittle. 1982. A simple method for displaying the hydropathic character of a protein. *J. Mol. Biol.* **157**:105-132.

Lawrence, C. E., S. F. Altschul, M. S. Boguski, J. S. Liu, A. F. Neuwald, and J. C. Wootton. 1993. Detecting subtle sequence signals: a Gibbs sampling strategy for multiple alignment. *Science* **262**:208-214.

Lipman, D. J., and W. R. Pearson. 1985. Rapid and sensitive protein similarity searches. *Science* **227**:1435-1441.

Lipman, D. J., S. F. Altschul, and J. D. Kececioglu. 1989. A tool for multiple sequence alignment. *Proc. Natl. Acad. Sci. USA* **86**:4412-4415.

Lipman, D. J., W. J. Wilbur, T. F. Smith, and M. S. Waterman. 1984. On the statistical significance of nucleic acid similarities. *Nucleic Acids Res.* **12**:215-226.

Livingstone, C. D., and G. J. Barton. 1996. Identification of functional residues and secondary structure from protein multiple sequence alignment. *Methods Enzymol.* **266**:497-512.

Mehta, P., J. Heringa, and P. Argos. 1995. A simple and fast approach to prediction of protein secondary structure from multiple aligned sequences with accuracy above 70% Prot. *Science* **4**:2517-2525.

Needleman, S. B., and C. D. Wunsch. 1970. A general method applicable to the search for similarities in the amino acid sequence of two proteins. *J. Mol. Biol.* **48**:443-453.

Neuwald, A. F., J. S. Liu, and C. E. Lawrence. 1995. Gibbs motif sampling: detection of bacterial outer membrane protein repeats. *Protein Sci.* **4**:1618-1621.

Pearson, W. R. 1988. Improved tools for biological sequence comparison. *Proc. Natl. Acad. Sci. USA* **85**:2444-2448.

Pearson, W. R. 1990. Rapid and sensitive sequence comparison with FASTP and FASTA. *Methods Enzymol.* **183**:63-98.

Pearson, W. R. 1991. Searching protein sequence libraries: comparison of the sensitivity and selectivity of the Smith-Waterman and FASTA algorithms. *Genomics* **11**:635-650.

Pearson, W. R., and D. J. Lipman. 1988. Improved tools for biological sequence comparison. *Proc. Natl. Acad. Sci. USA* **85**:2444-2448.

Pearson, W. R., and W. Miller. 1991. Searching protein sequence libraries: comparison of the sensitivity and selectivity of the Smith-Waterman and FASTA algorithms. *Genomics* **11**:635-650.

Pearson, W. R., and W. Miller. 1992. Dynamic programming algorithms for biological sequence comparison. *Methods Enzymol.* **210**:575-601.

Rost, B., and C. Sander. 1992. Jury returns on structure prediction. *Nature* (London) **360**:540.

Rost, B., and C. Sander. 1993. Improved prediction of protein secondary structure by use of sequence profiles and neural networks. *Proc. Natl. Acad. Sci. USA* **90**:7558-7562.

Rost, B., and C. Sander. 1993. Prediction of protein secondary structure at better than 70% accuracy. *J. Mol. Biol.* **232**:584-599.

Rost, B., and C. Sander. 1993. Secondary structure prediction of all-helical proteins in two states. *Protein Eng.* **6**:831-836.

Rost, B., and C. Sander. 1994. Combining evolutionary information and neural networks to predict protein secondary structure. *Proteins* **19**:55-77.

Rost, B., and C. Sander. 1994. Conservation and prediction of solvent accessibility in protein families. *Proteins* **20**:216-226.

Rost, B., C. Sander, and R. Schneider. 1993. Progress in protein structure prediction? *Trends Biochem. Sci.* **18**:120-123.

Rost, B., C. Sander, and R. Schneider. 1994. Redefining the goals of protein secondary structure prediction. *J. Mol. Biol.* **235**:13-26.

Rost, B., R. Casadio, P. Fariselli, and C. Sander. 1995. Prediction of helical transmembrane segments at 95% accuracy. *Prot. Sci.* **4**:521-533.

Rudd, K. E., W. Miller, J. Ostell, and A. Benson. 1990. Alignment of Escherichia coli K12 sequences to a genomic restriction map. *Nucleic Acids Res.* **18**:313-321.

Salamov, A. A., and V. V. Solovyev. 1995. Prediction of protein secondary structure by combining nearest-neighbor algorithms and multiple sequence alignments. *J. Mol. Biol.* **247**:11-15.

Sander, C., and R. Schneider. 1991. Database of homology-derived structures and the structural meaning of sequence alignment. *Proteins* **9**:56-68.

Sauer, R. T., R. R. Yocum, R. F. Doolittle, M. Lewis, and C.O. Pabo. 1982. Homology among DNA-binding proteins suggests use of a conserved super-secondary structure. *Nature* (London) **298**:447-451.

Schwartz, R. M., and M. O. Dayhoff. 1978. Origins of prokaryotes, eukaryotes, mitochondria, and chloroplasts. *Science* **199**:395-403.

Seely, O., D. F. Feng, D. W. Smith, D. Sulzbach, and R. F. Doolittle. 1990. Construction of a facsimile data set for large genome sequence analysis. *Genomics* **8**:71-82.

Smith, H. O., T. M. Annau, and S. Chandrasegaran. 1990. Finding sequence motifs in groups of functionally related proteins. *Proc. Natl. Acad. Sci. USA* **87**:826-830.

Smith, T. F., and M. S. Waterman. 1981a. Comparison of biosequences. *Adv. Appl. Math.* **2**:482-489.

Smith, T. F., and M. S. Waterman. 1981b. Identification of common molecular subsequences. *J. Mol. Biol.* **147**:195-197.

Smith, T. F., M. S. Waterman, and C. Burks. 1985. The statistical distribution of nucleic acid similarities. *Nucleic Acids Res.* **13**:645-656.

Solovyev, V. V., and A. A. Salamov. 1994. Predicting alpha-helix and beta-strand segments of globular proteins. *Comput. Appl. Biosci.* **10**:661-669.

Sonnhammer, E. L. L, and R. Durbin. 1990. A workbench for Large Scale Sequence Homology Analysis. *Comput. Appl. Biosci.* **10**:301-307.

Sonnhammer, E. L. L., and D. Kahn. 1994. Modular arrangement of proteins as inferred from analysis of homology. *Protein Sci.* **3**:482-492.

Stormo, G. D. 1990. Consensus patterns in DNA. *Methods Enzymol.* **183**:211-220.

Wahl, R., P. Rice, C. M. Rice, and M. Kroger. 1994. ECD–a totally integrated database of Escherichia coli K12. *Nucleic Acids Res.* **22**:3450-3455.

Wako, H., and T. L. Blundell. 1994. Use of amino-acid environment-dependent substitution tables and conformational propensities in structure prediction from aligned sequences of homologous proteins. 2. Secondary structures. *J. Mol. Biol.* **238**:693-708.

Wallace, J. C., and S. Henikoff. 1992. PATMAT: a searching and extraction program for sequence, pattern and block queries and databases. *Comput. Appl. Biosci.* **8**:249-524.

Waterman, M. S. 1984. Efficient sequence alignment algorithms. *J. Theor. Biol.* **108**:333-337.

Waterman, M. S. 1986. Multiple sequence alignment by consensus. *Nucleic Acids Res.* **14**:9095-9102.

Wilbur, W. J., and D. J. Lipman. 1983. Rapid similarity searches of nucleic acid and protein data banks. *Proc. Natl. Acad. Sci. USA* **80**:726-730.

Wu, T. T., and E. A. Kabat. 1970. An analysis of the sequences of the variable regions of Bence Jones proteins and myeloma light chains and their implications for antibody complementarity. *J. Exp. Med.* **132**:211-250.

Zvelebil, M. J. J. M., G. J. Barton, W. R. Taylor, and M. J. E. Sternberg. 1987. Prediction of protein secondary structure and active sites using the alignment of homologous sequences. *J. Mol. Biol.* **195**:957-961.

Books and Monographs

Adams, M. D., C. Fields, and J. C. Venter. 1994. *Automated DNA Sequencing and Analysis.* Academic Press, San Diego.

Barton, G. J. 1996. Protein sequence alignment and database scanning. *In* M. J. E. Sternberg (ed.), *Protein Structure Prediction: A Practical Approach.* IRL Press at Oxford University Press, Oxford.

Bishop, M. J., and C. J Rawlings (ed.). 1987. *Nucleic Acid and Protein Sequence Analysis - a Practical Approach,* p. 323-358. IRL Press, Oxford.

Dayhoff, M. O., R. M. Schwartz, and B. C. Orcutt. 1978. A model of evolutionary change in proteins: matrices for detecting distant relationships, p. 345-358. *In* M. O. Dayhoff (ed.), *Atlas of Protein Sequence and Structure*, vol. 5. National Biomedical Research Foundation, Washington, D.C.

Devereux, J. 1988. A rapid method for identifying sequences in large nucleotide sequence databases. Ph.D. thesis. University of Wisconsin, Madison.

Doolittle, R. F. 1986. *Of URFs and ORFs: a Primer on How To Analyze Derived Amino Acid Sequences.* University Science Books, Mill Valley, Calif.

Knecht, L., and G. H. Gonnet. 1992. *Alignment of Nucleotide with Peptide Sequences.* Report 184, Institute for Scientific Computing, ETH, Zurich.

Rost, B. 1995. Fitting 1D predictions into 3D structures, p. 132-151. *In* H. Bohr and S. Brunak (ed.), *Protein Folds: a Distance Based Approach.* CRC Press, Boca Raton, Fla.

Rost, B., and C. Sander. 1994. 1D secondary structure prediction through evolutionary profiles, p. 257-276. *In* H. Bohr and S. Brunak (ed.), *Protein Structure by Distance Analysis.* IOS Press, Amsterdam.

von Heijne, G. 1987. *Sequence Analysis in Molecular Biology: Treasure Trove or Trivial Pursuit?* Academic Press, San Diego.

9

Gopher and the Web

■ Navigating the Internet and the Web

■ Gopher holes and Web sites

Unlike other components of the Internet, Gopher and the Web do not have a well-defined function. The earliest tools, e-mail, FTP, and Telnet, all have clear and defined functions: e-mail handles electronic messages, FTP transfers files, and Telnet operates remote computers. Instead of being Internet components with singular roles, Gopher and the Web represent a new class of Internet components that function as information management systems. They allow the user to manage increasing amounts of information in a coordinated and logical fashion. They permit organized searches and queries for information. Most importantly, they are also integrated resources, serving to combine the functions of most if not all of the components that make up the Internet under a common and unifying interface.

Of these tools, the Web in particular has reshaped the Internet landscape. Prior to the Web, a resource on the Internet was primarily a tool of singular function: a server to analyze a sequence or to retrieve a specified piece of information. The display of relationships and linkages between different pieces of information was nearly impossible with the older components of the Internet. The Web not only allows these linkages, but encourages them. Further, it permits new classes of information to be displayed such as large centralized databanks and linked, distributed databanks.

Gopher has a lot to offer the research community. A vast range of molecular biology information and many sequence databases can be searched on the Internet with Gopher. As we discussed in Chapters 1 and 2, Gopher allows users to quickly navigate through layers of information and to search databases, thus enabling easy retrieval of information and connection to other services. Gopher was first distributed only a few years ago, yet there are well over 3,000 freely accessible Gopher directories on computers throughout the world, with many of these specific to molecular biology and sequence information. We list here the key advantages of Gopher.

■ There is a wealth of information available by Gopher, including many of the most critical resources for the genomic and molecular biology community.

■ The interface is consistent and computer-independent to allow seamless browsing of servers and databases.

■ Gopher offers the ability to browse the Internet and locate information that cannot be identified using earlier Internet components.

■ Gopher can locate information even when the user is unsure of what he/she is looking for.

- The use of bookmarks provides a user-defined index of resources of interest.

Even with all of these advantages, Gopher has been overshadowed by the Web. The Web offers all of the features of Gopher, has a more unified interface, and is simpler to use. As a result, many sites that offer both Gopher and Web access are phasing out Gopher servers, simply because of lack of demand. Clearly the Internet is again in transition, this time from Gopher to the Web.

Like the Internet and the genomic and molecular biology community, the Web is in transition as well. For scientists, three classes of Web sites are appearing: those that are focused on analysis tools, such as the sites that have been developed by NCBI and EBI; those that are focused on information and links to other sites, such as ExPASy or Pedro's Molecular Biology Tools; and a new class, the knowledge bases typified by smaller sites such as ECO2DBASE and EcoCyc or larger projects such as the Tree of Life.

The Web and Gopher sites that are described in the following pages are organized to reflect these three classes of sites, with one additional class: those that have been developed by organizations. First, we list five sites that have been developed by various science associations or organizations. These sites offer a mixture of information and news, along with important links to other sites on the Internet. Next, we list sites that focus on the analysis of sequence data. This is followed by informational and resource sites. Last, we will highlight four sites that represent the small but developing class of knowledge-based sites.

Navigating the Internet and the Web

While it is assumed that most users understand the basic principles of how to navigate on the Internet using a graphics-based Web browser (such as Mosaic, Netscape Navigator, or Internet Explorer), the following is a brief summary on how to use some of the most common features of the popular browser Netscape Navigator.

- **Back:** In a sequence of displayed pages, moves to previous screen.

- **Forward:** In a sequence of displayed pages, moves to next screen.

- **Home:** Returns to starting page.

- **Reload:** Sometimes when trying to link to a page, the system stalls. Reload tells the browser to try it again.

- **Print:** Prints the current screen page; however, a printer cannot always recreate a screen image properly.

- **Go:** By pulling down the menu, users can see a list of sites that have already been visited during the current session. By highlighting and clicking, the user can return to the highlighted site.

- **Bookmark:** A special electronic tag placed on a site that allows it to be selected from a list instead of either searching for it or reentering its URL.

■ **File..Mail To:** This browser option can be used to send e-mail to another Internet address.

You can perform searches on the Web in several different ways. Click on the "Net Search" button and a list of search engines will appear. Or, type in the URL of a search engine such as: Yahoo (http://www.yahoo.com), Alta Vista (http://www.altavista.com), Lycos (http://lycos.com), or Inktomi (http://inktomi.berkeley.edu). Many different search engines exist, and each has its own search commands and search methods.

Gopher holes and Web sites

Associations and Organizations

Web sites developed by associations and organizations are often advertisements for the developer rather than a true community resource. The sites listed below run counter to this rule. They are tightly focused resources that offer information and materials from their own proprietary files and they function quite nicely as gateways to related areas of interest on the Internet.

■ American Society for Microbiology

The ASM Web page is a convenient starting point to access microbiology-related information and a wide range of links to specific resources in genomic and molecular biology. Contents of the extensive list of ASM-published journals can be browsed as well. Most importantly, this site is well maintained and frequently updated. ASM has also established a Web site as a complement to this text, as discussed in the Preface. This site provides a compendium of active links to the resources described in this book. It will provide regular updates to the information listed here. Comments and suggestions directed to this address are encouraged.

Internet addresses

Web: http://www.asmusa.org   (ASM home page)

                   http://www.asmpress.org/isbn/1555811191/ (Book Web page)

■ ATCC (The American Type Culture Collection)

The American Type Culture Collection (ATCC) is an organization that provides biological products, technical services, and educational programs to organizations around the world. The mission of the ATCC is to serve as a curator and distributor of reference cultures, related biological materials, and associated data. This site offers access to the stock collections of the ATCC, which contain a wide range of bacterial, fungal, plant, and animal cell lines as well as virus stocks, DNAs, and antisera. The databases can be searched by text term. In addition, ordering forms can be downloaded and information on upcoming laboratory courses, workshops, and meetings can be accessed. The Web site is linked to several other related Internet sites in broad areas of culture collections and resources.

Internet addresses

Web:	http://www.atcc.org
Gopher:	gopher://culture.atcc.org

■ Cold Spring Harbor Laboratory

One of the oldest genetic and molecular biology laboratories in the world, Cold Spring Harbor Laboratory offers a well-developed Web site that contains historical information on the revolution in biology, announcements of courses and meetings, abstracts, and preprints from new books and manuals. In addition, a good assortment of broadly based links to other Internet resources is available. The site is current and frequently updated, making it a good choice for a starting place when browsing the Internet.

Internet address

Web:	http://www.cshl.org

■ Jackson Laboratory Server

Founded in 1929, the Jackson Laboratory has three functions in the scientific community: research in basic genetics and the role of genes in disease; education of the scientific and lay community; provider of genetically defined mice and other resources for science. The Web site reflects these three functions. This site offers the Mouse Genome Database (MGD), with genetic maps of each mouse chromosome. Searches for specific loci on these maps as well as for references and notes associated with loci can be performed. In addition, the site offers access to the extensive collection of strains in the Jackson Laboratory Stocks.

Internet address

Web:	http://www.jax.org

■ Protein Science's List of Links

This site belongs to the Protein Society, an organization of protein biochemists and molecular biologists that is based in San Diego, Calif. One particular strength of this site is its set of links to guides on using the Internet. As with the four other sites in this class, the Protein Society offers a Web site that contains announcements of courses and meetings, abstracts, and preprints from new books and manuals. In addition, a good assortment of broadly based links to protein databases and analysis tools on the Internet is available. As some of the links are little known but specialized and useful, this site is a good community resource. Further, it is well maintained and often updated.

 Internet addresses

Web: http://www.prosci.uci.edu

Gopher: gopher://gopher.prosci.uci.edu

Analysis Tools

This broad and loosely defined category contains sequence analysis tools and the necessary data to use them. With these sites, raw sequence data can be compared to the international databases, annotated, submitted for entry into the databases, and searched for domains and patterns. All sites offer, in addition to their own analysis tools, sets of complete links to other analysis tools.

■ BCD (Biological Computing Division) Web Biology Server

BCD is dedicated to both DNA and protein sequence analysis and offers the ability to browse through a number of databases in addition to some outstanding sequence analysis tools. These tools include a solid implementation of the Smith-Waterman algorithm for similarity analysis of both protein and nucleotide sequence data along with news and information. BCD also maintains a good set of links to other resources in the broad field of sequence analysis and databases.

 Internet address

Web: http://dapsas1.weizmann.ac.il

■ Biologist's Control Panel

The Biologist's Control Panel Web site is located at the Baylor College of Medicine. It is a well-constructed multi-purpose site offering analysis tools and an extensive range of links to databases (including those of NCBI, Genethon, GDB, PIR, SWISS-PROT, and others) along with links to library and literature resources (such as Cold Spring Harbor Laboratory Press, the Human Genome News, and others). It includes a multiple sequence alignment service and several useful human genome analysis resources. In addition, a guide to genetics and human genome resources can be found at this site.

 Internet address

Web: http://gc.bcm.tmc.edu:8088/bio

■ **European Bioinformatics Institute (EBI)**

While not as polished as the site from NCBI, the European Bioinformatics Institute (EBI) maintains an extensive and well-developed site. The EBI Gopher server uses a clear graphical interface to streamline access to the available network services, and it offers links to other genomic biology resources. The EBI Web site permits access to all of EBI's resources including submission of sequence data, database access, similarity analysis, database query and retrieval, and links to other Web resources worldwide. The resources include access to the primary and specialized databases maintained by EBI, collections of computer software and documentation, and sequence analysis tools that can be accessed via Gopher or the Web. Sequence data can be analyzed for similarity to other sequences by using BLITZ, FastA, and Mail-PROSITE. To retrieve sequence records from databases, EBI offers a mirror of the Retrieve system of NCBI and a newer, Web-specific tool called the Sequence Retrieval System (SRS). SRS is a tool to query the EBI/EMBL databases using keywords to retrieve specific sequences. Similar in design to Query, SRS is widely used in the EMBL community and can be found at many other server sites around the world.

Internet addresses

Web: http://www.ebi.ac.uk

Gopher: gopher://gopher.ebi.ac.uk

■ **GenomeNet**

GenomeNet is a Japanese computer network for genome research and related research areas in molecular and cellular biology. However, GenomeNet is not simply a network connection; it is an informatics infrastructure for biological sciences that services a wide range of databases and data analysis systems. GenomeNet may be accessed by a range of Internet protocols, but access through the Web is by far the most popular. A main feature of this server is the DBGET database system. Another useful option of the GenomeNet is the suite of sequence interpretation tools that include BLAST, MOTIF, and PSORT. The homology and motif search results are directly linked to the DBGET system to facilitate interpretation.

Internet address

Web: http://www.genome.ad.jp

■ **National Center for Biotechnology Information (NCBI)**

The NCBI's massive Web site is one of the strongest all-around tools and resources that the Internet has to offer. NCBI is arguably the best single site for genomic and molecular biology because of the depth of its databases and the power of its analysis tools. NCBI offers access to the BLAST, Query, and Retrieve servers, direct electronic submission of sequence data, and extensive links to other related Web sites, and it is home to the integrated database retrieval and search system Entrez. Entrez links the international nucleotide and protein sequence databases with literature references from MEDLINE, three-dimensional structure data, and complete genome data, making it a nearly indispensable resource for the community. Well maintained and frequently updated, the NCBI Web site also offers extensive links to other resources on the Internet.

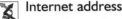 **Internet address**

Web: http://www.ncbi.nlm.nih.gov

■ National Institute of Genetics (NIG)

The NIG site is a counterpart to the sites offered by EBI and NCBI. While not as complete or well designed as those sites, it does offer a good set of analysis tools and links to the DDBJ sequence databases. The site is frequently updated and contains a set of well-maintained links to the outside.

Internet addresses

Web: http://www.nig.ac.jp/
Gopher: gopher://gopher.nig.ac.jp/

Information and Resources

The sites listed here are broadly defined. Most offer multiple functions. The primary focus of the sites listed in this section is the maintenance of comprehensive lists of molecular biology-related Internet resources such as databases, sequence analysis tools, and literature.

■ Agricultural Genome Information Server (AGIS)

A broad and well-designed Web site, AGIS contains a wide range of documentation and resources that focus on genomic and sequence analysis projects in the agricultural arena. Nearly every plant genomic database is either located or linked to this site. Along with the databases are links to analysis tools, documentation, and help files. AGIS is an outstanding gateway to plant molecular genetics and molecular biology resources on the Internet.

 Internet address

Web: http://probe.nalusda.gov:8000

■ Australian National Genomic Information Service (ANGIS)

ANGIS offers a wide range of documentation and resources on the Human Genome Project and a good set of links to related sites and tools. This site has sequence analysis tools, access to the international databases, and news on upcoming conferences and meetings. Well maintained and frequently updated, ANGIS offers a versatile regional site for sequence analysis.

Internet address

Web: http://morgan.angis.su.oz.au

■ CEPH-Genethon

The CEPH Genethon Web site offers various sources of information relating to the human genome and its analysis. It is also home of the experimental Web server GenomeView, which integrates access to a variety of human physical and genetic map databanks. From this site, the CEPH-Ginithon mapping data of the human genome can be searched (Ginithon is the French version of the word Genethon), including the YAC and STS data used to build a first-generation physical map of the human genome.

Internet addresses

Web: http://www.genethon.fr/genethon_fr.html

Gopher: gopher://gopher@genethon.fr

■ Cooperative Human Linkage Center (CHLC)

CHLC maintains detailed information on human genetic and integrated maps, combining data from several mapping groups. The site contains extensive sets of primary maps, collections of markers, examples of integrated maps, and genotype data. Extensive links are provided to other resources in the genetics community.

Internet addresses

Web: http://www.chlc.org

Gopher: gopher://gopher.chlc.org

■ ExPASy Molecular Biology Server

For the genomic and molecular biology community, ExPASy could be considered the Swiss Army knife of Web servers. It offers sequence analysis tools, access to nearly every molecular biology database, and extensive documentation and links to other resources on the Internet. This site contains recommendations of information sources for molecular biologists and computational biologists. Clearly one of the most popular molecular biology servers due to its content and multimedia capabilities, ExPASy is maintained by the Geneva University Hospital and the University of Geneva, Geneva, Switzerland.

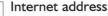 **Internet address**

Web: http://expasy.ch

■ GenLink

GenLink is a resource for human genetics that is currently under development. It provides linkage mapping information and software tools for the integration of physical and genetic linkage data to produce unified maps of the human genome. The site is composed of a genotypes database (GenotypesDB), a telomere database (TelDB), and a meiotic map database. GenotypesDB stores public genotype information including detailed meiotic map graphics with links to marker information at the Genome Data Base (GDB). The current GenotypesDB includes the CEPH reference family genotypes used to construct human genome maps that are available through GenLink. TelDB provides literature references and information about telomeric and subtelomeric regions of chromosomes from different organisms. WWW-TelDB includes a user interface to the searchable telomere literature database where users enter and/or select keywords to be used for their literature citations database query. Common queries are also presented in the form of links to results pages. TelDB will accept data submissions (or URL links) from telomere researchers wishing to share their data. The Human Meiotic Maps section contains detailed graphics and mapping information. The site has a concise set of links to other mapping and genetic resources available on the Internet.

Internet address

Web: http://www.genlink.wustl.edu

■ Human Chromosome Servers

With the increasing pace of genomic sequencing, several sites that are specific for a single human chromosome have appeared on the Web. These sites offer a single point of contact for a diverse range of information on each chromosome, including sequence and mapping data, genetic marker information, literature references, and links to other resources. While obviously evolving resources, they offer a good foundation to the informational needs of the Human Genome Project.

Internet addresses

Chromosome 1:   http://linkage.rockefeller.edu/chr1

Chromosome 3:   http://mars.uthscsa.edu/

Chromosome 7:   http://www.genet.sickkids.on.ca/chr7db/query.html

Chromosome 8:	http://gc.bcm.tmc.edu:8088/chr8/home.html
Chromosome 9:	http://www.gene.ucl.ac.uk/chr9/
Chromosome 10:	http://www.cric.com/htdocs/chr10-mapping/index.html
Chromosome 11:	http://chr11.bc.ic.ac.uk/
	http://mcdermott.swmed.edu/datapage
Chromosome 12:	http://paella.med.yale.edu/chr12/home.html
Chromosome 13:	http://genome1.ccc.columbia.edu/~genome
Chromosome 16:	http://www-ls.lanl.gov/dbqueries.html
	http://www.tigr.org/tdb/humgen/c16.html
Chromosome 17:	http://bioinformatics.weizmann.ac.il/
Chromosome 19:	http://www-bio.llnl.gov/bbrp/genome/genome.html
Chromosome 20:	http://www.expasy.ch/cgi-bin/lists?humchr20.txt
Chromosome 21:	http://www.expasy.ch/cgi-bin/lists?humchr21.txt
Chromosome 22:	http://www.cbil.upenn.edu/HGC22.html
	http://www.expasy.ch/cgi-bin/lists?humchr22.txt
	http://www.sanger.ac.uk/hum22
Chromosome X:	http://gc.bcm.tmc.edu:8088/chrX/home.html
	http://www.expasy.ch/cgi-bin/lists?humchrx.txt
Chromosome Y:	http://www.expasy.ch/cgi-bin/lists?humchry.txt
Mitochondrial chromosome:	http://infinity.gen.emory.edu/mitomap.html

■ Human Genome Management Information System (HGMIS)

HGMIS contains a wide range of documentation and resources on the Human Genome Project and extensive links to genetic-related sites and historical documents. This searchable, comprehensive site also has an electronic copy of the "Primer on Molecular Genetics," a popular resource for teachers, genetic counselors, and educational organizations.

Internet address

Web:	http://www.ornl.gov/hgmis

■ International Centre for Genetic Engineering and Biotechnology (ICGEB)

ICGEB Web and Gopher sites provide a comprehensive set of databases and analysis tools for biologists of European and the Middle Eastern member countries. Note that the

United States and Canada are not member countries. Analysis tools provided include the GCG and Intelligenetics packages and BLAST. The SRS system is provided for the retrieval of sequence information from the public databases. These databases include GenBank, new GenBank genome subsets, EMBL, PIR, SWISS-PROT, PROSITE, and several other specialized databases.

 **Internet addresses**

Web: http://www.icgeb.trieste.it

Gopher: gopher:// icgeb.trieste.it

■ Johns Hopkins Bioinformatics Server

The Johns Hopkins Bioinformatics Server is a set of databases and associated tools maintained by the Computational Biology group. A wide range of databases are linked to this site. A few of the more prominent ones are GenBank, GDB, OMIM, PDB, and the Encyclopedia of the Mouse Genome. Well organized and maintained, this site is a good place to start for database resource and a link to other Web sites.

Internet address

Web: http://www.gdb.org

■ Listing of Biology Internet Servers

This is a well-organized and well-designed list of servers that offer resources to the biology community. Updated on a frequent basis, this site has a wide range of servers. Each server or site listed is also linked to the list itself, making it easy to use this Web site as a starting point to explore the Web.

Internet address

Web: http://info.er.usgs.gov/network/science/biology/index.html

■ Pedro's BioMolecular Research Tools

Possibly the most complete list of biology sites on the Web, Pedro's BioMolecular Research Tools is a comprehensive list of links to information and services for the genomic and molecular biology community. The site is composed of three main parts: Molecular Biology Search and Analysis; Bibliographic, Text, and WWW Searches; and Guides, Tutorials, and Help Tools. The site offers links to information and services useful to molecular biologists, including extensive listings of molecular biology database search and analysis facilities, bibliographies, guides, tutorials, and journal contents. The site is offered at four locations around the globe to speed access. It suffers from one drawback: it is not frequently updated, and as a result some of its links to other sites are out of date.

Internet addresses

Web: http://www.public.iastate.edu/~pedro/research_tools.html

http://www.biophys.uniduesseldorf.de/bionet/research_tools.html

http://www.fmi.ch/biology/research_tools.html

http://www.peri.co.jp/Pedro/research_tools.html

■ Virtual Library: Biochemistry, Biophysics, and Molecular Biology

The Virtual Library is a searchable index of Internet resources cataloged under different subject headings. It is very useful for navigating to particular areas of the Internet because it contains a comprehensive listing of genetics resources on the Internet with links to each other. The site is updated frequently, and it is well maintained.

Internet address

Web: http://golgi.harvard.edu/biopages/biochem.html

Knowledge Bases

Knowledge bases are the newest, least developed, but most promising of all Internet tools available for genomic and molecular biology. With the informational content of one or more encyclopedias, they might at first seem like simple databases of information. However, they integrate this data from several distinct sources and offer a global view of life ranging from a single organism to the entire domain of life on the planet; they link diverse areas such as metabolism, functional properties, regulation, and genetic information. As such, all of these sites are works in progress and are incomplete. But each of them takes full advantage of the power of the Web to link information of many different types and allow analysis of the data in a freeform fashion.

■ EcoCyc

The encyclopedia of *Escherichia coli* genes and metabolism (EcoCyc) is a database that combines information about the genome and the intermediary metabolism of the bacterium *E. coli*. EcoCyc allows the query and exploration of the database using visualization tools to browse genomic maps and layouts of metabolic pathways. In this fashion it closes the gap between sequence data and functional analysis to permit the investigation of a broad range of questions. As a result, EcoCyc is both an encyclopedia of information on *E. coli* and a computational model of the organism.

While still in its infancy, EcoCyc offers enormous potential because of its approach to cellular physiology, metabolism, and regulation. This is one tool that encompasses the full power of the Web and could not exist as a community-based resource without the Web. A companion site, HincCyc, is being developed for the bacterium *Haemophilus influenzae*, the first free-living organism to be completely sequenced.

Internet address

Web: http://www.ai.sri.com/ecocyc/ecocyc.html

■ ECO2DBASE: Gene-Protein Database of *Escherichia coli*

The *E. coli* Gene-Protein Database (ECO2DBASE) is unique in that this database is configured on a global approach that allows the cell's total complement of polypeptides to be examined simultaneously. Two-dimensional polyacrylamide gel electrophoresis, or 2D gels, have been used to resolve the total protein content of the cell into discrete polypeptide species by two independent separation steps: isoelectric focusing and sodium dodecyl sulfate-polyacrylamide gel electrophoresis (SDS-PAGE). This approach to cellular physiology and regulation offers the possibility of constructing a genome expression map that displays the 2D gel location of every polypeptide encoded by the genome and matches each polypeptide spot to its gene. The purpose for creating this site goes beyond building the "master" 2D gel for *E. coli*. Instead, the overarching purpose is to provide investigators with physiological and regulatory data on the entire set of *E. coli* proteins. The goal is to monitor under what circumstances, to what level, and for what function each protein-encoding gene is expressed.

There are two components that serve as the foundation of the ECO2DBASE site. First, the Genome Expression Map Project (GEM) is designed to link each of the protein-encoding genes to a spot on the 2D gel. The second project, called the Response/Regulation Map Project (RRM), is focused on discovering the conditions under which each gene is expressed and on determining the underlying mechanism of regulation. In combination with the recently completed genome for *E. coli*, this site could become a paradigm for the functional analysis of cellular physiology, metabolism, and regulation.

Internet address

Web: http://pcsf.brcf.med.umich.edu/eco2dbase

■ Tree of Life

The Tree of Life is one of the most ambitious projects attempted on the Web or the Internet. Developed by David R. Maddison and Wayne P. Maddison at the University of

Arizona, this site is a distributed Internet project about phylogeny and biodiversity. It is designed to link information about the phylogenetic relationships and characteristics of organisms, to illustrate the diversity and unity of living organisms, and to link biological information available on the Internet in the form of a phylogenetic navigator. While still under development, the Tree of Life offers an unprecedented look at the power of genomic and molecular biology when integrated with the resources of the Internet through the Web.

Internet address

Web: http://phylogeny.arizona.edu/tree/phylogeny.html

■ TreeBASE

TreeBASE was initiated in 1996 by Michael Donoghue of Harvard University and Michael Sanderson of the University of California at Davis. The site focuses primarily on green plants. It is designed to serve as a repository and link to all morphological and genetic data for all phylogenetic trees and their supporting data. Links are maintained to the primary sequence data and mapping information as well as related resources such as the Web site Tree of Life, described above.

Internet addresses

Web: http://phylogeny.harvard.edu/treebase

                        http://herbaria.harvard.edu/treebase

Selected readings

Journal Articles

Blake, J. A., C. J. Bult, M. J. Donoghue, J. Humphries, and C. Fields. 1994. Interoperability of biological databases: a meeting report. *Syst. Biol.* **43**:585-589.

Cohen, D., I. Chumakov, and J. Weissenbach. 1993. A first-generation physical map of the human genome. *Nature* (London) **366**:698-701.

Koonin, E. V., R. L. Tatusov, and K. E. Rudd. 1995. Sequence similarity analysis of Escherichia coli proteins - functional and evolutionary implications. *Proc. Natl. Acad. Sci. USA* **92**:11921-11925.

Morell, V. 1996. TreeBASE: the roots of phylogeny. *Science* **273**:569.

Pennisi, E. 1996. From genes to genome biology. *Science* **272**:1736-1738.

Sanderson, M. J., M. J. Donoghue, W. Piel, and T. Eriksson. 1994. TreeBASE: a prototype database of phylogenetic analyses and an interactive tool for browsing the phylogeny of life. *Am. J. Bot.* **81**:183.

Smith, R. F. 1995. A brief guide to information resources supporting the Human Genome Project. *IEEE Eng. Med. Biol.* **14**:760-761.

Books and Monographs

Berlyn, M. B., K. B. Low, and K. E. Rudd. 1996. Integrated linkage map of *Escherichia coli* K-12, edition 9. *In* F. C. Neidhardt, R. Curtiss III, J. L. Ingraham, E. C. C. Lin, K. B. Low, B. Magasanik, W. Reznikoff, M. Riley, M. Schaechter, and H. E. Umbarger (ed.), *Escherichia coli and Salmonella: Cellular and Molecular Biology,* 2nd ed. ASM Press, Washington, D.C.

Hahn, H. 1996. *The Internet Complete Reference,* 2nd ed. Osbourne McGraw-Hill, Berkeley, Calif.

Swindell, S. R., R. R. Miller, and G. S. A. Myers (ed.). 1995. *Internet for the Molecular Biologist.* Horizon Scientific Press, Norfolk, England.

VanBogelen, R. A., K. Z. Abshire, A. Pertsemlidis, R. L. Clark, and F. C. Neidhardt. 1996. Gene-protein database of *Escherichia coli* K-12: edition 6. *In* F. C. Neidhardt, R. Curtiss III, J. L. Ingraham, E. C. C. Lin, K. B. Low, B. Magasanik, W. Reznikoff, M. Riley, M. Schaechter, and H. E. Umbarger (ed.), *Escherichia coli and Salmonella: Cellular and Molecular Biology,* 2nd ed. ASM Press, Washington, D.C.

Internet Publication

Hayden, D. 1994. *Guide to Molecular Biology Databases.* School of Library and Information Studies, University of Alberta, Calgary, Alberta, Canada.

CHAPTER 10 The Future of the New Biology and Its Internet Tools

More than ten years ago, the Internet was opened to the greater scientific and technological community when it made the transition from a Cold War instrument to a peacetime communication and information system. Almost simultaneously, the notion of determining the complete sequence of a free-living organism was seen as possible and the biological research community rose to explore this new frontier. Thus, the field of genomic biology and the resources of the Internet have grown up together along parallel paths that eventually became intertwined in the 1990s with the birth of the Web.

As a result of the convergence of these two communities, the entire biomedical research field now stands at a crossroads. Two questions are being asked. First, how will the Internet and its scientific analysis tools evolve? Second, how many genomes should be sequenced? Answers to both questions are also intertwined. The answer to the second question hinges primarily on what is learned from the first few completely sequenced genomes and what new questions will inevitably arise from their analysis. The answer to the first question is dependent on what questions the genomic biologists want to ask and what tools will be needed to address them.

More than five years ago, Nobel laureate Walter Gilbert urged the budding genomic biology community to pursue computer literacy and to tap into the tools of a worldwide network for both communication and database access. This recommendation has proven correct and farsighted. The key resources and informational exchanges that are vital for a worldwide cooperative approach to research reside on the Internet. Projects on the scale of sequencing entire genomes could not be designed, managed, and completed without access to its tools.

In many cases, Internet resources have supplanted or quickly will supplant existing locally based electronic archives and even printed texts. This was predicated based on several interrelated factors: the volume of data that has accumulated, the speed at which it accumulates, and the need to compare, analyze, and curate this data. A final selective pressure was the need for reliable, high-speed, if not nearly instantaneous access to this data.

As a result, sophisticated tools have grown up in this environment that have, in turn, spawned a user community that demands ever more sophisticated tools and an increasingly complex architecture of federated databases which share informational resources. The major sequence databases, for example, are so tightly linked that their informational content is nearly equivalent at all times.

With these thoughts in mind, what are the guiding principles for an Internet resource for the greater research community, be it Web site, database, or analysis tool? First: completeness. A resource should offer thorough coverage of its topic; the focus can be nar-

270

row, like ECO2DBASE, the gene-protein index of *Escherichia coli*, described in Chapter 9, or the analysis tool nnPredict, described in Chapter 6; or broad as in the major international sequence databases described in Chapter 4 or the ExPASy Web site covered in Chapter 9. The key is substance rather than style and appearance; the scientific community is looking for informational content when accessing a site or an analysis tool, not flashy graphics and a maze of links. A second requirement is ease of use and functionality. A site can contain useful resources, but they are of little use (and little used) if they are difficult to access. E-mail servers should offer straightforward directives and parameters based in part on those used by existing servers. Web sites should guide and link users to information in a coherent and directed fashion, rather than in a series of blind passageways that waste time and effort.

Third, the information and content should be curated and developed in concert with other resources. Simply put: don't reinvent the wheel; look to see what is available in the community. Determine what gaps exist, and design tools that fill those gaps and complement the existing resource base. A resource, particularly a database, is only as accurate as the accuracy of the raw information that it was built on. This means constant checking of informational content at several steps along the way: when the data is first being prepared for entry into a database (*not* after it has been posted); again through cross-checks with other databases and literature resources; and again by encouraging the user community to report problems and make suggestions.

The fourth requirement is lasting and current resources and tools. Data and the tools to access it are constantly expanding in complexity, volume, and utility. Resources must reflect this reality; they need to be updated regularly and to have a long lifespan in order to develop into a tool that is seen as a benchmark rather than a whim. Do not build a resource on the Internet unless the support infrastructure can be put in place to make it a lasting proposition.

The prospect of a new biology that can tap the growing resources of molecular sequence databanks and the analysis resources of the Internet and other computational tools is promising. Vast new sources of information are becoming available as a result of the increasing number of genomic sequencing projects. While this flood of information at first glance may appear overwhelming, it offers unprecedented possibilities for research and study. The current generation of informational and analysis servers we have described here offers a glimpse of what is currently possible and allows new minds to explore this biological world. Both current biomedical scientists and newcomers will be needed to nurture and shape this infant field. As computational tools and integrated information systems become more sophisticated and accessible, a new biology that crosses both experimental and theoretical arenas will take shape. Within this new biology it will be possible to grasp the complexity of an entire living organism and delve into the fundamental questions of what life is and how it functions.

Within this framework of a new biology, the central issues facing computational biology and sequence analysis might seem to be based chiefly on the relationship between molecular structure and function. Clearly, much of what is learned from the analysis of the first few complete genomes will have a direct and long-range impact on the essential questions of structure-function relationships. But, while structure-function relationships are certainly important, it must be emphasized that other scientific issues also can and will be addressed by computational analysis strategies. Continued improvements in sequence analysis at both the conceptual and application levels will be key in meeting these challenges.

With that in mind, what are the other central problems that will be addressed by genomic and molecular biology in the coming decade?

- **Elucidation of metabolic pathways and regulatory mechanisms.** A rapid estimation of the metabolic pathways present in microbial organisms can be achieved from a partial list of enzymes, which in turn can be acquired directly from sequence similarity data. To do this requires a fairly comprehensive collection of known pathways, along with protein sequences connected to the enzyme numbers. These now exist and are available. The related problem of determining a set of enzymes that must almost certainly exist, but for which no sequence has yet been identified, becomes a more manageable task in the presence of a large, organized body of known pathways. Connecting enzymes that must be present to unidentified sequences should be simplified by the presence of multiple complete genomes. By extension, the second part of this question, that of understanding the regulatory mechanisms involved in the control of metabolism, will likely prove more challenging. It seems likely that the essential components of such regulatory processes will be identified long before the exact structural basis and mechanism for the regulation can be fully understood.

- **Composition of protein families.** Even though a thorough understanding of structure-function relationships may be long in coming, an understanding of the composition and evolution of protein families will likely proceed rapidly. These insights will play a major supporting role in numerous specific problems. The clarification of how function evolved in specific cases will frequently be based initially on alignments, phylogenetic evidence, and correlation with known structural data. In addition, with the availability of a number of phylogenetically diverse, complete genomes it is now possible to rapidly identify operon structures and regulatory signals. Further, this information can be applied to assign functions to unidentified sequences from other organisms. The continued integration of these types of information into knowledge bases should speed the solutions of these questions.

While these two broad questions are both significant in their own right, they are also deeply interrelated. It is clear that advances on either question will have a direct impact on the other. For example, there are the obvious relationships between questions of metabolic organization and regulation as related to genetic regulation and genomic structure. Two other obvious avenues of study that can be more readily approached as information accumulates from genomic sequencing projects are questions concerning the structure of protein families and conservation in alignments and functional sites, as well as the phylogenic analysis of organisms and proteins. In each of these cases, comparative analysis will play a central role. Effective access to diverse categories of data will accelerate processes that now often require a number of discrete and cumbersome steps.

The last question that needs to be considered in the short term is how long it will take until a function is assigned to every gene in a single genome. The endpoint of genome analysis is not at the level of sequence determination. Instead, the endpoint lies with a functional understanding of the genome and its components. Sequence determination is a key step in that process, but it is only a step. Think about the impact such a base of knowledge would have on the study of other organisms. Broad inferences and generalizations could be made throughout and across phylogenetic trees. Although many questions would remain unanswered, this foundation would create a methodology for dissecting more complex genomes and gaining an infinitely deeper understanding of life.

Because of the broad progress being made in genomic and molecular biology, in the not too distant future it should be possible to merge the wealth of data on biological systems with the developing field of interactive multimedia to recreate specific cellular processes in real time. Using this concept as a starting point, it is a simple leap of the imagination to generate a virtual cell. In this vision of a new biological laboratory, basic cellular processes could be observed in a real-time fashion: proteins being synthesized, damaged DNA being repaired, cells replicating.

In this world, new types of questions could be asked and answered, questions that are simply too complex to ask in the current laboratory setting. For example, the effects of an alteration in the protein structure of an enzyme could be quickly tested merely by changing the desired amino acids in the stored protein sequence data. A computer would then generate a new three-dimensional model of the protein and introduce it into the cell. This mutated protein would then be integrated into the virtual cell, and both its immediate and long-term effects could be studied. Much of the current trial-and-error system of analyzing cellular function could be accomplished in a faster, far more interactive manner by using a virtual cell system. Traditional biological techniques would not be made obsolete by this virtual biology, but would work in concert with it. The virtual biologist would be able to test different scenarios for the real world, and the most promising would be implemented in actual systems.

At this point, the virtual cell is little more than words on paper. Unless the existing computational tools evolve and new, revolutionary systems emerge, the promises that the new biology holds will largely remain unfulfilled. These new tools will not emerge solely from the present biomedical research community. New minds from many different fields will need to contribute in order to make this new biology a reality.

Selected readings

Journal Articles

Boguski, M. S., and G. D. Schuler. 1995. Establishing a human transcript map. *Nature Genet.* **10**:369-371.

Collins, F., and D. Galas. 1993. A new five-year plan for the U.S. human genome project. *Science* **262**:43-46.

Gilbert, W. 1991. Towards a paradigm shift in biology. *Nature* (London) **349**:99.

Nierlich, D. P. 1996. Future directions for biomolecular databases. *ASM News* **62**:251-254.

Karp, P. 1995. A strategy for database interoperation. *J. Comput. Biol.* **2**:573-586.

Olson, M. V. 1995. A time to sequence. *Science* **270**:394-396.

Pennisi, E. 1996. From genes to genome biology. *Science* **272**:1736-1738.

Williamson, A. R., K. O. Elliston, and J. L. Sturchio. 1995. The Merck Gene Index: a public resource for genomics research. *J. NIH Res.* **7**:61-64.

Books and Monographs

U.S. Department of Health and Human Services and Department of Energy. 1990. *Understanding Our Genetic Inheritance: The U.S. Human Genome Project: The First Five Years.* National Technical Information Service, U.S. Department of Commerce, Springfield, Va.

Internet Publication

Hayden, D. 1994. *Guide to Molecular Biology Databases.* School of Library and Information Studies, University of Alberta, Calgary, Alberta, Canada.

Appendices

A. Internet addresses

B. Domains and usage

C. Computer hardware and software requirements

D. Nucleic acid and amino acid codes and properties

E. Glossary of terms

These five appendices serve as a quick reference guide both to the resources on the Internet discussed in the book and for molecular and genomic biology in general. Appendix A lists Internet addresses for all of the resources described in the chapters as well as a wide range of other useful servers, databases, and resources. This listing is divided by category and is meant to serve as a "cheat sheet" or Internet "phone book" that may be copied and kept alongside the laboratory computer. Appendix B lists the current top-level Internet domains and their usage. This can be useful in deciphering where a particular server or database is located geographically and in correcting addresses. Appendix C summarizes the minimum hardware and software requirements of a laboratory computer for genomic and molecular biology-related tasks and projects. Appendix D contains several related tables that summarize useful data on amino acids, nucleic acids, and genetic codes. Finally, Appendix E is a glossary of terms used in this text. It is designed for the computer and Internet novice as an aid to learning the terminology so that this vast electronic world will seem less intimidating and more inviting.

Appendix A. Internet Addresses

This listing of Internet addresses for servers, resources, and databases is not exhaustive, nor is it meant to be. There are literally thousands of biology-related sites on the Internet. Many, if not most, are at least somewhat relevant to genomic and molecular biology. If all of the individual laboratory Web sites were also folded into this list, it would be hopelessly unmanageable in scope and in sheer size. Instead, this appendix is designed to give a good cross-section of addresses for sequence analysis, pattern matching, information and news, and organizations. To remain as useful as possible and to reflect changes in this fast-evolving field, this listing will be updated frequently on the ASM Web site (http://www.asmpress.org/isbn/1555811191/).

■ Databases

Name	Internet Address		Purpose
DDBJ	*E-mail:*	ddbj@ddbj.nig.ac.jp	Information
(DNA DataBank of Japan)		ddbjsub@ddbj.nig.ac.jp	Data submission
		ddbjupdt@ddbj.nig.ac.jp	Data updates
		fasta@nig.ac.jp	FastA e-mail server
		blast@nig.ac.jp	BLAST e-mail server
		malign@nig.ac.jp	MAlign e-mail server
	Web:	http://www.ddbj.nig.ac.jp	Home page
EBI (European	*E-mail:*	datalib@ebi.ac.uk	Information
Bioinformatics Institute)		datasubs@ebi.ac.uk	Data submission
		netserv@ebi.ac.uk	E-mail fileserver
		nethelp@ebi.ac.uk	Help system
		update@ebi.ac.uk	Data updates, corrections
		blitz@ebi.ac.uk	BLITZ e-mail server
		fasta@ebi.ac.uk	FastA e-mail server
		prosite@ebi.ac.uk	PROSITE e-mail server
		retrieve@ebi.ac.uk	Retrieve e-mail server
	Gopher:	gopher.ebi.ac.uk	EBI Gopher server
	Web:	http://www.ebi.ac.uk	EBI home page
NCBI (National Center for	*E-mail:*	info@ncbi.nlm.nih.gov	Information
Biotechnology Information)		gb-sub@ncbi.nlm.nih.gov	Data submission
(GenBank)		update@ncbi.nlm.nih.gov	Data update
		blast@ncbi.nlm.nih.gov	BLAST server
		retrieve@ncbi.nlm.nih.gov	Retrieve server
	Web:	http://www.ncbi.nlm.nih.gov	NCBI Web site
		http://www3.ncbi.nlm.nih.gov/omim	
			OMIM Web site
GSDB (Genome Sequence	*E-mail:*	ncgr@ncgr.org	Information on NCGR
Database)		gsdbhelp@ncgr.org	Information on GSDB

Name	Internet Address		Purpose
GSDB (continued)	*E-mail:*	datasubs@gsdb.ncgr.org	Data submission
		update@gsdb.ncgr.org	Data updates
		websub@gsdb.ncgr.org	Help with Web submission
	Web:	http://www.ncgr.org/gsdb/	GSDB Web server
		http://www.ncgr.org/gsdb/websub.html	
			Data submission
		http://www.ncgr.org/gsdb/update.html	
			Data updates
Blocks	*E-mail:*	henikoff@howard.fhcrc.org	General help/information
		blocks@howard.fhcrc.org	Server address
		blockmaker@howard.fhcrc.org	Block construction server
	Web:	http://blocks.fhcrc.org	
GDB (Human Genome Database)	*E-mail:*	help@gdb.org	General help/information
		data@gdb.org	Data submission
		mailserv@gdb.org	GDB e-mail server
	Web:	http://gdbwww.gdb.org/	Home page (U.S.A.)
		http://morgan.angis.su.oz.au/gdb/gdbtop.html	
			Australia
		http://gdb.infobiogen.fr/	France
		http://gdbwww.dkfz-heidelberg.de/	
			Germany
		http://inherit1.weizman.ac.il/gdb/docs/gdbhome.html	
			Israel
		http://gdb.gdbnet.ad.jp/gdb/docs/gdbhome.html	
			Japan
		http://www-gdb.caos.kun.nl/gdb/gdbtop.html	
			Netherlands
		http://www.hgmp.mrc.ac.uk/gdb/docs/gdbhome.html	
			United Kingdom
		http://www.embnet.se/gdb/index.html	
			Sweden
PDB (Protein DataBank)	*E-mail:*	pdbhelp@bnl.gov	Information
		pdb@bnl.gov	Data submission

Name	Internet Address		Purpose
PDB (continued)	*E-mail:*	orders@pdb.pdb.bnl.gov	Orders
		errata@pdb.pdb.bnl.gov	Data updates
	Gopher:	gopher://pdb.pdb.bnl.gov	Gopher site
	Web:	http://www.pdb.bnl.gov	PDB Web site
PIR (Protein Identification Resource)	*E-mail:*	pirmail@nbrf.georgetown.edu	General help/information, Americas
		mewes@mips.embnet.org	General help/information, Europe and Africa
		tsugita@jipdalph.rb.noda.sut.ac.jp	General help/information, Asia and Australia
		pirsub@nbrf.georgetown.edu	Data submission
		fileserv@ nbrf.georgetown.edu	E-mail server
	Web:	http://www-nbrf.georgetown.edu	Americas
		http://www.mips.biochem.mpg.de	Europe and Africa
PROSITE	*E-mail:*	bairoch@cmu.unige.ch	General help/information
		prosite@ebi.ac.uk	PROSITE server
	Web:	http://expasy.ch	ExPASy server
REBASE	*E-mail:*	roberts@neb.com	Information/mailing lists
		macelis@neb.com	Information/mailing lists
	Web:	http://www.neb.com/rebase	Home page
SWISS-PROT	*E-mail:*	bairoch@cmu.unige.ch	Information
	Web:	http://expasy.ch	ExPASy server

■ Analysis Servers

Similarity and comparison

Name	Internet Address		Purpose
BLAST	*E-mail:*	blast@ncbi.nlm.nih.gov	NCBI server
		blast@ebi.ac.uk	EBI server
		blast-help@ncbi.nlm.nih.gov	Personal help

Name	Internet Address		Purpose
Blast (continued)	*Web:*	http://www.ncbi.nlm.nih.gov	NCBI home page
Bicserv	*E-mail:*	bicserv@sgbcd.weizmann.il	Server
		mail-server-comments@compugen.co.il	
			Personal help
BioSCAN	*E-mail:*	bioscan@cs.unc.edu	Server
		bioscan-info@cs.unc.edu	Personal help
	Web:	http://genome.cs.unc.edu/bioscan.html	
			Home page
BLITZ	*E-mail:*	blitz@ebi.ac.uk	Server
		nethelp@ebi.ac.uk	Personal help
		mpsrch_help@biocomp.ed.ac.uk	Help software
	Web:	http://www.ebi.ac.uk	EBI home page
DDBJ servers	*E-mail:*	fasta@nig.ac.jp	FastA server
		blast@nig.ac.jp	BLAST server
		malign@nig.ac.jp	MAlign server
		trouble@nig.ac.jp	Personal help
		ddbj@ddbj.nig.ac.jp	Information
	Web:	http://www.ddbj.nig.ac.jp	DDBJ home page
GenQuest	*E-mail:*	Q@ornl.gov	Server
		grailmail@ornl.gov	Personal help
	Web:	http://avalon.epm.ornl.gov/grail-bin/	
			GenQuest home page
Mail-FastA	*E-mail:*	fasta@ebi.ac.uk	EBI server
		mfasta@genius.embnet.dkfz-heidelberg.de	
			EMBnet server
		nethelp@ebi.ac.uk	Personal help
	Gopher:	gopher.ebi.ac.uk	EBI Gopher server
	Web:	http://www.ebi.ac.uk	EBI home page

Name	Internet Address		Purpose
Mail-QUICKSEARCH	*E-mail:*	quick@ebi.ac.uk	Server
		nethelp@ebi.ac.uk	Personal help
	Gopher:	gopher.ebi.ac.uk	EBI Gopher server
	Web:	http://www.ebi.ac.uk	EBI home page

Retrieval

Name	Internet Address		Purpose
GENIUSnet/NetServ	*E-mail:*	netserv@ebi.ac.uk	NetServ server
sequence servers		nethelp@ebi.ac.uk	Help system
		netserv@genius.embnet.dkfz-heidelberg.de	
			GENIUSnet server
	Web:	http://www.ebi.ac.uk	EBI home page
Query	*E-mail:*	query@ncbi.nlm.nih.gov	Server
		info@ncbi.nlm.nih.gov	Information
	Web:	http://www.ncbi.nlm.nih.gov	NCBI home page
Retrieve	*E-mail:*	retrieve@ncbi.nlm.nih.gov	NCBI server
		retrieve@ebi.ac.uk	EBI server
		retrieve-help@ncbi.nlm.nih.gov	Personal help
	Web:	http://www.ncbi.nlm.nih.gov	NCBI home page
STS and EST report servers	*E-mail:*	retrieve@ncbi.nlm.nih.gov	NCBI server
		retrieve@ebi.ac.uk	EBI server
		retrieve-help@ncbi.nlm.nih.gov	Personal help
	Web:	http://www.ncbi.nlm.nih.gov	NCBI home page

Domain, pattern, and profile analysis

Name	Internet Address		Purpose
Blocks and Block Maker	*E-mail:*	blocks@howard.fhcrc.org	Blocks server
		blockmaker@howard.fhcrc.org	Block Maker server
		henikoff@howard.fhcrc.org	Personal help
	Web:	http://blocks.fhcrc.org	Home page
CBRG Servers	*E-mail:*	cbrg@inf.ethz.ch	Server
		korosten@inf.ethz.ch	Personal help
	Web:	http://cbrg.inf.ethz.ch	Home page
Domain	*E-mail:*	domain@hubi.abc.hu	Server
		hegyi@hubi.abc.hu	Personal help
	Web:	http://www.abc.hu/blast.html	Home page
MotifFinder	*E-mail:*	motiffinder@genome.ad.jp	Server
		motif-manager@genome.ad.jp	Personal help
	Web:	http://www.genome.ad.jp/SIT/MOTIF.html	
			Home page
ProDom Domain server	*E-mail:*	prodom@toulouse.inra.fr	Server
		proquest@toulouse.inra.fr	Personal help
		Daniel.Kahn@toulouse.inra.fr	Daniel Kahn
		Jerome.Gouzy@toulouse.inra.fr	Jerome Gouzy
		Florence.Corpet@toulouse.inra.fr	Florence Corpet
	Web:	http://protein.toulouse.inra.fr/prodom.html	
			Home page, France
PSORT	E-mail:	psort@nibb.ac.jp	Server
		nakai@nibb.ac.jp	Personal help
	Web:	http://psort.nibb.ac.jp	Home page

Name	Internet Address		Purpose
SBASE	*E-mail:*	sbase@icgeb.trieste.it	Server
		comment@icgeb.trieste.it	Personal help
	Gopher:	icgeb.trieste.it	Gopher server
	Web:	http://www.icgeb.trieste.it	Home page

Gene identification

Name	Internet Address		Purpose
EcoParse	*E-mail:*	ecoparse@cse.ucsc.edu	Server
GeneID	*E-mail:*	geneid@darwin.bu.edu	Server
		graf@darwin.bu.edu	Personal help: GeneID
		engel@virus.fki.dth.dk	Personal help: NetGene
GeneMark	*E-mail:*	genemark@ford.gatech.edu	Server, U.S.A.
		genemark@embl-ebi.ac.uk	Server, EBI
		mb56@prism.gatech.edu	Personal help, U.S.A.
		gt1619a@prism.gatech.edu	Personal help, U.S.A.
		nethelp@embl-ebi.ac.uk	Personal help, EBI
	Web:	http://amber.biology.gatech.edu/~genemark	
			Information and news
NetGene	*E-mail:*	netgene@cbs.dtu.dk	Server
		rapacki@cbs.dtu.dk	Personal help

Secondary structure

Name	Internet Address		Purpose
nnPredict	*E-mail:*	nnpredict@celeste.ucsf.edu	Server
		nnpredict-request@celeste.ucsf.edu	
			Personal help
	Web:	http://www.cmpharm.ucsf.edu/~nomi/nnpredict.html	
			Home page

Name	Internet Address		Purpose
PredictProtein	*E-mail:*	predictprotein@embl-heidelberg.de	Server
		predict-help@embl-heidelberg.de	Personal help
	Web:	http://www.embl-heidelberg.de/predictprotein	Home page
SOPMA (Self Optimized Prediction Method from Alignments)	*E-mail:*	deleage@ibcp.fr	Server
	Web:	http://www.ibcp.fr	Home page

Miscellaneous

Name	Internet Address		Purpose
BIOSCI Newsgroup Network	*E-mail:*	biosci-server@daresbury.ac.uk	News and information site for Europe, Africa, Asia
		biosci-server@net.bio.net	News and information site for Americas, Pacific Rim
		biosci-help@net.bio.net	Personal help
	Web:	http://www.bio.net	Home page
		http://www.bio.net/archives.html	Newsgroups archive page
		http://www.bio.net/bio-journals.html	Journals archive page
		http://www.bio.net/hypermail/employment/	Employment archive pages
		http://www.bio.net/addrsearch.html	Address database search page
		http://www.bio.net/hypermail/methds-reagnts/	Methods newsgroup archive pages

■ Web and Gopher Organizations

Name	Internet Address
American Society for Microbiology	http://www.asmusa.org
ATCC: American Type Culture Collection	http://www.atcc.org gopher://culture.atcc.org
Cell Press	http://www.cell.com
Coli Genetic Stock Center (CGSC)	http://cgsc.biology.yale.edu/
Cold Spring Harbor Laboratory	http://www.cshl.org
Genetics Computer Group (GCG)	http://www.gcg.com
Jackson Laboratory Server	http://www.jax.org
Molecular Biology Related Resources	http://expasy.ch/cgi-bin/listdoc
Nature Journal	http://www.nature.com
New England BioLabs	http://www.neb.com
Protein Science's List of Links	http://www.prosci.uci.edu gopher://gopher.prosci.uci.edu
Science Magazine	http://www.sciencemag.org

Information and resources

Name	Internet Address
Biological Computing Division (BCD) Web Biology Server	http://dapsas1.weizmann.ac.il
Biologist's Control Panel	http://gc.bcm.tmc.edu:8088/bio
GenomeNet	http://www.genome.ad.jp
National Institute of Genetics (NIG)	http://www.nig.ac.jp/ gopher://gopher.nig.ac.jp/
Agricultural Genome Information Server (AGIS)	http://probe.nalusda.gov:8000
Australian National Genomic Information Service (ANGIS)	http://morgan.angis.su.oz.au
CEPH-Genethon	http://www.genethon.fr/genethon_fr.html gopher://gopher@genethon.fr
Cooperative Human Linkage Center (CHLC)	http://www.chlc.org gopher://gopher.chlc.org
ExPASy Molecular Biology Server	http://expasy.ch

Name	Internet Address
GenLink	http://www.genlink.wustl.edu
Human Genome Management Information System (HGMIS)	http://www.ornl.gov/hgmis
International Centre for Genetic Engineering and Biotechnology (ICGEB)	http://www.icgeb.trieste.it gopher://icgeb.trieste.it
Johns Hopkins Bioinformatics Server	http://www.gdb.org
Listing of biology internet servers	http://info.er.usgs.gov/network/science/biology/index.html
Pedro's BioMolecular Research Tools	http://www.public.iastate.edu/~pedro/research_tools.html
	http://www.biophys.uniduesseldorf.de/bionet/research_tools.html
	http://www.fmi.ch/biology/research_tools.html http://www.peri.co.jp/Pedro/research_tools.html
Virtual Library: Biochemistry, Biophysics, and Molecular Biology	http://golgi.harvard.edu/biopages/biochem.html

Knowledge bases and databases

Name	Internet Address
16S Ribosomal RNA Mutation Database (16SMDB) and the 23S Ribosomal RNA Mutation Database (23SMDB)	http://www.fandm.edu/Departments/Biology/Databases/RNA.html
C. elegans Genome Project	http://www.sanger.ac.uk http://genome.wustl.edu
Cholinesterase Gene Server Database (ESTHER)	http://www.montpellier.inra.fr:70/cholinesterase
Codon Usage Tabulated from GenBank (CUTG)	http://tisun4a.lab.nig.ac.jp/codon/CUTG.html http://www.dna.affrc.go.jp/~nakamura/codon.html
EcoCyc	http://www.ai.sri.com/ecocyc/ecocyc.html
E. coli Database (ECD)	http://susi.bio.unigiessen.de/usr/local/www/html/ecdc.html
ENZYME Databank	http://expasy.ch/sprot/enzyme.html
FlyBase	http://cbbridges.harvard.edu:7081 (U.S.A.-Harvard) http://flybase.bio.indiana.edu:82 (U.S.A.-Indiana) http://www.embl-ebi.ac.uk/flybase (U.K.) http://astorg.u-strasbg.fr:7081 (France) http://www.angis.su.oz.au:7081 (Australia) http://shigen.lab.nig.ac.jp:7081 (Japan)

Name	Internet Address
Gene-Protein Database of E. coli (ECO2DBASE)	http://pcsf.brcf.med.umich.edu/eco2dbase
GenProtEc: database of *E. coli* gene products	http://mbl.edu/html/ecoli.html
Glucocorticoid Receptor Resource	http://nrr.georgetown.edu/grr/GRR/GRR.html
GRBase: Gene Regulation Database	http://www.access.digex.net/~regulate/trevgrb.html
Haemophilia A Mutation Search Test and Resource Site (HAMSTeRS)	http://www.hamsters.rpms.ac.uk
Haemophilus influenzae genome	http://www.tigr.org/

Human chromosome servers

Chromosome 1	http://linkage.rockefeller.edu/chr1
Chromosome 3	http://mars.uthscsa.edu/
Chromosome 7	http://www.genet.sick kids.on.ca/chr7db/query.html
Chromosome 8	http://gc.bcm.tmc.edu:8088/chr8/home.html
Chromosome 9	http://www.gene.ucl.ac.uk/chr9/
Chromosome 10	http://www.cric.com/htdocs/chr10-mapping/index.html
Chromosome 11	http://chr11.bc.ic.ac.uk/ http://mcdermott.swmed.edu/datapage
Chromosome 12	http://paella.med.yale.edu/chr12/home.html
Chromosome 13	http://genome1.ccc.columbia.edu/~genome
Chromosome 16	http://www-ls.lanl.gov/dbqueries.html http://www.tigr.org/tdb/humgen/c16.html
Chromosome 17	http://bioinformatics.weizmann.ac.il/
Chromosome 19	http://www-bio.llnl.gov/bbrp/genome/genome.html
Chromosome 20	http://www.expasy.ch/cgi-bin/lists?humchr20.txt
Chromosome 21	http://expasy.ch/cgi-bin/lists?humchr21.txt
Chromosome 22	http://www.cbil.upenn.edu/HGC22.html http://www.expasy.ch/cgi-bin/lists?humchr22.txt http://www.sanger.ac.uk/hum22
Chromosome X	http://gc.bcm.tmc.edu:8088/chrX/home.html http://www.expasy.ch/cgi-bin/lists?humchrx.txt
Chromosome Y	http://www.expasy.ch/cgi-bin/lists?humchry.txt
Mitochondrial chromosome	http://infinity.gen.emory.edu/mitomap.html
Large Ribosomal Subunit RNA Structure Database	http://rrna.uia.ac.be/rrna/lsuform.html

Name	Internet Address
LIST (Yeast genetic database)	http://www.ch.embnet.org/
MIPS Protein Database/Yeast Genome Database	http://www.mips.biochem.mpg.de
Molecular Probe Database (MPDB)	http://www.ist.unige.it/interlab/mpdb.html
NRSub: *Bacillus subtilis* genome database	http://ddbjs4h.genes.nig.ac.jp/ http://acnuc.univ-lyon1.fr//nrsub/nrsub.html
PAH Mutation Analysis Consortium database	http://www.mcgill.ca/pahdb
PRINTS Protein Fingerprint Database	http://www.biochem.ucl.ac.uk/bsm/dbbrowser/PRINTS/ PRINTS.html
Ribosomal Database Project (RDP)	http://rdpwww.life.uiuc.edi
RNA Modification Database	http://www-medlib.med.utah.edu/RNAmods/RNAmods.html
QUEST Protein Database Center	http://siva.cshl.org
Saccharomyces Genome Database (SGD)	http://genome-www.stanford.edu/Saccharomyces
Signal Recognition Particle Database (SRPDB)	http://pegasus.uthct.edu/SRPDB/SRPDB.html
Small Ribosomal Subunit RNA Structure Database	http://rrna.uia.ac.be/rrna/ssuform.html
Small RNA Database	http://mbcr.bcm.tmc.edu/smallRNA/smallrna.html
SWISS-2DPAGE	http://www.expasy.ch
TransTerm: a database of translation factors	http://biochem.otago.ac.nz.800/Transterm/home_page.html
Tree of Life	http://phylogeny.arizona.edu/tree/phylogeny.html
TreeBASE	http://phylogeny.harvard.edu/treebase http://herbaria.harvard.edu/treebase
uRNA Database	http://pegasus.uthct.edu/uRNADB/uRNADB.html
Yeast Protein Database (YPD)	http://www.proteome.com/YPDhome.html

Appendix B. Domains and Usage

The following table lists common top-level Internet domains and their associated country and/or type of institution. This list can help identify what country a server is located in as well as its institutional or organizational affiliation.

Domain Definition	Domain
Commercial (primarily U.S.A.)	.com
Educational (primarily U.S.A.)	.edu
Government (primarily U.S.A.)	.gov
Network	.net
Non-commercial organization	.org
Old-style ARPA net	.arpa
Old-style international organization	.int
Military (U.S.A. only)	.mil
Algeria	.dz
Antarctica	.ag
Argentina	.ar
Armenia	.am
Australia	.au
Austria	.at
Azerbaijan	.az
Belarus	.by
Belgium	.be
Bermuda	.bm
Brazil	.br
Bulgaria	.bg
Cameroon	.cm
Canada	.ca
Chile	.cl
China	.cn
Colombia	.co
Costa Rica	.cr
Croatia	.hr

Domain Definition	Domain
Cyprus	.cy
Czech Republic	.cz
Denmark	.dk
Ecuador	.ec
Egypt	.eg
Estonia	.ee
Faroe Islands	.fo
Fiji	.fj
Finland	.fi
France	.fr
Germany	.de
Great Britain	.gb
Greece	.gr
Greenland	.gl
Guam (U.S.A.)	.gu
Guinea	.gn
Hong Kong	.hk
Hungary	.hu
Iceland	.is
India	.in
Indonesia	.id
Iran	.ir
Ireland	.ie
Israel	.il
Italy	.it
Jamaica	.jm
Japan	.jp
Kazakstan	.kz
Kuwait	.kw
Latvia	.lv
Lebanon	.lb
Liechtenstein	.li

Domain Definition	Domain
Lithuania	.lt
Luxembourg	.lu
Malaysia	.my
Mexico	.mx
Moldavia	.md
Morocco	.ma
Namibia	.na
Netherlands	.nl
New Zealand	.nz
Nicaragua	.ni
Norway	.no
Panama	.pa
Peru	.pe
Philippines	.ph
Poland	.pl
Portugal	.pt
Puerto Rico	.pr
Reunion (French)	.re
Romania	.ro
Russian Federation	.ru
Saudi Arabia	.sa
Singapore	.sg
Slovak Republic	.sk
Slovenia	.si
South Africa	.za
South Korea	.kr
Soviet Union	.su
Spain	.es
Sri Lanka	.lk
Svalbard and Jan Mayen Islands	.sj
Sweden	.se
Switzerland	.ch

Domain Definition	Domain
Taiwan	.tw
Thailand	.th
Trinidad and Tobago	.tt
Tunisia	.tn
Turkey	.tr
Ukraine	.ua
United Arab Emirates	.ae
United Kingdom	.uk
United States	.us
Uruguay	.uy
Venezuela	.ve
Zambia	.zm
Zimbabwe	.zw

If the host computer is not located in the United States, then the type of institution running a server on the Internet is often indicated by adding either .co for commercial, .or for non-profit organization, or .ac for academic before the country domain. For example, an academic institution in Chile would have

.ac.cl

as the top-level domains, while a non-profit organization in the United Kingdom would use the following top-level domains:

.or.uk

and in Greece a commercial business would have an address with the top-level domains

.co.gr

Appendix C. Computer Hardware and Software Requirements

In the early years of sequence analysis, only a few specialized centers had access to the necessary computing facilities and programming expertise to perform the needed work. This meant that most analyses were distributed; i.e., they were performed at specialized sites, with the results returned to the user. By the early to mid-1980s, the wide availability of personal computers and software that could perform useful analyses resulted in a shift from distributed analysis to local analysis. As a result, it became more efficient and effective to perform the analyses on a local computer within a laboratory. With the growth of the analysis tools and resources on the Internet, the pendulum is swinging back toward a distributed model for some aspects of sequence analysis. In this revised model, the primary and specialized databases are maintained at centralized resource centers, with fast network connections and more powerful personal computers allowing researchers to use both local software and remote computers with equal ease.

The advantages that this model for sequence analysis offers are several. First, a laboratory needs to have only a moderate-priced computer and a high-speed network connection to perform complex manipulations of sequence data. Second, the databases are maintained by centralized resources, reducing the amount of information storage capacity that is needed on the local computer. There are two principal drawbacks with this type of system, however. First, analyses are dependent on distant servers and the limited set of algorithms that they offer. Second, if the network goes down, it is impossible to perform any analysis of data. It should be noted that network reliability has improved dramatically over the past decade and, further, that every year there is an increasing choice of algorithms and options for them.

With these comments as a foundation, what is the optimal personal computer for performing sequence analysis? Unfortunately, there is no clear-cut answer to that question. Because of the continual and rapid pace of improvements in personal computer technology, it is impossible to make specific and lasting recommendations about what system to purchase for sequence analysis. Considering that, on average, computer processing power improves about one- to twofold per year, this means that in 4 or 5 years the same money spent on a system now will purchase a system that is ten times more powerful. So, instead of purchasing a specific computer based on a set of performance specifications, look first to what tasks the computer system must accomplish in the laboratory.

A laboratory computer will be used for more tasks than sequence analysis. Word-processing functions such as reports, manuscripts, and other writing needs are usually among the most common types of work. Graphing of data, the preparation of figures, and other graphics-intensive tasks are also commonly required in the laboratory. In addition, many laboratories will need or want some type of local sequence analysis capability, ranging from simple restriction maps to more demanding sequence assembly, to such computationally intensive operations as molecular modeling. These needs will dictate how powerful a computer is required.

Thus, for basic software a laboratory typically needs a word processor, a drawing program, a spreadsheet program or database, and perhaps a simple sequence analysis package such as DNA Strider. To meet the first three needs an integrated software package such as ClarisWorks is often useful. It is cross-platform, opens most documents quickly, and does not require a great deal of computer "muscle" to operate well. Another obvious consideration for laboratory software is a communications package. There are a wide range of "Internet in a box" software kits that simplify setting up a client system connected to the Internet. At a minimum this type of software package should include a Web browser and e-mail capability.

The next consideration in the acquisition of a laboratory computer is cost: what funds are available to purchase a system (hardware and software) and an Internet connection. When budgeting for a laboratory computer, two rules apply. First, always purchase the fastest processor that is affordable, and second, get twice as much hard drive capacity as needed.

Another important consideration is system compatibility, although this is not as critical as it has been in the past. Within the personal computer world there are two basic systems: DOS/Windows machines, based on microprocessors from Intel, and Macintosh computers, based on microprocessors from Motorola, Apple Computer, and IBM. Some departments and institutions have standardized on one type of computer system to simplify operations. Others use a mixture of computers. It is also important to take into account what computer is being used at home as well: a compatible system is a necessity. Lastly, do not purchase a system based on an obsolescent microprocessor. If purchasing a new DOS/Windows system, do not consider anything less than one with a Pentium processor. If purchasing a Macintosh system from either Apple Computer or one of the growing number of Macintosh clone manufacturers, get a system based on a PowerPC processor.

With all of those factors in mind, the following are some general purchase guidelines for computer systems in a laboratory setting:

■ **Basic system: $2,500.00** maximum

PowerPC 603/100 MHz (Mac) or Pentium/100 Mhz (Windows)
1 gigabyte (Gb) hard drive
16 megabytes (Mb) RAM
14–15" monitor
28.8 Kbaud or faster modem
Integrated software package and Internet communications software
Laser or ink jet printer

■ **Mid-range system: $4,000.00** maximum

PowerPC 603/200 Mhz (Mac) or Pentium/200 Mhz (Windows)
1 Gb hard drive
24 Mb RAM
17" monitor
Back-up storage system such as a Zip, SysQuest, Jaz, or magneto-optical drive
28.8 Kbaud or faster modem
Integrated software package and Internet communications software
Laser or ink jet printer

■ High-end system: $6,000.00 and up

> PowerPC-class/250 Mhz (Mac) or PentiumPro-class/250 Mhz (Windows)
> 2–5 Gb hard drive
> 32 Mb RAM or more
> 17–21" monitor
> Back-up storage system such as a Zip, SysQuest, Jaz, or magneto-optical drive
> 28.8 Kbaud or faster modem or a dedicated high-speed connection
> Integrated software package, Internet communications software, and specialized
>   analysis software
> Laser or ink jet printer

Appendix D. Nucleic Acid and Amino Acid Codes and Properties

Nucleic Acids: Codes

Code	Bases	Mnemonic
A	A	A-denine
C	C	C-ytosine
G	G	guanine
T (or U)	T	T-hymine (or U-racil)
R	A or G	pu-R-ine
Y	C or T	p-Y-rimidine
S	G or C	S-trong (3 H-bonds)
W	A or T	W-eak (2 H-bonds)
K	G or T	K-eto
M	A or C	a-M-ino
B	C or G or T	not-A
D	A or G or T	not-C
H	A or C or T	not-G
V	A or C or G	not-(T or U)
N (or X)	Any base	a-N-y (or unknown)
	Gap	

Nucleic Acids: Properties

Nucleic Acid	Molecular Weight
ATP	507.2
CTP	483.2
GTP	523.2
UTP	484.2
dATP	491.2
dCTP	467.2
dGTP	507.2
TTP	482.2

Amino Acids: Codes and Properties

Amino Acid	Letter Codes		Molecular Weight, pH 7
	1 letter	3 letter	
Alanine	A	Ala	89.09
Cysteine	C	Cys	121.16
Aspartate	C	Asp	132.12
Glutamate	C	Glu	146.13
Phenylalanine	F	Phe	165.19
Glycine	G	Gly	75.07
Histidine	H	His	155.16
Isoleucine	I	Ile	131.17
Lysine	K	Lys	147.19
Leucine	L	Leu	131.17
Methionine	M	Met	149.21
Asparagine	N	Asn	132.12
Proline	P	Pro	115.13
Glutamine	Q	Gln	146.15
Arginine	R	Arg	175.20
Serine	S	Ser	105.09
Threonine	T	Thr	119.02
Valine	V	Val	117.15

Amino Acid	Letter Codes		Molecular Weight, pH 7
	1 letter	3 letter	
Tryptophan	W	Trp	204.22
Tyrosine	Y	Tyr	181.19
Water			18.01
Alpha-amino group			8.56
Alpha-carboxyl group			3.56

Standard Genetic Code

First Position (5´)	Second Position								Third Position (3´)
	U		C		A		G		
U	UUU	Phe	UCU	Ser	UAU	Tyr	UGU	Cys	U
	UUC	Phe	UCC	Ser	UAC	Tyr	UGC	Cys	C
	UUA	Leu	UCA	Ser	UAA	Stop	UGA	Stop	A
	UUG	Leu	UCG	Ser	UAG	Stop	UGG	Trp	G
C	CUU	Leu	CCU	Pro	CAU	His	CGU	Arg	U
	CUC	Leu	CCC	Pro	CAC	His	CGC	Arg	C
	CUA	Leu	CCA	Pro	CAA	Gln	CGA	Arg	A
	CUG	Leu	CCG	Pro	CAG	Gln	CGG	Arg	G
A	AUU	Ile	ACU	Thr	AAU	Asn	ADU	Ser	U
	AUC	Ile	ACC	Thr	AAC	Asn	AGC	Ser	C
	AUA	Ile	ACA	Thr	AAA	Lys	AGA	Ser	A
	AUG	Met	ACG	Thr	AAG	Lys	AGG	Arg	G
G	GUU	Val	GCU	Ala	GAU	Asp	GGU	Gly	U
	GUC	Val	GCC	Ala	GAC	Asp	GGC	Gly	C
	GUA	Val	GCA	Ala	GAA	Glu	GGA	Gly	A
	GUG	Val	GCG	Ala	GAG	Glu	GGG	Gly	G

Alternative Genetic Codes

The genetic code, originally thought to be universal in all organisms, is now known to be in a state of evolution. In 1981 it was first discovered that mammalian mitochondria did not use the standard genetic code, but an alternative one. Since that time, many other alternative genetic codes have been discovered, some in mitochondria or other organelles, others in organisms such as bacteria, protozoa, algae, and yeasts. The following listing of alternative genetic codes is adapted from a variety of sources, but primarily from an excellent compilation posted on the Web by Andrzej Elzanowski and James Ostell at the National Center for Biotechnology Information. To maintain consistency in nomenclature, the name that NCBI has given each of these alternative codes is used in this adapted listing. For further information please refer to the NCBI Web site or the selected references listed at the end of this section

Several genetic codes are summarized in this listing:

- Standard Code (Universal)
- Vertebrate Mitochondrial Code
- Yeast Mitochondrial Code
- Mold, Protozoan, and Coelenterate Mitochondrial Code and Mycoplasma/Spiroplasma Code
- Invertebrate Mitochondrial Code
- Ciliate, Dasycladacean, and Hexamita Nuclear Code (Ciliate Macronuclear)
- Echinoderm Mitochondrial Code (Echinodermate Mitochondrial)
- Euplotid Nuclear Code (Alternative Ciliate Macronuclear)
- Bacterial Code (Eubacterial)
- Alternative Yeast Nuclear Code (Alternative Yeast)
- Ascidian Mitochondrial Code
- Flatworm Mitochondrial Code
- Blepharisma Nuclear Code

Each code is displayed as a five-line table. The first line lists the amino acids, using the single letter code. The second line lists the amino acid(s) that can be used for protein initiation. The next three lines give the first, second, and third bases of the codons that encode each amino acid in the first line. Finally, other differences from the standard or universal genetic code are summarized below the table: differences in codons, initiation codons, and species range for the genetic code, with other comments.

■ Standard Code

```
Amino acid   FFLLSSSSYY**CC*WLLLLPPPPHHQQRRRRIIIMTTTTNNKKSSRRVVVVAAAADDEEGGGG
Initiation   ---------------------------------M------------------------------
Base 1       TTTTTTTTTTTTTTTTCCCCCCCCCCCCCCCCAAAAAAAAAAAAAAAAGGGGGGGGGGGGGGGG
Base 2       TTTTCCCCAAAAGGGGTTTTCCCCAAAAGGGGTTTTCCCCAAAAGGGGTTTTCCCCAAAAGGGG
Base 3       TCAGTCAGTCAGTCAGTCAGTCAGTCAGTCAGTCAGTCAGTCAGTCAGTCAGTCAGTCAGTCAG
```

Initiation codon
AUG

■ Vertebrate Mitochondrial Code

```
Amino acid   FFLLSSSSYY**CCWWLLLLPPPPHHQQRRRRIIMMTTTTNNKKSS**VVVVAAAADDEEGGGG
Initiation   -------------------------------MMMM--------------M-----------
Base 1       TTTTTTTTTTTTTTTTCCCCCCCCCCCCCCCCAAAAAAAAAAAAAAAAGGGGGGGGGGGGGGGG
Base 2       TTTTCCCCAAAAGGGGTTTTCCCCAAAAGGGGTTTTCCCCAAAAGGGGTTTTCCCCAAAAGGGG
Base 3       TCAGTCAGTCAGTCAGTCAGTCAGTCAGTCAGTCAGTCAGTCAGTCAGTCAGTCAGTCAGTCAG
```

Differences from the standard code

	Alternate	Standard
AGA	Stop	Arg
AGG	Stop	Arg
AUA	Met	Ile
UGA	Trp	Stop

Alternative initiation codons

Bos: AUA
Homo: AUA, AUU
Mus: AUA, AUU, AUC
Coturnix, Gallus: GUG

Systematic range

Vertebrata

Comment

The transcripts of several vertebrate mitochondrial genes end in U or UA. Upon polyadenylation these become the termination codon UAA.

■ Yeast Mitochondrial Code

```
Amino acid   FFLLSSSSYY**CCWWTTTTPPPPHHQQRRRRIIMMTTTTNNKKSSRRVVVVAAAADDEEGGGG
Initiation   ----------------------------------M-----------------------------
Base 1       TTTTTTTTTTTTTTTTCCCCCCCCCCCCCCCCAAAAAAAAAAAAAAAAGGGGGGGGGGGGGGGG
Base 2       TTTTCCCCAAAAGGGGTTTTCCCCAAAAGGGGTTTTCCCCAAAAGGGGTTTTCCCCAAAAGGGG
Base 3       TCAGTCAGTCAGTCAGTCAGTCAGTCAGTCAGTCAGTCAGTCAGTCAGTCAGTCAGTCAGTCAG
```

Differences from the standard code

	Alternate	Standard
AUA	Met	Ile
CUU	Thr	Leu
CUC	Thr	Leu
CUA	Thr	Leu
CUG	Thr	Leu
UGA	Trp	Stop
CGA	Absent	Arg
CGC	Absent	Arg

Systematic Range

Saccharomyces cerevisiae, Candida glabrata, Hansenula saturnus, Schizosaccharomyces pombe, and *Kluyveromyces thermotolerans.*

Comments

1. Other CGN codons are rare in *Saccharomyces cerevisiae* and they have not been found in *Candida glabrata.*

2. While the AUA codon is common in the gene *var1,* coding for the only mitochondrial ribosomal protein, it appears to be rare in genes encoding enzymes. The coding assignments for the codons AUA and CUU are uncertain in *Hansenula saturnus.* AUA is believed to code for Met or Ile. CUU may code for Leu, but probably does not code for Thr.

3. The termination codon UGA is rarely used in *Schizosaccharomyces pombe.* It appears to code for Trp when it is present. There is no evidence for other nonstandard codon assignments in this species.

4. In *Kluyveromyces thermotolerans,* the coding assignment of CUN to Thr has not been determined.

■ Mold, Protozoan, and Coelenterate Mitochondrial Code and Mycoplasma/Spiroplasma Code

```
Amino acid   FFLLSSSSYY**CCWWLLLLPPPPHHQQRRRRIIIMTTTTNNKKSSRRVVVVAAAADDEEGGGG
Initiation   --MM--------------M-----------MMMM--------------M-----------
Base 1       TTTTTTTTTTTTTTTTCCCCCCCCCCCCCCCCAAAAAAAAAAAAAAAAGGGGGGGGGGGGGGGG
Base 2       TTTTCCCCAAAAGGGGTTTTCCCCAAAAGGGGTTTTCCCCAAAAGGGGTTTTCCCCAAAAGGGG
Base 3       TCAGTCAGTCAGTCAGTCAGTCAGTCAGTCAGTCAGTCAGTCAGTCAGTCAGTCAGTCAGTCAG
```

Differences from the standard code

The codon UGA encodes the amino acid Trp instead of serving as a termination codon.

Alternative initiation codons

Trypanosoma: UUA, UUG, CUG

Leishmania: AUU, AUA

Tetrahymena: AUU, AUA, AUG

Paramecium: AUU, AUA, AUG, AUC, GUG, GUA(?)

Systematic range

Mycoplasmatales: *Mycoplasma, Spiroplasma*

Fungi: *Acremonium, Candida parapsilosis, Dekkera/Brettanomyces, Eeniella, Emericella nidulans, Neurospora crassa, Podospora anserina, Trichophyton rubrum,* and probably *Ascobolus immersus, Aspergillus amstelodami, Claviceps purpurea,* and *Cochliobolus heterostrophus*

Protozoa: *Leishmania tarentolae, Paramecium tetraurelia, Tetrahymena pyriformis, Trypanosoma brucei,* and probably *Plasmodium gallinaceum*

Metazoa: Coelenterata: *Ctenophora and Cnidaria*

Comments

1. This genetic code is also used for the kinetoplast DNA. Kinetoplasts are modified mitochondria that are more commonly called maxicircles and minicircles.
2. This genetic code is not used in the *Acholeplasmataceae* and plant-pathogenic varieties of mycoplasma-like organisms (MLO).

■ Invertebrate Mitochondrial Code

```
Amino acid  FFLLSSSSYY**CCWWLLLLPPPPHHQQRRRRIIMMTTTTNNKKSSSSVVVVAAAADDEEGGGG
Initiation  ---M----------------------------M-MM--------------M------------
Base 1      TTTTTTTTTTTTTTTTCCCCCCCCCCCCCCCCAAAAAAAAAAAAAAAAGGGGGGGGGGGGGGGG
Base 2      TTTTCCCCAAAAGGGGTTTTCCCCAAAAGGGGTTTTCCCCAAAAGGGGTTTTCCCCAAAAGGGG
Base 3      TCAGTCAGTCAGTCAGTCAGTCAGTCAGTCAGTCAGTCAGTCAGTCAGTCAGTCAGTCAGTCAG
```

Comment

The codon AGG is absent in *Drosophila*.

Differences from the standard code

	Alternate	Standard
AGA	Ser	Arg
AGG	Ser	Arg
AUA	Met	Ile
UGA	Trp	Stop

Alternative initiation codons
Apis: AUA, AUU, AUC
Polyplacophora: GUG
Ascaris, Caenorhabditis: UUG

Systematic range
Nematoda: *Ascaris, Caenorhabditis*
Mollusca: *Bivalvia, Polyplacophora*
Arthropoda/Crustacea: *Artemia*
Arthropoda/Insecta: *Drosophila, Locusta migratoria, Apis mellifera*

Comments

1. The codon GUG may function as an initiation codon in *Drosophila*.
2. The codon AUU is not used as an initiation codon in *Mytilus*.

■ Ciliate, Dasycladacean, and Hexamita Nuclear Code

```
Amino acid   FFLLSSSSYYQQCC*WLLLLPPPPHHQQRRRRIIIMTTTTNNKKSSRRVVVVAAAADDEEGGGG
Initiation   --------------------------------M---------------------------
Base 1       TTTTTTTTTTTTTTTTCCCCCCCCCCCCCCCCAAAAAAAAAAAAAAAAGGGGGGGGGGGGGGGG
Base 2       TTTTCCCCAAAAGGGGTTTTCCCCAAAAGGGGTTTTCCCCAAAAGGGGTTTTCCCCAAAAGGGG
Base 3       TCAGTCAGTCAGTCAGTCAGTCAGTCAGTCAGTCAGTCAGTCAGTCAGTCAGTCAGTCAGTCAG
```

Differences from the standard code

The codons UAA and UAG encode the amino acid Gln instead of serving as termination codons.

Systematic range

Ciliata: *Oxytricha,* Oxytrichidae, *Paramecium, Stylonychia, Tetrahymena,* and probably *Glaucoma chattoni*

Dasycladaceae: *Acetabularia* and *Batophora*

Diplomonadida: *Diplomonadida* and *Hexamita inflata*

Comment

    1. Note that the complete ciliate macronuclear code has not been determined.

    2. The codon UAA is known to code for Gln only in the Oxytrichidae.

■ Echinoderm Mitochondrial Code

```
Amino acid   FFLLSSSSYY**CCWWLLLLPPPPHHQQRRRRIIIMTTTTNNNKSSSSVVVVAAAADDEEGGGG
Initiation   --------------------------------M---------------------------
Base 1       TTTTTTTTTTTTTTTTCCCCCCCCCCCCCCCCAAAAAAAAAAAAAAAAGGGGGGGGGGGGGGGG
Base 2       TTTTCCCCAAAAGGGGTTTTCCCCAAAAGGGGTTTTCCCCAAAAGGGGTTTTCCCCAAAAGGGG
Base 3       TCAGTCAGTCAGTCAGTCAGTCAGTCAGTCAGTCAGTCAGTCAGTCAGTCAGTCAGTCAGTCAG
```

Differences from the standard code

	Alternate	Standard
AAA	Asn	Lys
AGA	Ser	Arg
AGG	Ser	Arg
UGA	Trp	Stop

Systematic range

Asterozoa (starfishes) and Echinozoa (sea urchins)

■ **Euplotid Nuclear Code**

```
Amino acid   FFLLSSSSYY**CCCWLLLLPPPPHHQQRRRRIIIMTTTTNNKKSSRRVVVVAAAADDEEGGGG
Initiation   ---------------------------------M----------------------------
Base 1       TTTTTTTTTTTTTTTTCCCCCCCCCCCCCCCCAAAAAAAAAAAAAAAAGGGGGGGGGGGGGGGG
Base 2       TTTTCCCCAAAAGGGGTTTTCCCCAAAAGGGGTTTTCCCCAAAAGGGGTTTTCCCCAAAAGGGG
Base 3       TCAGTCAGTCAGTCAGTCAGTCAGTCAGTCAGTCAGTCAGTCAGTCAGTCAGTCAGTCAGTCAG
```

Differences from the standard code
The codon UGA encodes the amino acid Cys instead of serving as a termination codon.

Systematic range
Ciliata: Euplotidae

■ **Bacterial Code**

```
Amino acid   FFLLSSSSYY**CC*WLLLLPPPPHHQQRRRRIIIMTTTTNNKKSSRRVVVVAAAADDEEGGGG
Initiation   ---M--------------M------------M--M--------------M------------
Base 1       TTTTTTTTTTTTTTTTCCCCCCCCCCCCCCCCAAAAAAAAAAAAAAAAGGGGGGGGGGGGGGGG
Base 2       TTTTCCCCAAAAGGGGTTTTCCCCAAAAGGGGTTTTCCCCAAAAGGGGTTTTCCCCAAAAGGGG
Base 3       TCAGTCAGTCAGTCAGTCAGTCAGTCAGTCAGTCAGTCAGTCAGTCAGTCAGTCAGTCAGTCAG
```

Differences from the standard code
None

Alternative initiation codons
GUG, UUG, AUU, and CUG

Systematic range
All bacteria including the eubacteria and archaebacteria.

Comments
1. The codon UGA codes at low efficiency for Trp in *Bacillus subtilis* and possibly in *Escherichia coli*.
2. The codon CUG is known to function as an initiation codon for RepA, a plasmid-encoded protein in *Escherichia coli* and possibly in *Methanobacterium thermoautotrophicum*.
3. There is no evidence for the codon AUU serving as an initiation codon in the archaebacteria.

■ **Alternative Yeast Nuclear Code**

```
Amino acid   FFLLSSSSYY**CC*WLLLSPPPPHHQQRRRRIIIMTTTTNNKKSSRRVVVVAAAADDEEGGGG
Initiation   -------------------M---------------M----------------------------
Base 1       TTTTTTTTTTTTTTTTCCCCCCCCCCCCCCCCAAAAAAAAAAAAAAAAGGGGGGGGGGGGGGGG
Base 2       TTTTCCCCAAAAGGGGTTTTCCCCAAAAGGGGTTTTCCCCAAAAGGGGTTTTCCCCAAAAGGGG
Base 3       TCAGTCAGTCAGTCAGTCAGTCAGTCAGTCAGTCAGTCAGTCAGTCAGTCAGTCAGTCAGTCAG
```

Differences from the standard code
The codon CUG encodes the amino acid Ser instead of Leu.

Alternative initiation codons
The codon CAG may be used in *Candida albicans*.

Systematic range
Endomycetales (yeasts), specifically *Candida albicans, Candida cylindracea, Candida melibiosica, Candida parapsilosis,* and *Candida rugosa.*

Comments
All other yeasts, including *Saccharomyces cerevisiae, Candida azyma, Candida diversa, Candida magnoliae, Candida rugopelliculosa, Yarrowia lipolytica,* and *Zygoascus hellenicus,* use the standard genetic code.

■ Ascidian Mitochondrial Code

```
Amino acid   FFLLSSSSYY**CCWWLLLLPPPPHHQQRRRRIIMMTTTTNNKKSSGGVVVVAAAADDEEGGGG
Initiation   ----------------------------------M-----------------------------
Base 1       TTTTTTTTTTTTTTTTCCCCCCCCCCCCCCCCAAAAAAAAAAAAAAAAGGGGGGGGGGGGGGGG
Base 2       TTTTCCCCAAAAGGGGTTTTCCCCAAAAGGGGTTTTCCCCAAAAGGGGTTTTCCCCAAAAGGGG
Base 3       TCAGTCAGTCAGTCAGTCAGTCAGTCAGTCAGTCAGTCAGTCAGTCAGTCAGTCAGTCAGTCAG
```

Differences from the standard code

	Alternative	Standard
AGA	Gly	Arg
AGG	Gly	Arg
AUA	Met	Ile
UGA	Trp	Stop

Systematic range
Ascidiaceae (sea squirts), specifically the Pyuridae.

■ Flatworm Mitochondrial Code

```
Amino acid   FFLLSSSSYYYY*CCWWLLLLPPPPHHQQRRRRIIIMTTTTNNNKSSSSVVVVAAAADDEEGGGG
Initiation   -----------------------------------M----------------------------
Base 1       TTTTTTTTTTTTTTTTCCCCCCCCCCCCCCCCAAAAAAAAAAAAAAAAGGGGGGGGGGGGGGGG
Base 2       TTTTCCCCAAAAGGGGTTTTCCCCAAAAGGGGTTTTCCCCAAAAGGGGTTTTCCCCAAAAGGGG
Base 3       TCAGTCAGTCAGTCAGTCAGTCAGTCAGTCAGTCAGTCAGTCAGTCAGTCAGTCAGTCAGTCAG
```

Differences from the standard code

	Alternative	Standard
AAA	Asn	Lys
AGA	Ser	Arg
AGG	Se	Arg
UAA	Tyr	Stop
UGA	Trp	Stop

Systematic Range
Platyhelminthes (flatworms)

■ Blepharisma Nuclear Code

```
Amino acid  FFLLSSSSYY*QCC*WLLLLPPPPHHQQRRRRIIIMTTTTNNKKSSRRVVVVAAAADDEEGGGG
Initiation  ----------------------------------M-----------------------------
Base 1      TTTTTTTTTTTTTTTTTCCCCCCCCCCCCCCCCAAAAAAAAAAAAAAAAGGGGGGGGGGGGGGGG
Base 2      TTTTCCCCAAAAGGGGTTTTCCCCAAAAGGGGTTTTCCCCAAAAGGGGTTTTCCCCAAAAGGGG
Base 3      TCAGTCAGTCAGTCAGTCAGTCAGTCAGTCAGTCAGTCAGTCAGTCAGTCAGTCAGTCAGTCAG
```

Differences from the standard code
The codon UAG encodes the amino acid Gln instead of serving as a termination codon.

Systematic range
Blepharisma

■ Other alternative initiation codons

Several other codons either are used to initiate translation or have been postulated to function in initiation as follows.

1. The codons GUG, UUG, and possibly CUG are used in the archaebacteria.
2. The codons AUA, GUG, UUG, and AUC or AAG may be used by the yeasts.
3. The codon ACG initiates translation of certain proteins in adeno-associated virus type 2, the phage T7 mutant CR17, Sendai virus, and rice chloroplasts. Also, this codon is the most effective non-AUG initiation codon in mammalian cells.
4. Finally, the codon CUG has been shown to be the initiation codon for one of the two alternative products of the human c-myc gene. Its role as an initiation codon for other transcripts and in other systems is unknown.

Selected readings

Journal Articles

Jukes, T. H., and S. Osawa. 1993. Evolutionary changes in the genetic code. *Comp. Biochem. Physiol.* **106**:489-494.

Osawa, S., T. H. Jukes, K. Watanabe, and A. Muto. 1992. Recent evidence for evolution of the genetic code. *Microbiol. Rev.* **56**:229-264.

Internet Publications

Elzanowski, A., and J. Ostell. 1996. *The Genetic Codes.* National Center for Biotechnology Information, National Library of Medicine, National Institutes of Health, Bethesda, Md.

Appendix E. Glossary of Terms

Advanced Research Projects Agency (ARPA)   A research organization that was originally part of the United States Department of Defense. Also known as DARPA for Defense Advanced Research Projects Agency.

American Standard Code for Information Interchange (ASCII)   The American Standard Code for Information Interchange sets a unique numeric definition for each of the most commonly used characters in Western language, including numerals, letters (both upper- and lowercase), and some symbols. An ASCII set is these characters and their numbers. In the context of a file, an ASCII file is one that contains only text characters: numbers, letters, and standard punctuation.

Anonymous FTP   A file transfer protocol (FTP) that allows the retrieval of files from public sites (*see* File Transfer Protocol).

Archie   An Internet service tool that searches anonymous FTP sites for files based on user-defined parameters.

Archive   A collection of data, text, programs, or other electronic information stored for other parties to access or retrieve, usually without charge.

ARPA   *See* Advanced Research Projects Agency.

ASCII   *See* American Standard Code for Information Interchange.

Backbone   The backbone, in network terms, is that portion of the network structure that carries the majority of the network traffic. In a properly designed network, the backbone is the highest-capacity and fastest section of the total network, often with redundant routes to ensure reliability.

Bandwidth   The capacity of a network to carry data and transfer it.

Baud   A measure of modem speed that is equal to one signal or bit per second: 300 baud equals 300 bits per second (300 bps).

BITNET   Acronym for Because It's Time Net. The BITNET is an academic collective of networks. It is also the forerunner to, and is now a component of, the Internet.

Bookmark   When you visit a Web site that you want to return to, you can save the site's location in your browser's bookmark file. When you are ready to return to the site, pull down the Bookmarks menu and click on the site you want to jump to.

Browser   Client software that reads HTML-formatted documents and displays them on your PC screen. Most browsers track where you've been to help you navigate in the Web, and also allow you to save documents. Most also support the use of bookmarks so that you can mark your favorite sites and return to them later.

CERN   European Center for Particle Physics, Switzerland.

Client   In its simplest form, a client is a software program that an individual uses to send requests to a server. In a broader definition it can mean a computer or computer program that requests a service of a host computer or program.

DARPA   Acronym for the Defense Advanced Research Projects Agency, part of the United States Department of Defense. Also known as ARPA for Advanced Research Projects Agency.

Domain   A network or portion thereof that operates under a single administrative umbrella such as an institution, a group of institutions, a geographical region, or a country.

EARN   European Academic Research Network.

Electronic mail (e-mail)   Messages that travel through computer networks rather than being committed to paper and being sent via the traditional postal services.

EMBnet   European Molecular Biology Network.

Ethernet   A standard for physical and data link layer specifications.

FAQ   Acronym for Frequently Asked Question. Often the staff of a server will publish a list of FAQs and their answers.

Fileserver   A computer that provides files to other computers via a network.

File Transfer Protocol (FTP)   An Internet protocol that defines a common method of transferring files across networks and between remote computers.

Freeware   A computer program or software that is available free of charge. Remember, just because a program or piece of software is free, that doesn't mean it is in the public domain. Usually the author or developer retains the copyright to it.

FTP   *See* File Transfer Protocol.

Gateway   A computer that is connected to two or more independent networks and can transfer files between them, even if the networks cannot directly process data from each other.

Gopher   A versatile menu-driven information service that allows access to various types of data and sites within the Internet. Gopher supports searching a site for documents based on their content.

Home page   The entry point of a Web site. Typically it will include information about the site, instructions on how to use the site, and links to the contents of the site.

Host   A computer that allows users to communicate with other computers or hosts on a network.

Hotlink   A link between two documents on the Web, or from one location to another within a document. Synonymous with hypertext.

Hotlist   A hypertext list of Web sites arranged according to the focus of the site or the fancy of the hotlist's creator. Very useful when well researched and maintained, but very time-consuming when lists link to lists that link to others and so on.

HTML   *See* Hypertext Markup Language.

HTTP   *See* Hypertext Transfer Protocol.

Hypermedia   Hypertext that includes links to other forms of media.

Hypertext   The linking of one document to another document, or to another location within the same document. This allows the document creator to highlight words that, when clicked on, cause the browser to load another URL.

Hypertext Markup Language (HTML)   The standard language used for creating hypermedia documents within the World Wide Web.

Hypertext Transfer Protocol (HTP)   The standard language that World Wide Web clients and servers use to communicate.

Integrated Services Digital Network (ISDN)   A system for combining voice and data communications.

Internet   A global collective of computer networks running TCP/IP.

ISDN   *See* Integrated Services Digital Network.

LAN   *See* Local Area Network.

Listserv   Software that manages mailing lists on the Internet.

Local Area Network (LAN)   Two or more computers connected together via cabling. LANs are typically found connecting computers within a laboratory, building, or institution with one connection to the Internet.

Login   The process by which a user identifies him/herself to the host computer, usually in a two-step process composed of a userid and a password.

Mailing list   A set of Internet e-mail addresses that belong to the same discussion group. An e-mail message sent to one address on the list is sent to all of the other addresses.

MEMEX   A conceptual machine that could show the trails of information that its users viewed.

Menu bar   A common element in graphic computer interfaces that allows users to select options from menus.

Mosaic   A mouse-driven graphic interface to the World Wide Web developed by the NCSA.

National Center for Supercomputing Applications (NCSA)   A federally funded organization whose mission is to develop and research high-technology resources for the scientific community.

NCSA   *See* National Center for Supercomputing Applications.

News Network Transfer Protocol (NNTP)   A method of transferring news articles.

Newsgroup   A single message area on the UseNet covering a specific topic area.

NNTP   *See* News Network Transfer Protocol.

Offline   Describes actions or tasks performed when not connected to another computer via a network.

Online   Describes actions or tasks performed when connected to another computer via a network.

Packet   A network message that is composed of a header, addressing information, and data.

Protocol   A set of instructions or rules that govern how two computers exchange information.

Public domain   Software that can be freely used, copied, distributed, and modified. *Also see* Freeware *and* Shareware.

Route   A listing of computers that a message passes through from its origin to its final destination.

Router   A device that directs or routes data between networks or different parts of a network.

Scroll bar   A graphic computer interface element that allows the user to scroll through electronic documents on the computer screen.

Server   A combination of both computer hardware and software which supplies a service to requests submitted by client computers. The server processes the requests and then returns the results to the client. In short, it is a computer that makes services available on a network. For example, a fileserver makes files available.

Shareware   A method of software distribution in which the software may be freely distributed and may be tested prior to purchasing it.

Simple Mail Transfer Protocol (SMTP)   SMTP defines the manner by which e-mail is transferred between computers on the Internet.

SMTP   *See* Simple Mail Transfer Protocol.

Snail mail   Mail sent via the traditional postal service.

TCP/IP   *See* Transmission Control Protocol/Internet Protocol.

Telnet   An Internet standard program which allows users to remotely use computers across networks.

Transmission Control Protocol/Internet Protocol (TCP/IP)   The software communication protocol of the Internet. One computer communicates with another computer through the Internet using the TCP/IP format.

Uniform Resource Locator (URL)   A standardized way of representing different documents, media, and network services on the World Wide Web. It gives every document of the Internet its own unique address. URLs are most commonly associated with Web sites but also apply to e-mail servers, Gopher servers, and any other computer with an Internet address.

UNIX   A computer operating system developed by Bell Laboratories.

URL   *See* Uniform Resource Locator.

UseNet   A collection of computer systems that routinely exchange posted e-mail messages as part of a newsgroup, allowing widespread discussion of topics. In other words, a global news-reading network.

Userid   A name used to log on to a host computer.

Veronica   A network service that allows users to search Gopher systems for documents.

WAIS   *See* Wide Area Information Server.

WAN   *See* Wide Area Network.

Webmaster   The administrator responsible for the management and often design of a World Wide Web site.

Wide Area Network (WAN)   A group of geographically separated computers connected via dedicated lines or satellite links.

Wide Area Information Server (WAIS)   A service which allows users to intelligently search for information among databases distributed throughout the Internet.

World Wide Web (WWW)   The initiative to create a universal, hypermedia-based method to access information and resources on the Internet. Also used (incorrectly) to refer to the Internet.

WWW   *See* World Wide Web.

Index